Springer Geophysics

The Springer Geophysics series seeks to publish a broad portfolio of scientific books, aiming at researchers, students, and everyone interested in geophysics. The series includes peer-reviewed monographs, edited volumes, textbooks, and conference proceedings. It covers the entire research area including, but not limited to, geodesy, planetology, geodynamics, geomagnetism, paleomagnetism, seismology, and tectonophysics.

More information about this series at http://www.springer.com/series/10173

Jun Yao · Zhao-Qin Huang

Fractured Vuggy Carbonate Reservoir Simulation

Second Edition

Jun Yao
School of Petroleum Engineering
China University of Petroleum
Qingdao, Shandong
China

Zhao-Qin Huang
China University of Petroleum
Qingdao, Shandong
China

Springer Geophysics
ISBN 978-3-662-55031-1   ISBN 978-3-662-55032-8   (eBook)
DOI 10.1007/978-3-662-55032-8

Jointly published with Petroleum Industry Press, Beijing, China.

The print edition is not for sale in China Mainland. Customers from China Mainland please order the
print book from Petroleum Industry Press.

Library of Congress Control Number: 2016944323

This Springer imprint is published by Springer Nature
The registered company is Springer-Verlag GmbH Germany
The registered company address is: Heidelberger Platz 3, 14197 Berlin, Germany

Preface

This book focuses on the numerical methods of the two-phase fluid flow and the displacement in fractured vuggy porous carbonate reservoirs as well as quantitative approaches for describing such multi-scale physical processes. The book is intended to complement the existing literature by presenting new advances and updated developments in the two-phase fluid flow in fractured porous media, especially for fractured vuggy carbonate reservoirs. The material of this book is based primarily on (1) a series of peer-reviewed papers, published by our research group, (2) the technical reports that we have done during the research projects, including National Program on Key Basic Research Project (973 Program) and National Key Technologies R & D Program of China, and (3) the course notes that we used to teach undergraduate and graduate courses on advanced multiphase fluid flow in porous media at the China University of Petroleum (East China). The publications that this book is based on are related to the research on the subject of two-phase fluid flows in fractured vuggy porous media, which we have carried out or been involved with since the late 2000s.

This book can be used as a textbook or reference for senior undergraduate and graduate students in petroleum engineering, hydrogeology or groundwater hydrology, soil sciences, and other related engineering fields, such as civil and environmental engineering. It can also serve as a reference book for petroleum reservoir engineers, and other engineers and scientists working in the area of flow and transport in fractured porous media, especially in fractured karstic/vuggy media.

The contents of the book are organized to cover fundamentals of two-phase fluid flow in fractured and fractured vuggy porous media based on the discrete medium concepts and its corresponding applications. It discussed the multi-physical processes and principles governing coupling two-phase free flow and porous flow by using Navier–Stokes equations and Darcy's law. The book starts from the discrete fracture model, and then various numerical approaches are introduced to model the immiscible two-phase fluid flow in fractured reservoirs. Specifically, we proposed the discrete fracture-vug network model (DFVN) to analyze immiscible two-phase flow in fractured vuggy porous media. The DFVN model is an extension of discrete fracture model for fractured vuggy porous media. Based on these discrete models, an efficient equivalent medium numerical simulation is also developed and presented, which is more suitable for practical applications. In addition, the book reviews the hybrid and multi-scale concepts, approaches, and developments for modeling two-phase flow in fractured vuggy porous media. In an effort to include the new developments, the book also presents mathematical formulations and numerical modeling approaches for two-phase flow by using multi-scale finite element methods.

Qingdao, China   Jun Yao

Acknowledgements

We would like to thank Dr. Yue-Ying Wang and Dr. Na Zhang in the China University of Petroleum (East China) for their tremendous help in preparing the manuscript. In addition, they have contributed the major materials of Chaps. 5 and 6 for the hybrid model and multi-scale method in the book. I could not have completed this book without their support and help.

I would also like to thank my colleague at China University of Petroleum (East China), Prof. Ai-Fen Li, for her thorough technical review of the manuscript and sharing her expertise in reservoir engineering. I am highly indebted to Dr. Yang Li at the SINOPEC company for his review of this manuscript and meaningful suggestions, especially for his discussions during our collaboration during the research projects (including National Program on Key Basic Research Project (973 Program) and National Key Technologies R & D Program of China).

I would like to take this opportunity to thank many of my current and former colleagues and students in the China University of Petroleum (East China) who have made this book possible. Specifically, I would like to thank the editorial and production staff for their work and professionalism.

Contents

Chapter 1
Introduction

Abstract This chapter reviews background and progress made in development and application of the numerical theory and methods of fluid flow in fractured vuggy carbonate reservoirs. It discusses the characteristics of fractured vuggy porous media and the importance of quantifying flow processes in fractured vuggy porous media to scientific understandings and engineering applications. In addition, this chapter points out the need in further studies of the physics of complicated multiphase fluid flow in fractured vuggy porous media, driven by recent development in several frontiers of energy and natural resources. Then, this chapter is concluded by introducing the contents of the remaining five chapters.

Keywords Fractured vuggy carbonate reservoirs · Porous medium flow · Free-fluid flow · Discrete fracture-vug networks model · Multi-scale numerical simulation

1.1 Background

With the rapid development of economics, China has become the second largest fuel-consuming country. The demand for oil has increased from 3.2×10^8 t in 2005 to 4.98×10^8 t in 2013. The annual growth rate has reached 7.43 %. Since 1993, China became an oil-importer country, and the oil and gas import have increased from 1.567×10^7 t in 1993 to 2.89×10^8 t in 2013. The ratio of hydrocarbon import resources has reached to 60 %, and it still has an increasing trend (Xu 2013). To ensure the sustainable development of our economics, petroleum industry of China confronts with several challenges.

Hydrocarbon mainly reserved in terrestrial detrital reservoir-sand marine-carbonate reservoirs. The latter can be further subdivided into three types: (1) pore type carbonate reservoir, (2) fractured carbonate reservoir, and (3) fractured vuggy carbonate reservoir. The theory of oil and gas development of terrestrial detrital reservoirs, and water-flooding theory and applications have been the fundamentals of the rapid and steady development of our oil industry in the last half

© Petroleum Industry Press and Springer-Verlag Berlin Heidelberg 2017
J. Yao and Z.-Q. Huang, *Fractured Vuggy Carbonate Reservoir Simulation*,
Springer Geophysics, DOI 10.1007/978-3-662-55032-8_1

century. With the increasingly high degree to exploitation and development in terrestrial detrital reservoirs, the challenge has become much larger. Our aim has transferred from east to west. At present, the reserves in carbonate formation are 50 % of known economic reserves all over the world, and the production is higher than 60 %. The total oil and gas resources in marine-carbonate formation are larger than 300×10^8 t oil equivalent quantity. The oil mass is 150×10^8 t, mainly distributed in Tarimu Basin and HuaBei area. But the proved reserves are only about 11 %, among which fractured vuggy reservoir accounts for 2/3 of the proved reserves. So it will be the main exploitation and development region in the future in China (Yao and Zisheng 2007).

There have been more than 25-years history for the development of fractured vuggy carbonate reservoir in China. Zhuangxi area of Shengli oil field was developed in 1986, and Tahe oil field has been also developed for more than 15 years. In Tahe oil field, natural energy plays the main role in exploitation and development. Both a single well-water injection for displacing oil and fractured vuggy unit water injection development are adopted, which gain a relatively good effect of increasing recovery of oil. However the overall development needs to be further enhanced. Up to now, the recovery efficiency of such kind of reservoir is only 13–15 %, and the utilization rate is very low which is less than half that of terrestrial detrital reservoirs. The annual declining rate of production is more than 25 %. One of the main reasons is lack of understanding for fluids flow in fractured vuggy carbonate reservoir.

Fractured vuggy carbonate reservoir usually experiences the processes of multi-period tectonic movement, multi-period Karst superposition reconstruction and multi-period hydrocarbon accumulation processes. Reservoir porous media possesses strong heterogeneity and anisotropy. There are many types of reservoir porosities, and the relationship between oil and water is complicated. Multi-type fluid flow coexist in such carbonate reservoirs, including the free-fluid flow in large-scale cavities and the porous flow in porous matrix. The flow regimes may range from laminar flow (in low-permeability matrix) to turbulent flow (in cavities or fractures drilled with a well). It is a kind of complex coupling fluid flow (Li 2013). Therefore, the conventional theory and fundamentals of fluid flow through porous media is not completely appropriate for the study on this kind of reservoir. There is an urgent need to establish a new set of fluid flow mathematical model and numerical simulation methods to describe and model the fluid flow processes in fractured vuggy carbonate reservoirs.

1.2   Characteristics of Fractured Vuggy Carbonate Reservoirs

Different from conventional clastic rock reservoir, fractured vuggy carbonate reservoir has a strong anisotropy and multi-scale characteristics:

(1) The storage categories of medium is various, including matrix, fractures, vugs, and cavities.
(2) The range of scale variation is huge, especially for the space size of fractures and vugs may range from mm-scale to the m-scale.
(3) Influenced by tectonic movement in latter stage, fractures, and solution cavities are filled seriously, which aggravates the heterogeneous of formation. There are not only mechanical and physical filling with sand and mud but also chemical filling such as siliceous and calcite.

The above three characteristics lead to the extreme complexity of the fluid flow in fractured vuggy carbonate reservoir. How to model such complex fluid flow and make a prediction of production is always the major challenge in the development and management of fractured vuggy carbonate reservoir.

(1) Triple-porosity model
For the current reservoir simulation of fractured vuggy carbonate reservoirs, the concept models and modeling approaches still use or refer to the research methods which are applied in fractured reservoirs, such as double-porosity model, triple-porosity model, and their extended models. These methods still belong to the scope of seepage mechanisms used in conventional porous media. Double-porosity model was first established to model the matrix-fracture flow system in 1960s by Barrenblatt et al. (1960). Afterwards, Warren and Root build a more complete Warren-Root double-porosity model (Warren and Root 1963), and it has been widely used in fractured reservoir simulation nowadays. Then, the study on double-porosity model mainly focused on the calculation of the exchange flow function between matrix and fracture systems (Coats 1989; Kazemi et al. 1976; Saidi 1983; Thomas et al. 1983; Ueda et al. 1989). Recently, based on the classical double-porosity model, Pruess et al. make a subdivision on matrix rock gridding, and proposed the MINC (Multiple INteraction Continua) model (Pruess and Narasimhan 1985; Wu and Pruess 1988).

During the development process of fractured vuggy carbonate reservoirs, however, there are some special fluid flow phenomena and problems, which cannot been explained by the current double-porosity models. For example, the circular exploitation and development in Tahe oil field show that there exists the third storage space, i.e., vug system which cannot be neglected in fractured vuggy carbonate reservoirs. To this end, researchers proposed the triple-porosity model in terms of the idea on double-porosity model. Currently, the corresponding studies on triple-porosity model mainly focus on well-testing analysis region. We can distinguish whether the reservoir has triple-porosity characteristics by well-testing analysis curves (Chang et al. 2004; Yao and Zisheng 2007; Yao et al. 2004). Recently, Kang (2010) and Wu et al. (2011) extend triple-porosity model to the study on reservoir simulation of fractured vuggy carbonate reservoirs.

This triple-porosity model can describe the phenomenon of preferential flow in fractured vuggy reservoirs in some degree, and it simultaneously considers the mass exchange between fracture, matrix, and vug systems, which is much closer to reality. However, this model is based on assumption that matrix and vug are divided into the same size-shape medium by fractures; assumption is too simple and cannot fully describe the discontinuity and multi-scale characteristics of fractures and vugs. Moreover, there is no corresponding theory and method to determine the mass exchange function coefficient between matrix, fractures, and vugs systems. Especially for two-phase and multiphase fluid flow, the difficulty of introducing the effects of gravity and wettability to the mass exchange function is nontrivial. In addition, the triple-porosity model still belongs to the conventional continuous medium model. The triple continua assumption is feasible only under the condition that there is a high fractured degree and connectivity of fractures and vugs. Furthermore, the triple-porosity model cannot describe the multi-scale coupling flow characteristic. As a result, in many practice applications, the triple-porosity model will result in a relative big error.

(2) Equivalent medium model
Different from double-and triple-porosity models, equivalent medium model regards the overall fractured vuggy reservoir as a continuous porous system. We can represent its heterogeneity by the corresponding equivalent parameters. This model has a high calculation efficiency and a simple requirement for parameters. It has gained a far-reaching development in rock hydraulics (Zhang 2005; Zhou and Wang 2004). Currently, the study on this model mainly focuses on the single phase flow in fractured porous media. For two-phase or multiple phase flow and fractured vuggy reservoirs, there is no mature theory and method to calculate the corresponding equivalent parameters, such as the equivalent relative permeabilities and equivalent capillary curve (Wang et al. 2011; Zhou and Wang 2004). Actually, the theoretical basis of equivalent medium model is up-scale theory, and its mathematical essence is to reduce the differential orders of the flow equations on fine scale. Thus, the up-scale model is macroscopic scale, and we can smooth the effect of heterogeneity and multi-scale characteristics of fractures and vugs.

Recently, Huang et al. developed an oversampling technique to describe the macroscopic heterogeneity of fractured vuggy medium and the connectivity of fractures between coarse scale grid blocks. The results of single flow is much more accurate than the conventional equivalent medium model (Yan et al. 2013). But there is still huge error in two phases flow and multiple phases flow problems. The main reason is that such equivalent medium model is largely simplified on macroscopic scale, and it cannot capture the fine-scale characteristic of fractures and vugs.

(3) Discrete medium model
After a long-term geologic process of carbonate formation, it will generate a discrete surface with different type, scale and mechanical property including joint, fracture, and fault. Meanwhile, due to the karst and washout in different periods, it will generate discrete vug system. Thus, all the rock system growing with fractures and vugs are discrete. They belong to discrete medium concept. If an accurate discrete fracture or fracture-vug network model can be obtained, we can describe the fluid flow more accurately in fractured vuggy medium. Because the corresponding REV (Representative Elementary Volume) does not exist in such fractured vuggy carbonate reservoir, the above two types of continua medium model

are not effective any more, and the discrete medium model will have an obvious advantage and should be applied.

The concept of discrete fracture is first proposed by Snow for rock hydraulic problem (Snow 1968). The discrete fracture model used in nowadays reservoir simulation is developed by Noorishad and Mehran (1982). They developed a finite element numerical scheme for 2-D solute diffusion-convection problem in discrete fractured porous medium by using the upstream method. During the process of calculation, matrix rock adopts 2-D surface elements, and the fractures are discretized as 1-D line element. These two different dimension elements are coupled by using superposition principle. Since the double-porosity model and finite difference method were popular on reservoir numerical simulation, this model has not been paid too much attention by the petroleum industry.

Until 1999, Kim and Deo applied the discrete model to simulate the two-phase flow in fractured reservoir (Kim and Deo 1999, 2000). In the last 15 years, the discrete fracture model has a considerable development in reservoir numerical simulation, and various numerical formulations have been proposed, including finite difference method, Galerkin finite element method, control-volume finite element method, finite volume method, mixed finite element method, and mimic finite difference method. Based on the concept and methods used in fractured porous medium, Yao et al. proposed discrete fracture-vug network (DFVN) model (Yao et al. 2010a, b). These model add vugs system into the classical discrete fracture model, which can be considered as an extension of the discrete fracture model.

Discrete medium model makes an explicit description of fractures or vugs in reservoirs. And it use flux equivalent principal to regard flow in fracture as seepage flow. This model has a high precision and possesses a good reality. Meanwhile this model can be used to solve the relevant equivalent parameters of double-porosity, triple-porosity, and equivalent medium models. In recent years, with the development of geological model building technique, we can establish the detailed multi-scale discrete fracture or fracture-vug geological model. However, the corresponding fluid flow model and numerical simulation method are still not mature.

1.3 Purpose and Scope

In the past decade, the authors have made a deep and systematic research on discrete medium model, and developed a complete system of fractured vuggy carbonate reservoir simulation based on discrete medium models. The objectives of this book are to discuss the discrete concept models and the corresponding numerical simulation of fracture vuggy carbonate reservoirs. The remaining chapters of the book are organized as follows.

Chapter 2 focuses on the classical discrete fracture model and the related numerical schemes. First, a brief introduction and review of its principal and development has been done. Then, different numerical schemes are presented to model the fluid flow in discrete fracture model, including finite element method, finite volume method, and mimic finite difference method. Due to the present challenge raised from the unstructured gridding of such complex discrete models, we have developed an efficient embedded discrete fracture model based on mimic finite difference method. The new model can use the current mature finite difference simulator. This model can be appropriate for the complicated shape and there is no need for complex unstructured gridding process. It is a non-matched grid model which can reduce the computational resources and improve the computational efficiency.

Chapter 3 describes the discrete fracture-vug network model and its corresponding numerical simulation. By adding vug system into the classical discrete fracture model, we propose the discrete fracture-vug network model. This model divide fractured vuggy media into three flow systems: (1) the matrix system (including microfractures and matrix porosity), (2) the macroscopic fractures system and (3) vugs system, among which matrix and fractures system belong to seepage region and vug system is the free-fluid flow region. First, a basic model for coupling two-phase free flow with porous flow is developed. It is valid on the representative elementary volume (REV) scale and accounts for mass and momentum transfer across the fluid-porous interface. The development is based on a two-step up-scaling approach, in which the volume averaging method is applied. The comparisons between analytical solutions and Beavers-Joseph experimental data indicate that the new fits are more in line with the experimental data than the previous studies. Then, the Galerkin finite element method has been used to model the fluid flow in the porous region based on the discrete fracture model. For the free-flow region, the upstream Petrov-Galerkin finite element method has been applied to discretize the average two-fluid model based on operator splitting method. And then an alternate solution scheme is used to couple such two regions.

In Chapter 4, an efficient numerical model has been developed for immiscible two-phase flow in fractured karst reservoirs based on the idea of equivalent continuum representation, which is suitable to the field-scale reservoir simulation. Based on the discrete fracture-vug model and homogenization theory, the equivalent absolute permeability tensors for each grid blocks are calculated. Then an analytical procedure to obtain a pseudo-relative permeability curves for a grid block-containing fractures and cavities has been successfully implemented. Next, a full-tensor simulator has been designed based on a hybrid numerical method (combining mixed finite element method and finite volume method). Some numerical examples have been used to validate the method. Summing up, an efficient fluid flow model and its modeling theory have been developed in this dissertation, which can be applied to the fractured vuggy carbonate reservoirs.

Chapter 5 focuses on the hybrid models for fractured porous media and the corresponding numerical simulation methods. The fractured reservoir with complex fracture networks has strong heterogeneity and various scale fractures. To this end, we develop a hybrid fracture model and discuss the corresponding numerical method and technique. At first, the criterion and classification of fractures are discussed. Based on this, different flow mathematical models are applied to describe the fluid flow in different scale fracture systems. And then, the finite element numerical schemes are used for the hybrid model.

In Chap. 6, we have presented our recent research results of the multi-scale simulation methods used in fractured vuggy reservoirs. The difficulty in analyzing multiphase fluid flow in real reservoirs is mainly caused by the strong heterogeneity of the reservoirs. The multiple scales in reservoirs may span several orders of magnitude. It takes a long time to calculate multi-scale problem by utilizing conventional numerical method. Multi-scale method incorporates the small-scale information into the base functions; therefore, multiple scale method has exclusive advantages when it is applied to reservoir numerical simulation. The multi-scale methods only need to carry out the coarse mesh on the macroscale. The multi-scale basis function, constructed by solving the partial differential equations on the coarse mesh, could capture the small-scale information. It aims at reducing the computational amount and capturing the small-scale characteristics. Besides, the efficiency can be further improved by applying parallel computation. In this chapter, we present some applications of multi-scale methods to fluid flows in carbonate reservoirs. We discuss multi-scale methods for transport equations and their coupling to flow equations which are solved using MsFEMs.

References

Barenblatt GI, Zheltov IP, Kochina IN (1960) Basic concepts in the theory of seepage of homogeneous liquids in fissured rocks [strata]. J Appl Math Mech 24:1286–1303

Chang X, Yao J, Dai W, Wang Z (2004) The study of well test interpretation method for a triple medium reservoir. J Hydrodyn 19:339–346

Coats KH (1989) Implicit compositional simulation of single-porosity and dual-porosity reservoirs. In: SPE symposium on reservoir simulation. Society of Petroleum Engineers, Houston, Texas

Kang Z (2010) Mathematic model for flow coupling of crevice-cave type carbonate reservoir. Pet Geol Oilf Dev Daqing 29:29–32

Kazemi H, Merrill LS Jr, Porterfield KL, Zeman PR (1976) Numerical simulation of water-oil flow in naturally fractured reservoirs. Soc Pet Eng J 16:317–326

Kim J, Deo MD (2000) Finite element, discrete-fracture model for multiphase flow in porous media. AIChE J 46:1120–1130

Kim JG, Deo MD (1999) Comparison of the performance of a discrete fracture multiphase model with those using conventional methods. In: SPE symposium on reservoir simulation, pp 359–371

Li Y (2013) The theory and method for development of carbonate fractured-vuggy reservoirs in Tahe oilfield. Acta Pet Sin 34:115–121. doi:10.7623/syxb201301013

Noorishad J, Mehran M (1982) An upstream finite element method for solution of transient transport equation in fractured porous media. Water Resour Res 18:588–596

Pruess K, Narasimhan NT (1985) A practical method for modeling fluid and heat flow in fractured porous media. Soc Pet Eng J 25:14–26

Saidi AM (1983) Simulation of naturally fractured reservoirs. In: SPE reservoir simulation symposium. Society of Petroleum Engineers

Snow DT (1968) Rock fracture spacings, openings, and porosities. J Soil Mech Found Div 94:73–92

Thomas LK, Dixon TN, Pierson RG (1983) Fractured reservoir simulation. Soc Pet Eng J 23:42–54

Ueda Y, Murata S, Watanabe Y, Funatsu K (1989) Investigation of the shape factor used in the dual-porosity reservoir simulator. In: SPE Asia-Pacific conference. Society of Petroleum Engineers, Sydney, Australia

Wang Y, Yao J, Huang Z (2011) Review on fluid flow models through fractured rock. J Daqing Pet Inst 35:42–48

Warren JE, Root PJ (1963) The behavior of naturally fractured reservoirs. Soc Pet Eng J 3:245–255

Wu Y-S, Pruess K (1988) A multiple-porosity method for simulation of naturally fractured petroleum reservoirs. SPE Reserv Eng 3:327–336

Wu Y-S, Di Y, Kang Z, Fakcharoenphol P (2011) A multiple-continuum model for simulating single-phase and multiphase flow in naturally fractured vuggy reservoirs. J Pet Sci Eng 78:13–22

Xu B (2013) 2013 annual report of oil and gas industry in the world. Beijing

Yan X, Huang Z, Sun H, Yao J (2013) An novel calculation method of equivalent permeability of fractured reservoir based on oversampling technique. In: The 12th national conference of fluid flow through porous media, China, pp 90–93

Yao J, Zisheng W (2007) Theory and method for well test interpretation in fractured-vuggy carbonate reservoirs. China University of Petroleum Press, Shandong Dongying

Yao J, Dai W, Wang Z (2004) Well test interpretation method for triple medium reservoir with variable wellbore storage. J Univ Pet China 28:46–51

Yao J, Huang Z, Wang Z, Lv X (2010a) Mathematical model of fluid flow in fractured vuggy reservoirs based on discrete fracture-vug network model. Acta Pet Sin 31:815–819

Yao J, Huang Z-Q, Li Y-J, Wang C-C, Lv X-R (2010b) Discrete fracture-vug network model for modeling fluid flow in fractured vuggy porous media. In: International oil and gas conference and exhibition in China

Zhang Y (2005) Rock hydraulics and engineering. China WaterPower Press, Beijing

Zhou Z, Wang J (2004) Dynamics of fluids in fractured media. China WaterPower Press, Beijing

Chapter 2
Discrete Fracture Model

Abstract This chapter introduces the concept of discrete fracture model. It started by reviewing the background and the state of the art of discrete fracture model. It then explains three numerical methods to solve discrete fracture model including the Galerkin finite element method, the control volume method, and the mimetic finite difference method. In this chapter, detailed process of the establishment of mathematical model and the corresponding solution for the three numerical methods are explained. Then these numerical methods are applied to some examples. By the end of the chapter, the embedded discrete fracture model is introduced. A full discussion of the establishment and solving for the embedded discrete fracture mathematical model is included.

Keywords Discrete fracture model · Numerical simulation · Galerkin finite element method · Control volume method · Mimetic finite difference method · Embedded discrete fracture model

2.1 Background and the State of the Art

Fracture, the smallest geological structure (Van Golf-Racht 1982), is any break or fracture occurring in the rock caused by the deformation or the physical diagenesis in the rock. All rocks in the earth's crust are fractured to some extent. In groundwater dynamics, the rock masses with well-developed fractures are known as fractured porous media. The problems of two-phase flow in fractured porous media widely exist in different engineering practices, such as the oil and gas field development, the prevention and control of groundwater pollution, and the disposal of underground nuclear wastes (Slough et al. 1999; Yuan et al. 2004).

In fracture porous media, two types of media may be distinguished: fractured media of single porosity and fractured media of double porosity. Both media are composed of a network of fractures surrounding rock blocks, but what differentiates the two types of reservoirs are the porosity and permeability of the rock blocks. In the first case, the rock blocks are practically impervious while in the second case the

J. Yao and Z.-Q. Huang, *Fractured Vuggy Carbonate Reservoir Simulation*,
Springer Geophysics, DOI 10.1007/978-3-662-55032-8_2

porosity and permeability are quite significant. Fractures distribute very randomly and display multiscale features (Zhang 2005; Zhou and Wang 2004; Rutqvist et al. 2002) (as shown in Fig. 2.1). Therefore, it is difficult to establish an accurate and effective mathematical flow model and the corresponding numerical simulation method, which is the current research focus of petroleum industry and rock hydraulics (Feng et al. 2009; Huang et al. 2010; Reichenberger et al. 2006; Yao et al. 2010; Zhang and Wu 2010; Zhou 2007).

Since the 1980s, discrete fracture model (DFM) has been a great deal of development. The major characteristic of DFM is explicit expression and reducing dimension. For DEM, fractures are viewed as entity and the relationship between fractures and matrix can be established without interporosity flow function. This kind of model keeps their computational accuracy while the data volume and the cost of computation are reduced. Meanwhile, the model considers the permeability of matrix, namely fluid flow in matrix as well as in fractures. For this reason, DFM can not only accurately describe the flow character within fractures, but also describe the inhomogeneous character and seepage character of fracture media.

Originally, Noorishad and Mehran (1982) put forward DFM to solve single-phase flow problem in 2-D porous media. In this model, fractures are viewed as 1-D entity and finite element method is used to solve transient transport equation.

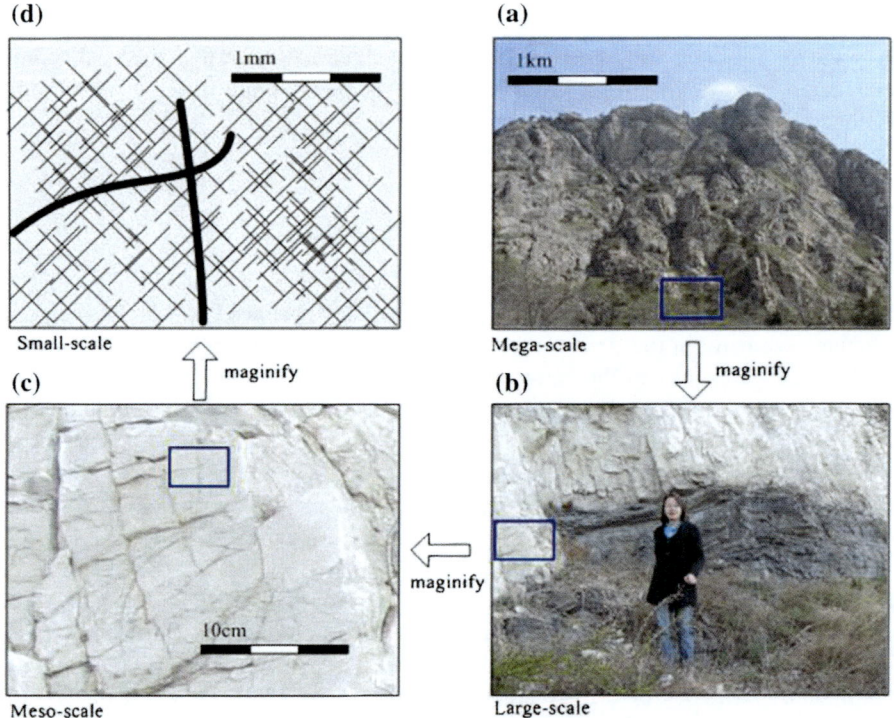

Fig. 2.1 Fractured porous media of different scales

While Kim and Deo (2000) adopt finite element to do discretization to discrete fracture model and combine matrix and fractures according to superposition principle. For nonlinear partial differential equations, pressure and saturation fully implicit scheme and Newton's method can solve it.

In 2003, Karimi-Fard and Firoozabadi (2003) adopt discrete fracture model to solve the two-phase flow problem of fractured media. As shown in Fig. 2.2, the model used line element to do discretization for fractures and used irregular mesh element such as triangle to do discretization for matrix. Furthermore, Galerkin finite element method is used to process numerical simulation based on implicit pressure–explicit saturation equation. The method greatly simplifies the problem so it can be applied to any complex structure in fractured media. There is an excellent match between the result of this method and traditional numerical simulation method that is based on single-porosity model. On this basis, Jun Yao et al. did further research, and the validity of model and algorithm has been verified by computation examples (Yao et al. 2010). By analyzing the impacts of fractures on the water flooding development effect, the discrete fracture model is regarded as the method which has good applicability for reservoir of low development degree of fractures, especially when the reservoir has several large fractures that control the direction of flow.

Lange et al. (2004) put forward a new discretization method of discrete fracture model. Based on the concept of dual media, the model does discretization to complex fractures according to minimum principia of calculation amount based on the geologic model and confirms fracture pressure at real fractures. Fractures are discretized by the model in each horizontal plane of formation, which means confirm compute nodes at every intersection and the end points of fractures and combine rock blocks with every fracture elements by rapid processing algorithms according to minimum principia of distance from fracture mesh, as shown in Fig. 2.3.

Above scholars mostly use finite element method when solving the model while finite element method cannot ensure locally mass conserving, so some scholars apply finite volume method which is based on physical conservation to discrete fracture model. P. Bastian et al. performed two-phase flow numerical simulations of fractured media to discrete fracture model by finite volume method and developed corresponding simulator (Bastian et al. 2000). In 2004, S. Geiger et al. applied

Fig. 2.2 Schematic of mesh generation in discrete fracture model

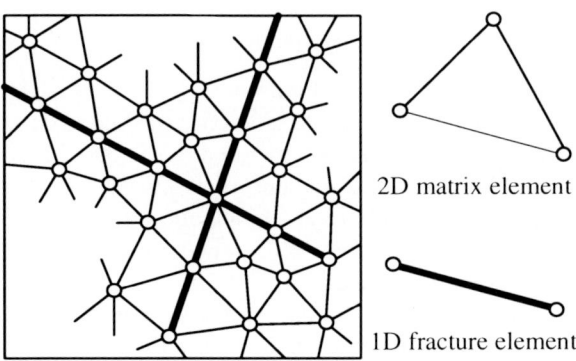

2D matrix element

1D fracture element

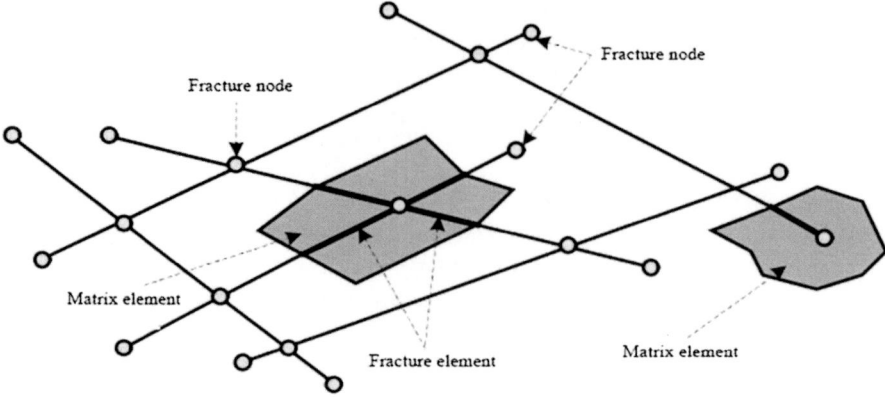

Fig. 2.3 Schematic of discrete fracture model

control volume method to solve flow potential equations while solving saturation equations by finite volume method (Geiger et al. 2004).

In recent years, discrete fracture model is getting more and more attention interiorly and becoming a hot research topic with the development of unconventional resources such as shale gas, tight reservoirs, and fractured reservoir. Further research has been done by Huang et al. (2011) in two-phase flow simulation of fractured reservoirs based on discrete fracture model. Combining the appropriate unstructured mesh generation technique, discrete fracture model can keep the arbitrary of development and distribution for fractures very well, describes the heterogeneity, anisotropy, and discontinuity of fractured media, and depicts the unique flow characteristic in fractures. Lv et al. (2012) did some research of discrete fracture mesh flow simulation based on control volume method. The high efficiency of calculation and the validity of flow simulation theory and algorithm for discrete fracture model based on finite volume method have been verified by examples (Lv 2010; Lv et al. 2012).

In recent 15 years, discrete fracture model has had a great development in fractured reservoir numerical simulation and several numerical discrete forms have sprung up, which include finite difference method, Galerkin finite element method, control volume method, finite volume method, mixed finite element method, mimetic finite difference method, etc.

(1) Finite difference method
Slough et al. applied finite difference method to do some research about multiphase flow problem for discrete fractured media based on discrete fracture model (Slough et al. 1999). Discrete fracture model is discretized into regular structured mesh to adapt to finite difference computation format in this study, while the discrete fracture always has complex geometry in practical problems. So this method has not been widely promoted.

After that, Lee et al. first proposed an embedded discrete fracture model to make full use of the existing mature finite difference reservoir numerical simulator and

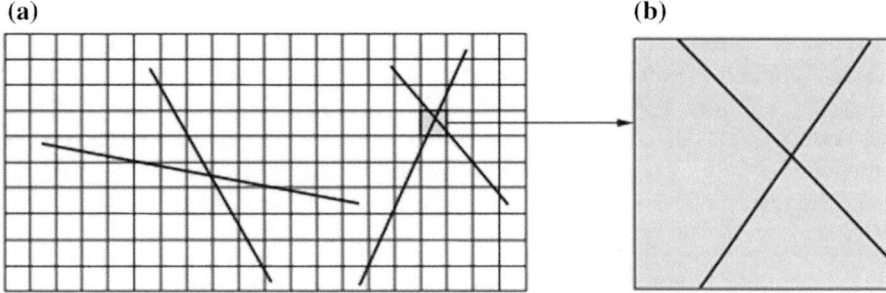

Fig. 2.4 Schematic of embedded discrete fracture model

adapt to the complex geometry of discrete fracture model (Lee et al. 2001). As shown in Fig. 2.4a, this model is typical nonmatching grid. In recent years, Li, Moinfar, Panfili, and Zhou et al. did further promotion and improvement to the model (Li and Lee 2008; Moinfar et al. 2012; Panfili et al. 2013; Zhou et al. 2014). Recently, Xia Yan et al. established a new embedded discrete fracture numerical computation format based on mimetic finite difference method to adapt to the condition of full tensor permeability (Yan et al. 2014).

(2) Galerkin finite element method
Based on the work of Kim and Deo (1999, 2000), Karimi-Fard and Firoozabadi used Galerkin finite element method to study water flooding numerical simulation in discrete fracture model (Karimi-Fard et al. 2003). Considering the influence of different wettability and comparing with the numerical result of single-porosity model (fractures are viewed as narrow high permeable zone), the validity of discrete fracture model can be verified. However, although Galerkin finite element method has whole conservation, the local conservation of elements cannot be guaranteed, especially on the singularities such as injection and production well. The oscillation of solutions is existent even when applied in the upstream format. Toward this, Zhang et al. put forward local conservation Galerkin finite element method (Zhang et al. 2013). In essence, the method meets the quantity of flow continuity condition at boundary of elements by the post-processing of element node to guarantee the local conservation of elements, which is similar to mixed finite element method. The method has not been given strict mathematic proof, so whether the method can extend to discrete fracture model remains to be studied.

(3) Control volume method
Based on control volume method, Monteagudo and Firoozabadi established a discrete fracture numerical computation format that has good local conservation to make up the short of Galerkin finite element method (Monteagudo and Firoozabadi 2007). Furthermore, they did some research about 3-D two-phase immiscible flow problem of fractured media. After that, Matthai et al. did further study about the method. A mixed mesh computation format was established to improve applicability of this method to the discrete fracture model (Matthäi and Belayneh 2004).

Reichenberger et al. developed a fully implicit numerical computation format based on control volume method (Reichenberger et al. 2006). Two sets of mesh are needed in control volume computation to solve the control volume of element node in every mesh: one is the initial mesh based on element nodes; another one is the auxiliary mesh system based on the central point of element. Consequently, the computational amount of this method will be increased compared with Galerkin finite element method.

(4) Finite volume method

Based on finite volume method, Granet et al. established a set of new discrete fracture numerical computation format and studied the 2-D incompressible two-phase flow (Granet et al. 1998). After that, Karimi-Fard et al. extended the method to 3-D multiphase flow problem based on GPRS reservoir numerical simulator of Stanford University (Karimi-Fard et al. 2003, 2004). Above computation format belongs to two-point flux approximation (TPFA). Accordingly, it cannot adapt to the condition of full tensor permeability. Sandve et al. deduced multipoint flux approximation (MPFA) of discrete fracture model to solve the problem (Sandve et al. 2012).

Finite volume method has good local conservation and little computation compared with finite element method and been widely used in reservoir numerical simulation. However, finite volume method is not as convenient as finite element method when it deal with cross fracture in discrete fracture model. To solve this problem, Karimi-Fard et al. put forward Delta–Star method to deal with cross fracture with the experience of resistance analysis method in cross circuit (Karimi-Fard et al. 2004). For single-phase flow, the method has high computation precision; for two-phase flow, Karimi-Fard et al. indicate that the computation error can meet the requirements only when the fracture densities are small, while the applicability and validity have not been verified for large-scale computation in reservoir.

(5) Mixed finite element method

In 1970s, Raviart and Thomas successfully applied the mixed finite method to reservoir numerical simulation and put forward the famous low-order RT_0 mixed finite element computation format (Raviart and Thomas 1977). Mixed finite element method is viewed as the finite volume method in finite element method for the good local conservation. Recently, Hoteit and Firroozabadi studied incompressible two-phase flow problem in discrete fracture model by combining mixed finite element method and discontinuous Galerkin finite element method (Hoteit and Firoozabadi 2006). They put forward an upstream weighted computation format that has high computational accuracy when they deal with cross fracture. For mixed finite element method, the point is the structure of the pressure and velocity basis function. For triangle, quadrangle, and regular hexahedron, the structure of the basis function has matured theory and method. The matured universal method for 3-D unstructured mesh such as tetrahedron and irregular polyhedron element has not been developed, which restricts the development and application of mixed finite element method in discrete fracture reservoir numerical simulation to some extent.

(6) Mimetic finite difference method
Huang et al. deduced a new discrete fracture numerical computation format based on mimetic finite difference method and studied incompressible two-phase flow problem (Huang et al. 2014). The mimetic finite difference method which is put forward by Breezi et al. (2005) has been widely used in research such as computational fluid dynamics (Lie et al. 2012; Lipnikov et al. 2014), electromagnetism, reservoir numerical simulation, etc., for the good local conservation and application to complicate mesh. The method is treated as the mixed finite element method in finite volume method, which means the computation format is similar to mixed finite element method and the difference is structure of element computation format. Mimetic finite difference method can structure computation format only based on single mesh element, so it can adapt to arbitrary complicate mesh system even the concave mesh. Because of reducing requirement to the mesh, mimetic finite difference method is more applicable than mixed finite element method for the flow simulation of complicate discrete fracture model.

2.2   Galerkin Finite Element Numerical Simulation

2.2.1   Discrete Fractured Model

Affected by the generation environment (stress, deposition, erosion, effloresce, etc.), fractures have complicated geometric configuration. It is necessary to simplify the fractures for convenience. Usually fractures are simplified into a parallel plate model inside which flow follows Navier–Stokes equation. For laminar flow conditions, velocity distribution along the fracture aperture can be obtained. Rewriting the quantity of flow in the form of equivalent Darcy's law gives the fractures' equivalent permeability. Evidently, the flow parameters and correlative physical quantities keep constant along the direction of the fracture in aperture, so reducing the dimension of the fracture in aperture direction is feasible. Fractures are simplified into 1-D line element for 2-D problem, and 2-D surface area element for 3-D problem (Fig. 2.5). Such simplification is the fundamental concept of the discrete fractured model.

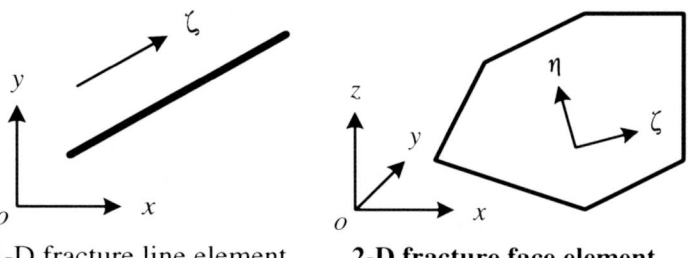

1-D fracture line element   **2-D fracture face element**

Fig. 2.5 Schematic of simplified fracture

The matrix system comprising microfissure and rock mass are regarded as equivalent porous continuum and the macroscopic fractures are represented manifestly as discrete fractures. As shown in Fig. 2.1, fractures occur at a variety of scales, from microscopic to field scale. Therefore, the division of microfissure and macroscopic fissure should comply with the specific research problem and the required precision of numerical simulation. Generally, a large fracture should be longer than a mesh of numerical simulation.

Therefore, the whole fracture porous media consists of matrix system and fracture system. The research region is $\Omega = \Omega_m + \sum a_i \times (\Omega_f)_i$, where m represents matrix, f represents fracture, and a is the aperture of the i-th fracture. Assuming the representative element volumes of both matrix and fracture system exist, the two-phase flow equations FEQ (Flow Equations) are applicable to the entire research area. Then for the discrete fractured model, the integral form of the flow equation can be expressed as

$$\int_\Omega \text{FEQ } d\Omega = \int_{\Omega_m} \text{FEQ } d\Omega_m + \sum_i a_i \times \int_{(\Omega_f)_i} \text{FEQ } d(\Omega_f)_i \qquad (2.1)$$

When neglecting the storage and seepage ability of matrix system, the research area only include fracture system and the above model degenerates into DFN (Discrete Fracture Network) model. When we consider the storage and seepage ability of the matrix system, the above model is the discrete fracture model. If the fractures are treated as microfissure, the above model only includes matrix system and changes into classical porous media flow model.

2.2.2 Two-Phase Flow Mathematical Model

For simplicity, we only consider isothermal flow of impressible fluid, which is similar to the analysis of other flow problem. Flow equations include mass conservation equation, generalized Darcy's law, saturation auxiliary equation, and capillary pressure relationship. Specific equations are as follows:

$$\phi \frac{\partial S_\alpha}{\partial t} + \nabla \cdot \mathbf{v}_\alpha = q_\alpha, \quad \alpha = \text{w, n} \qquad (2.2)$$

$$\mathbf{v}_\alpha = -\mathbf{K} \frac{k_{r\alpha}}{\mu_\alpha} (\nabla p_\alpha + \rho_\alpha g \nabla z), \quad \alpha = \text{w, n} \qquad (2.3)$$

$$S_w + S_n = 1 \qquad (2.4)$$

$$p_c(S_w) = p_n - p_w, \qquad (2.5)$$

where ϕ is porosity; S_l is saturation; v_i is seepage velocity, (m/s); D is Hamilton operator; q_l is source term, (1/s); w, n denote wetting phase and non-wetting phase, respectively; K is permeability tensor, (m^2); k_{rl} is relative permeability; μ_l is fluid viscosity, (Pa s); p_l is fluid pressure, (Pa); ρ_l is fluid density, (kg/m^3); g is acceleration of gravity; z denotes highness, positive on the upward side, (m); p_c is capillary pressure, (Pa). Herein, we define the flow potential Φ_l as follows:

$$\Phi_\alpha = p_\alpha + \rho_\alpha gz \tag{2.6}$$

and the corresponding capillary force potential Φ_c as

$$\Phi_c = \Phi_n - \Phi_w = p_c + (\rho_n - \rho_w)gz \tag{2.7}$$

Based on the above definitions, flow Eqs. (2.2), (2.3), and (2.5) can be written as

$$\phi\frac{\partial S_w}{\partial t} + \nabla \cdot (-K\lambda_w\nabla\Phi_w) = q_w \tag{2.8}$$

$$\phi\frac{\partial S_n}{\partial t} + \nabla \cdot (-K\lambda_n\nabla\Phi_n) = q_n \tag{2.9}$$

$$\Phi_c = \Phi_n - \Phi_w, \tag{2.10}$$

where

$$\lambda_w = \frac{k_{rw}}{\mu_w}, \quad \lambda_n = \frac{k_{rn}}{\mu_n} \tag{2.11}$$

denote the mobility coefficient of wetting phase and non-wetting phase, respectively.

Substituting Eqs. (2.4) and (2.10) into Eqs. (2.8) and (2.9) leads to the flow potential equation and phase saturation equation of wetting phase, written in the form of matrix:

$$\begin{bmatrix} 0 & 0 \\ 0 & \phi \end{bmatrix}\frac{\partial}{\partial t}\begin{bmatrix} \Phi_w \\ S_w \end{bmatrix} + \nabla \cdot \left\{ -\begin{bmatrix} K(\lambda_w + \lambda_n) & K\lambda_n p_c' \\ K\lambda_w & 0 \end{bmatrix}\nabla\begin{bmatrix} \Phi_w \\ S_w \end{bmatrix} \right\} = \begin{bmatrix} q_n + q_w \\ q_w \end{bmatrix}, \tag{2.12}$$

where

$$p_c'\nabla S_w = \nabla\Phi_c = \frac{d\Phi_c}{dS_w}\nabla S_w = \frac{dp_c}{dS_w}\nabla S_w \tag{2.13}$$

The initial condition and boundary conditions are stated below:
(1) Initial conditions

$$\Phi_\alpha(x, 0) = \Phi_\alpha(x), S_\alpha(x, 0) = S_\alpha(x), \quad \text{at } t = 0 \tag{2.14}$$

(2) Dirichlet boundary conditions

$$\Phi_\alpha(x, t) = \Phi_\alpha, S_\alpha(x, t) = S_\alpha, \quad \text{on } \Gamma_D \tag{2.15}$$

(3) Neumann boundary conditions (the outer boundary is impermeable), i.e.,

$$\begin{cases} v_\alpha \cdot n = (-K\lambda_\alpha \nabla \Phi_\alpha) \cdot n = 0 \\ \nabla S_\alpha \cdot n = 0 \end{cases}, \quad \text{on } \Gamma_N, \tag{2.16}$$

where n is the outer normal unit vector of outer boundary. N points to the outer normal direction of the interface of fracture line and outer boundary for 2-D problem and of fracture surface and outer boundary for 3-D problem.
(4) Internal impermeable boundary conditions mainly refer to the impermeable internal boundaries such as fault and fracture filled with mud, etc.

$$v_l \cdot n = (-K\lambda_\alpha \nabla \Phi_\alpha) \cdot n = 0, \quad \text{on } \Gamma_F, \tag{2.17}$$

where n is normal unit vector of internal boundary.

By substituting Eq. (2.12) as two-phase flow equation FEQ of fractured porous media into Eq. (2.1) and with the above initial conditions and boundary conditions, the complete mathematical model of the discrete fractured model can be developed.

2.2.3   Finite Element Numerical Formula

The discrete fractured model usually has complex fracture network structure and fractures distribute randomly. So in the numerical calculations, unstructured meshes are often used to adapt to its complex geometrical configuration. Therefore, the finite element method is applied to solve the numerical problem. The Galerkin weighted residual method is used to deduce the finite element calculation formula of Eq. (2.12). For convenience, the flow potential equation and phase saturation equation of wetting phase in Eq. (2.12) are derived separately, and the corresponding weight functions are variations of flow potential and saturation, respectively. Specific equations are as follows:

(1) Flow potential equation.

$$\int_{\Omega} \nabla \cdot [-\boldsymbol{K}(\lambda_{\mathrm{w}} + \lambda_{\mathrm{n}}) \nabla \Phi_{\mathrm{w}}] \delta \Phi_{\mathrm{w}} \mathrm{d}\Omega + \int_{\Omega} \nabla \cdot \left(-\boldsymbol{K} \lambda_{\mathrm{n}} p_{\mathrm{c}}' \nabla S_{\mathrm{w}}\right) \delta \Phi_{\mathrm{w}} \mathrm{d}\Omega$$
$$= \int_{\Omega} (q_{\mathrm{n}} + q_{\mathrm{w}}) \delta \Phi_{\mathrm{w}} \mathrm{d}\Omega \tag{2.18}$$

(2) Saturation equation.

$$\int_{\Omega} \phi \frac{\partial S_{\mathrm{w}}}{\partial t} \delta S_{\mathrm{w}} \mathrm{d}\Omega + \int_{\Omega} \nabla \cdot (-\boldsymbol{K} \lambda_{\mathrm{w}} \nabla \Phi_{\mathrm{w}}) \delta S_{\mathrm{w}} \mathrm{d}\Omega = \int_{\Omega} q_{\mathrm{w}} \delta S_{\mathrm{w}} \mathrm{d}\Omega \tag{2.19}$$

After integration by parts, combining impermeable boundary conditions (2.16), we obtain

$$\int_{\Omega} [\boldsymbol{K}(\lambda_{\mathrm{w}} + \lambda_{\mathrm{n}}) \nabla \Phi_{\mathrm{w}}] \nabla(\delta \Phi_{\mathrm{w}}) \mathrm{d}\Omega + \int_{\Omega} \left(\boldsymbol{K} \lambda_{\mathrm{n}} p_{\mathrm{c}}' \nabla S_{\mathrm{w}}\right) \nabla(\delta \Phi_{\mathrm{w}}) \mathrm{d}\Omega$$
$$= \int_{\Omega} (q_{\mathrm{n}} + q_{\mathrm{w}}) \delta \Phi_{\mathrm{w}} \mathrm{d}\Omega \tag{2.20}$$

$$\int_{\Omega} \phi \frac{\partial S_{\mathrm{w}}}{\partial t} \delta S_{\mathrm{w}} \mathrm{d}\Omega + \int_{\Omega} \left(\boldsymbol{K} \lambda_{\mathrm{w}} \nabla \Phi_{\mathrm{w}}\right) \nabla(\delta S_{\mathrm{w}}) \mathrm{d}\Omega = \int_{\Omega} q_{\mathrm{w}} \delta S_{\mathrm{w}} \mathrm{d}\Omega \tag{2.21}$$

For 2-D problem, Delaunay triangular mesh is employed to subdivide the whole research region and 1-D line element is employed to represent fracture. For 3-D problem, Delaunay triangular mesh is used to subdivide the fracture surface; the entire research region is subdivided by relevant tetrahedron or hexahedron, as shown in Fig. 2.6.

In each element, finite element approximation of flow potential and saturation is

$$\Phi_{\mathrm{w}} \approx \sum_{i=1}^{m} N_i(\Phi_{\mathrm{w}})_i = \boldsymbol{N}(\boldsymbol{x}) \Phi_{\mathrm{w}}(t) \; S_{\mathrm{w}} \approx \sum_{i=1}^{m} N_i(S_{\mathrm{w}})_i = \boldsymbol{N}(\boldsymbol{x}) \boldsymbol{S}_{\mathrm{w}}(t), \tag{2.22}$$

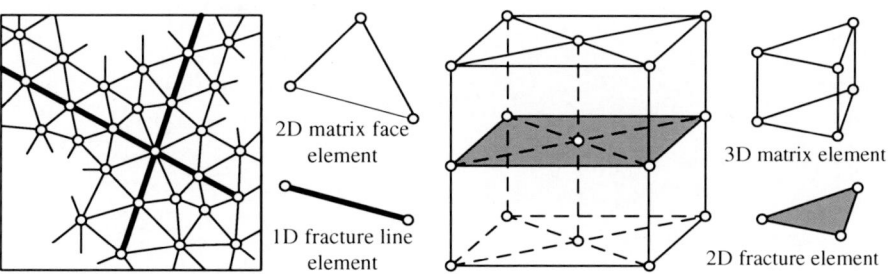

Fig. 2.6 Mesh schematics of discrete fractured model. **a** 2-D problem; **b** 3-D problem

where m is the number of element nodes; $N = [N_1, \ldots, N_m]$ is shape function; $\boldsymbol{\Phi}_{\mathrm{w}} = \left[(\Phi_{\mathrm{w}})_1, \ldots, (\Phi_{\mathrm{w}})_m\right]^{\mathrm{T}}$ is flow potential value of wetting phase at element nodes; $\boldsymbol{S}_{\mathrm{w}} = \left[(S_{\mathrm{w}})_1, \ldots, (S_{\mathrm{w}})_m\right]^{\mathrm{T}}$ is saturation value of wetting phase at element nodes.

Substituting Eq. (2.22) into Eqs. (2.20) and (2.21) and considering the arbitrariness of variation results in the following equation:

$$\begin{bmatrix} 0 & 0 \\ 0 & \boldsymbol{M}_{\mathrm{S}} \end{bmatrix} \begin{bmatrix} \dot{\boldsymbol{\Phi}}_{\mathrm{w}} \\ \dot{\boldsymbol{S}}_{\mathrm{w}} \end{bmatrix} + \begin{bmatrix} \boldsymbol{B}_{\Phi 1} & \boldsymbol{B}_{\Phi 2} \\ \boldsymbol{B}_{\mathrm{S1}} & \boldsymbol{B}_{\mathrm{S2}} \end{bmatrix} \begin{bmatrix} \boldsymbol{\Phi}_{\mathrm{w}} \\ \boldsymbol{S}_{\mathrm{w}} \end{bmatrix} = \begin{bmatrix} \boldsymbol{Q}_{\Phi} \\ \boldsymbol{Q}_{\mathrm{S}} \end{bmatrix}, \tag{2.23}$$

where

$$\boldsymbol{B}_{\Phi 1} = \sum_e \boldsymbol{B}_{\Phi 1}^e = \sum_e \int_{\Omega^e} \nabla^{\mathrm{T}} N \left[K(\lambda_{\mathrm{w}} + \lambda_{\mathrm{n}})\right] \nabla N \, \mathrm{d}\Omega^e;$$
$$\boldsymbol{B}_{\Phi 2} = \sum_e \boldsymbol{B}_{\Phi 2}^e = \sum_e \int_{\Omega^e} \nabla^{\mathrm{T}} N \left(K \lambda_{\mathrm{n}} p_{\mathrm{c}}'\right) \nabla N \mathrm{d}\Omega^e;$$
$$\boldsymbol{Q}_{\Phi} = \sum_e \boldsymbol{Q}_{\Phi}^e = \sum_e \int_{\Omega^e} \nabla^{\mathrm{T}} N (q_{\mathrm{n}} + q_{\mathrm{w}}) \mathrm{d}\Omega^e; \quad \boldsymbol{M}_{\mathrm{S}} = \sum_e \boldsymbol{M}_{\mathrm{S}}^e \int_{\Omega^e} N^{\mathrm{T}} \phi N \mathrm{d}\Omega^e;$$
$$\boldsymbol{B}_{\mathrm{S1}} = \sum_e \boldsymbol{B}_{\mathrm{S1}}^e = \sum_e \int_{\Omega^e} \nabla^{\mathrm{T}} N (K \lambda_{\mathrm{w}}) \nabla N \mathrm{d}\Omega^e; \quad \boldsymbol{B}_{\mathrm{S2}} = 0; \quad \boldsymbol{Q}_{\mathrm{S}} = \sum_e \boldsymbol{Q}_{\mathrm{S}}^e = \sum_e \int_{\Omega^e} \nabla^{\mathrm{T}} N q_{\mathrm{w}} \mathrm{d}\Omega^e,$$

where e denotes the elements set.

The interface of fractures and the matrix system needs special numerical calculation, for the reducing dimension of fractures in the discrete fractured model. As shown in Fig. 2.6, the nodes of fracture element and the matrix element are coincident at the interface. Assuming the water is wetting phase, the pressure of water phase at interface of fracture and the matrix is continuous. Therefore, the flow potential of water is continuous. After calculating the fracture element and the matrix element, respectively, and using superposition principle, the completely matrix equation can be developed, as shown in Fig. 2.7.

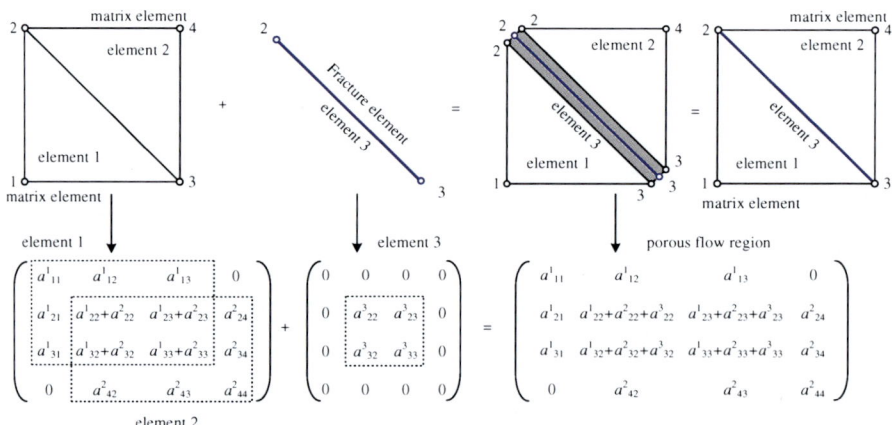

Fig. 2.7 Schematic of processing fracture and matrix element

Fig. 2.8 Capillary pressure curve for fracture and matrix

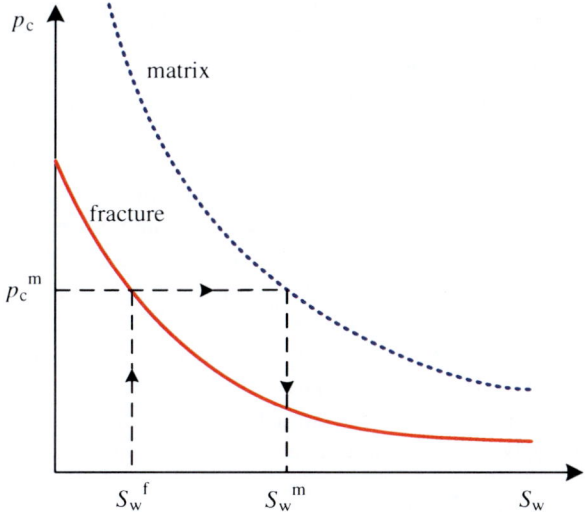

Generally, the capillary pressure curves of fractures and the matrix system are different. The saturation at interface of fractures and the matrix system is not always continuous, as shown in Fig. 2.8. Herein, the water phase saturation equation of fractures needs special treatment in above superposing process. For the flow potential of water is continuous, the capillary force potential and the capillary pressure are both continuous. Combining Fig. 2.8, the saturation of fractures and the matrix system meet the equation as follows:

$$\lambda^{up} = \begin{cases} \lambda_i & \text{if} \quad \Phi_i \geq \Phi_j \\ \lambda_j & \text{if} \quad \Phi_i < \Phi_j \end{cases} \tag{2.24}$$

For fractures, Eq. (2.12) can be written as

$$\begin{bmatrix} 0 & 0 \\ 0 & 0 \end{bmatrix} \frac{\partial}{\partial t} \begin{bmatrix} \Phi_w^f \\ S_w^f \end{bmatrix} + \nabla \cdot \left\{ -\begin{bmatrix} \mathbf{K}^f \left(\lambda_w^f + \lambda_n^f \right) & \mathbf{K}^f \lambda_n^f \left(p_c^f \right)' \\ \mathbf{K}^f \lambda_w^f & 0 \end{bmatrix} \nabla \begin{bmatrix} \Phi_w^f \\ S_w^f \end{bmatrix} \right\} = \begin{bmatrix} q_w^f + q_n^f \\ q_w^f \end{bmatrix} \tag{2.25}$$

Substituting the water phase flow potential continuous conditions and Eq. (2.24) into Eq. (2.25), we obtain

$$\begin{bmatrix} 0 & 0 \\ 0 & \phi^f \frac{ds_w^f}{ds_w^m} \end{bmatrix} \frac{\partial}{\partial t} \begin{bmatrix} \Phi_w^m \\ S_w^m \end{bmatrix} + \nabla \cdot \left\{ -\begin{bmatrix} \mathbf{K}^f \left(\lambda_w^f + \lambda_n^f \right) & \mathbf{K}^f \lambda_n^f \left(p_c^f \right)' \frac{ds_w^f}{ds_w^m} \\ \mathbf{K}^f \lambda_w^f & 0 \end{bmatrix} \nabla \begin{bmatrix} \Phi_w^m \\ S_w^m \end{bmatrix} \right\}$$
$$= \begin{bmatrix} q_w^f + q_n^f \\ q_w^f \end{bmatrix} \tag{2.26}$$

As shown in Eq. (2.26), only when $\frac{\mathrm{d}s_w^f}{\mathrm{d}s_w^m} = 1$, $S_w^f = S_w^m$ can be derived, the saturation continuous, i.e.,

obtaining the matrix equations by combining equations of fractures and the matrix system according to Eq. (2.1) and Fig. 2.7. For each node, the corresponding network element at the interface of fractures is not always in the same fracture line or fracture surface. Therefore, the global coordinates change into local coordinates when we get the specific property matrix. After superposing the above specific property matrix, the algebra equations can be derived. For the time term, the backward difference will be used to solve the problem. Then, the flow potential and distribution of matrix saturation of water phase can be developed. The saturation value of water phase in fractures can be deduced by Eq. (2.24). The mass matrix Ms of water phase saturation equation is usually off-diagonal consistent mass matrix. The row-sum lumping method is used to get lumped mass matrix (Reddy 1993), and specific equation in e element as follows:

$$[\hat{\boldsymbol{M}}]_{ii}^e = \sum_{j=1}^{m} \int_{\Omega} N_i^e \phi N_j^e \mathrm{d}\Omega, \quad [\hat{\boldsymbol{M}}]_{ij}^e = 0 \qquad (2.27)$$

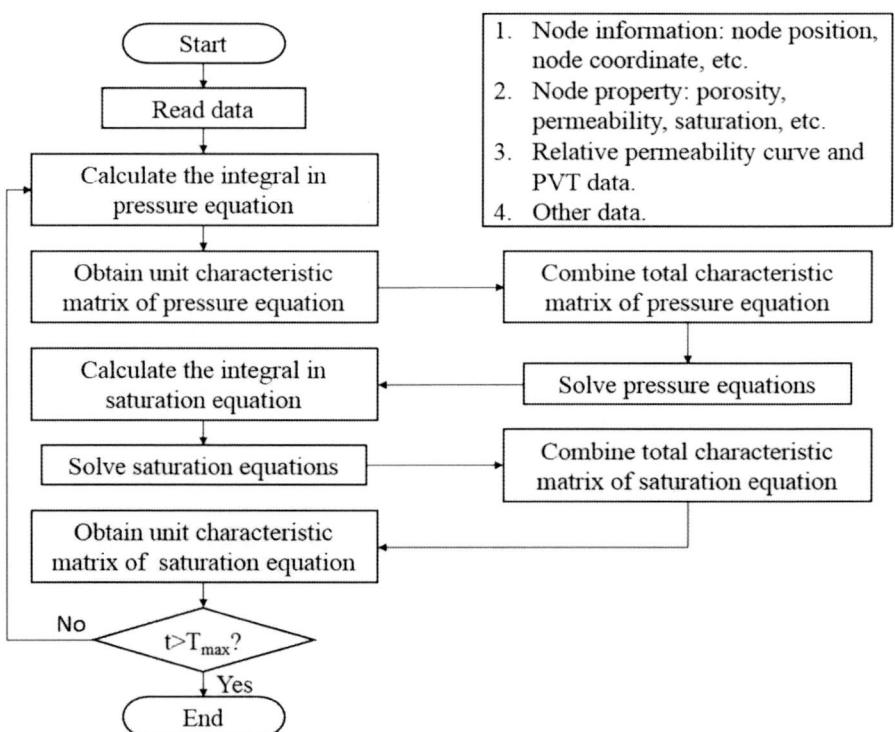

Fig. 2.9 Schematic of discrete fracture finite element numerical calculation process

Based on above numerical calculation formula, the MATLAB programming language is used to program corresponding finite element numerical calculation program of two-phase discrete fracture. The specific process is shown in Fig. 2.9.

It is possible that numerical oscillation occurs if convection dominated, when the standard Galerkin finite method is used to solve two-phase flow problem. Therefore, the upwind Galerkin calculation formula is used to solve the equations, specific equation as follows:

$$\lambda^{up} = \begin{cases} \lambda_i, & \Phi_i \geq \Phi_j \\ \lambda_j, & \Phi_i < \Phi_j \end{cases}, \tag{2.28}$$

where the fluidity coefficient is defined at every nodes of elements. The upwind Galerkin calculation formula has well stability and convergence. Corresponding analysis refers to references (Dalen 1979; Helmig and Huber 1998).

2.2.4 Numerical Examples and Applications

(1) Single fracture model

First, consider a two-phase flow problem in a single fracture. For search convenience, the wetting phase is always water and the non-wetting phase is always oil in the following context. Assume that fracture is filled with oil initially, water is injected from the left end in a constant speed and the pressure in the right end keeps the initial pressure. Make further assumption as follows: the fracture is partially filled and the length is 100 m, porosity $\phi = 0.25$, aperture is 1 mm, absolute permeability $K = 1\,\mu m^2$, viscosity of water $\mu_w = 1\,mPa\,s$, viscosity of oil $\mu_o = 5\,mPa\,s$, both irreducible water saturation and residual oil saturation are zero, water phase relative permeability $k_{rw} = S_w^2$, oil phase relative permeability $k_m = (1 - S_w)^2$, initial pressure is 10 MPa, and injection rate $q = 6.0 \times 10^{-6}\,m/s$.

Uniform mesh is applied, and the number of nodes is 251; the elements are quadratic; neglect the gravity and capillary pressure; the fluid is incompressible. This example is a typical Buckley–Leverett problem and its analytical solution is expressed as

$$x = \frac{f'_w(S_w)}{\phi A} \int_0^t q_{in} dt, \tag{2.29}$$

where A is the cross-sectional area of fracture, (m^2); $f_w = \lambda_w/(\lambda_w + \lambda_o)$ is water cut, $f'_w(S_w) = df_w/dS_w$. Figure 2.10 shows that there is an excellent match between numerical solution and analytical solution.

Fig. 2.10 Comparison between numerical and analytical solution

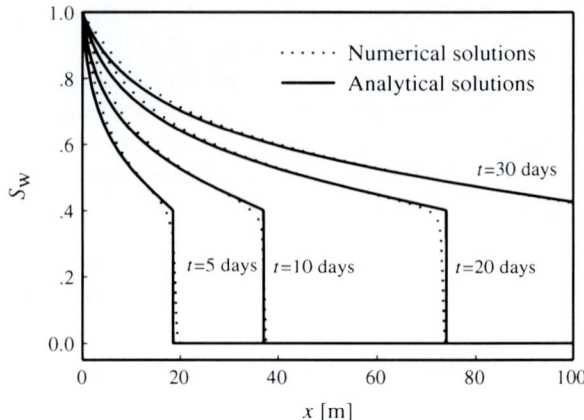

Fig. 2.11 Fractured reservoir with complex fractures

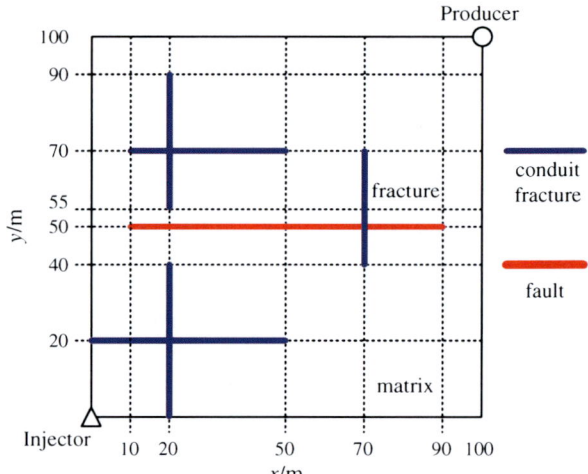

(2) Double porosity example

Consider the complex fracture reservoir model in Fig. 2.11, which includes diversion fracture and fault. The thickness of the reservoir is 10 m. Porosity of homogeneous isotropic matrix $\phi = 0.2$, permeability $K_m = 1000\,\mu m^2$; fracture aperture $a = 1\,mm$, permeability $K_f = a^2/12 = 8.33 \times 10^4\,\mu m^2$. Viscosity of water $\mu_w = 1\,mPa\,s$ viscosity of oil $\mu_o = 5\,mPa\,s$, irreducible water saturation is $S_{wr} = 0$, residual oil saturation is $S_{or} = 0.2$.

Water phase relative permeability $k_{rw} = S_w^2$, oil phase relative permeability $k_m = (1 - S_w)^2$, initial pressure is 10 MPa, both injection and production rates are

Fig. 2.12 Finite element
meshes

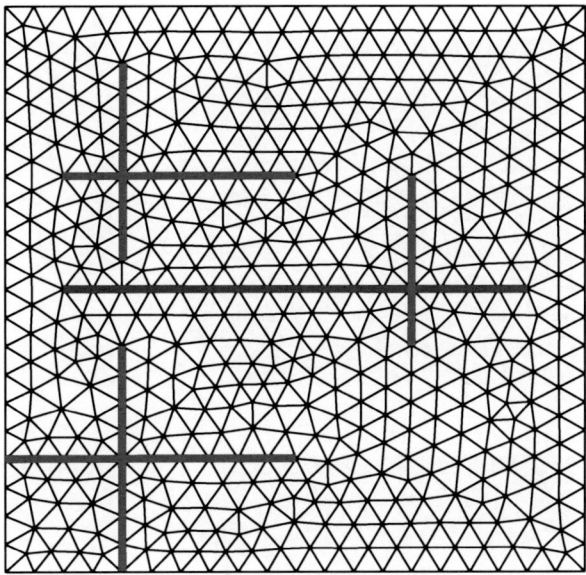

$q = 30\,\mathrm{m}^3/\mathrm{d}$. Assume that the reservoir model is water wetting, and the capillary pressure curve follows the Brooks–Corey capillary pressure function, and that both the matrix's capillary pressure and the fracture's pressure are considered.

$$p_c(S_w) = p_d\left(\frac{S_w - S_{wr}}{1 - S_{wr} - S_{or}}\right)^{-\frac{1}{\lambda}}, \ 0.2 \le \lambda \le 3.0 \qquad (2.30)$$

For matrix, threshold value $p_c = 10000\,\mathrm{Pa}$, λ equals to 2.0. For fractures, threshold value $p_d = 1000\,\mathrm{Pa}$, λ equals to 1.0.

As illustrated in Fig. 2.12, the finite element meshes consist of 532 nodes and 982 elements. Figure 2.13 is representing the water saturation distributions at different times. The results indicate that the injected water displaces oil down the matrix and then moves forward rapidly along fractures when the oil/water front encounters conduit fractures, as suggested by Fig. 2.13a. At the same time, fault acts as a flow barrier, which forces the fluid flow along the extension direction of the fault, as shown in Fig. 2.13b; conduit fractures connected with fault can rapidly introduce lower fluid into upper reservoir across the fault, as shown in Fig. 2.13c, d. Evidently, the existence of fractures results in strong heterogeneity and anisotropy, which have a great influence in the water flooding development. Owing to the presence of capillary pressure, the recovery of water flooding development is improved because of the expansion of sweep area; but the entire development effect is still controlled by fractures.

(a)   (b)

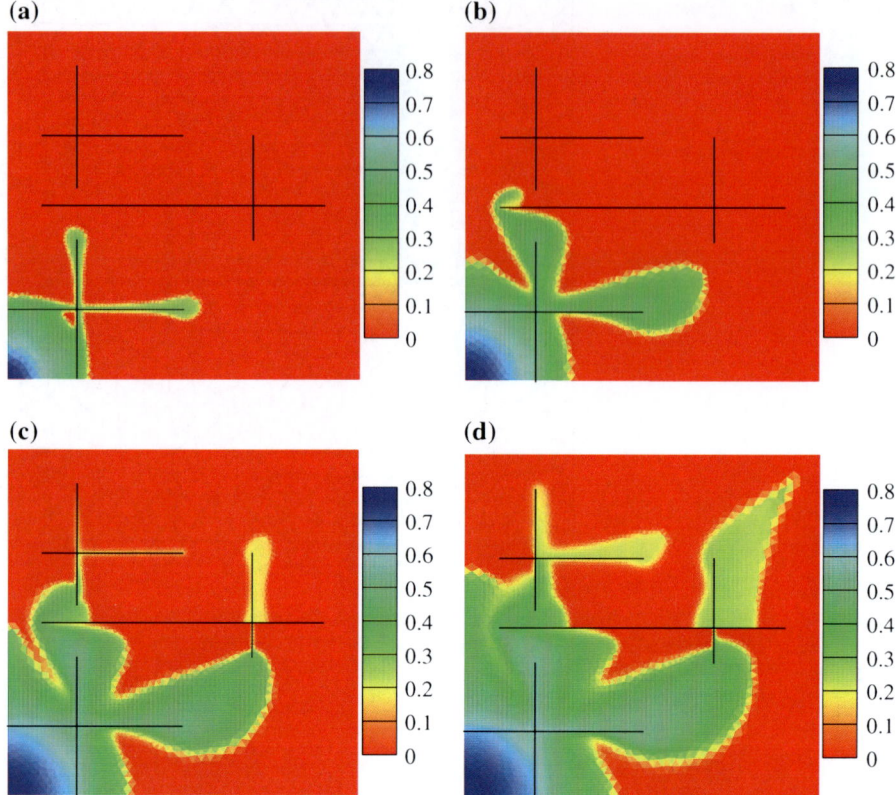

(c)   (d)

Fig. 2.13 Water saturation distributions at different times. **a** After 20 days; **b** after 50 days; **c** after 80 days; **d** after 120 days

2.3   Control Volume Method Numerical Simulation

First, two-phase flow control equations are developed. Then results in discrete fracture model according to the equivalence principle of quantity of flow for single fracture are presented, accompanied with the saturation relationship at the interface of inhomogeneous media. The fundamental assumptions of the reservoir model are as follows:

(1) The flow in this reservoir model is isothermal flow;
(2) Considering the existence of two-phase: water and oil, which cannot dissolve and react with each other, and their flow both follow the Darcy law;
(3) The fluid in matrix and fractures is slight compressible;
(4) Neglecting the compressibility of rock mass;
(5) Neglecting the effect of gravity, considering the effect of capillary pressure.

2.3.1 Two-Phase Flow Control Equations

The two-phase flow control equations comprise mass conservation equation, the Darcy law, state equation, saturation equation, capillary pressure relationship, etc. Considering the gravity and capillary pressure of fluid, the mathematic model which could describe the two-phase flow of slightly compressible fluid in reservoir can be established.

(1) Mass conservation equation.
Based on the principle of mass conservation, the continuity equation of oil phase and water phase can be established, respectively.
For oil phase:

$$-\nabla \cdot (\rho_o \cdot \mathbf{v}_o) + Q_o = \frac{\partial(\phi \rho_o S_o)}{\partial t} \tag{2.31}$$

For water phase:

$$-\nabla \cdot (\rho_w \cdot \mathbf{v}_w) + Q_w = \frac{\partial(\phi \rho_w S_w)}{\partial t}, \tag{2.32}$$

where o represents oil, w represents water; ρ_i is fluid density, kg/m^3; v_i is fluid velocity, m/s; ϕ is formation porosity; S_i fluid saturation; Q_i is source term which represents mass change in unit time and unit volume, and Q_i equals to positive value for injection well and Q_i equals to negative value for producing well, $kg/(m^3 s)$.

(2) Momentum equation.
When the fluid in reservoir follows the Darcy law, flow velocity can be expressed as follows:
For oil phase:

$$\mathbf{v}_o = -\frac{k_{ro}\mathbf{K}}{\mu_o}(\nabla p_o + \rho_o g \nabla z) \tag{2.33}$$

For water phase:

$$\mathbf{v}_w = -\frac{k_{rw}\mathbf{K}}{\mu_w}(\nabla p_w + \rho_w g \nabla z), \tag{2.34}$$

where μ_i is fluid viscosity, Pa s; p_i is fluid pressure; g is acceleration of gravity, m/s^2; z is highness from a reference plane, positive on the upward side, m; K_{rl} is relative permeability; $\mathbf{K}$ is permeability tensor, which changes into scalar K in isotropic matrix, m^2.

For 2-D problem, the permeability tensor K is defined as

$$K = \begin{bmatrix} K_{xx} & K_{xy} \\ K_{yx} & K_{yy} \end{bmatrix}$$

For 3-D problem, the permeability tensor K is defined as

$$K = \begin{bmatrix} K_{xx} & K_{xy} & K_{xz} \\ K_{yx} & K_{yy} & K_{yz} \\ K_{zx} & K_{zy} & K_{zz} \end{bmatrix}$$

If the feature vector of K coincides with the direction of coordinate axis, K degenerates into the tensor in diagonal form:

$$K = \begin{bmatrix} K_{xx} & & \\ & K_{yy} & \\ & & K_{zz} \end{bmatrix}$$

The direction of permeability tensor is always different from coordinate axis, especially for the complex reservoir. Therefore, full permeability tensor is needed for making up the deviation. But most of the simulator cannot simulate this kind of permeability now (Durlofsky 1993).

(3) State equation.
Considering the compressibility of oil and water, we obtain
    For oil phase:

$$C_o = \frac{1}{\rho_o} \frac{d\rho_o}{dp_o} \tag{2.35}$$

    For water phase:

$$C_w = \frac{1}{\rho_w} \frac{d\rho_w}{dp_w}, \tag{2.36}$$

where ρ_i is fluid density, kg/m^3; p_i is fluid pressure, Pa; C_l is elastic compression coefficient of fluid, Pa^{-1}.

(4) Auxiliary equation.
Saturation equation:

$$S_o + S_w = 1 \tag{2.37}$$

Capillary pressure equation:

$$p_o - p_w = p_c(S_w), \qquad (2.38)$$

where p_c is capillary pressure, Pa.

Substituting above momentum equations of oil phase and water phase into continuity equation, respectively, we obtain

For oil phase:

$$\nabla \cdot \left(\rho_o \cdot \frac{k_{ro}\boldsymbol{K}}{\mu_o}(\nabla p_o + \rho_o g \nabla z) \right) + Q_o = \frac{\partial(\phi\rho_o S_o)}{\partial t} \qquad (2.39)$$

For water phase:

$$\nabla \cdot \left(\rho_w \cdot \frac{k_{rw}\boldsymbol{K}}{\mu_w}(\nabla p_w + \rho_w g \nabla z) \right) + Q_w = \frac{\partial(\phi\rho_w S_w)}{\partial t} \qquad (2.40)$$

Equations (2.39) and (2.40) are simplified by compound function derivation law, we obtain

For oil phase:

$$\nabla \cdot \left(\rho_o \cdot \frac{k_{ro}\overline{\overline{\boldsymbol{k}}}}{\mu_o}(\nabla p_o + \rho_o g \nabla z) \right) + Q_o = \phi\left(\rho_o \frac{\partial S_o}{\partial t} + S_o \frac{\partial \rho_o}{\partial t} \right) \qquad (2.41)$$

For water phase:

$$\nabla \cdot \left(\rho_w \cdot \frac{k_{rw}\overline{\overline{\boldsymbol{k}}}}{\mu_w}(\nabla p_w + \rho_w g \nabla z) \right) + Q_w = \phi\left(\rho_w \frac{\partial S_w}{\partial t} + S_w \frac{\partial \rho_w}{\partial t} \right) \qquad (2.42)$$

Assuming condition as follows:

$$\frac{\partial \rho_i}{\partial t} = \frac{\partial \rho_i}{\partial p_i}\frac{\partial p_i}{\partial t} l = (o,w) \qquad (2.43)$$

Then Eqs. (2.41) and (2.42) can be expressed as

For oil phase:

$$\nabla \cdot \left(\rho_o \cdot \frac{k_{ro}\boldsymbol{K}}{\mu_o}(\nabla p_o + \rho_o g \nabla z) \right) + Q_o = \phi\rho_o\left(\frac{\partial S_o}{\partial t} + S_o \frac{1}{\rho_o}\frac{\partial \rho_o}{\partial p_o}\frac{\partial p_o}{\partial t} \right) \qquad (2.44)$$

For water phase:

$$\nabla \cdot \left(\rho_w \cdot \frac{k_{rw}\boldsymbol{K}}{\mu_w} (\nabla p_w + \rho_w g \nabla z) \right) + Q_w = \phi \rho_w \left(\frac{\partial S_w}{\partial t} + S_w \frac{1}{\rho_w} \frac{\partial \rho_w}{\partial p_w} \frac{\partial p_w}{\partial t} \right) \quad (2.45)$$

Substituting state equation into Eqs. (2.44) and (2.45), we obtain
For oil phase:

$$\nabla \cdot \left(\rho_o \cdot \frac{k_{ro}\boldsymbol{K}}{\mu_o} (\nabla p_o + \rho_o g \nabla z) \right) + Q_o = \phi \rho_o \frac{\partial S_o}{\partial t} + \phi \rho_o S_o C_o \frac{\partial p_o}{\partial t} \quad (2.46)$$

For water phase:

$$\nabla \cdot \left(\rho_w \cdot \frac{k_{rw}\boldsymbol{K}}{\mu_w} (\nabla p_w + \rho_w g \nabla z) \right) + Q_w = \phi \rho_w \frac{\partial S_w}{\partial t} + \phi \rho_w S_w C_w \frac{\partial p_w}{\partial t} \quad (2.47)$$

Divide Eq. (2.46) and Eq. (2.47) by fluid density ρ_i ($i = \{$ o, w $\}$), and standard control equation can be developed, which can describe immiscible displacement of two-phase slightly compressible fluid.
For oil phase:

$$\phi \frac{\partial S_o}{\partial t} + \phi S_o C_o \frac{\partial p_o}{\partial t} - \nabla \cdot \left(\frac{k_{ro}\boldsymbol{K}}{\mu_o} (\nabla p_o + \rho_o g \nabla z) \right) - q_o = 0 \quad (2.48)$$

For water phase:

$$\phi \frac{\partial S_w}{\partial t} + \phi S_w C_w \frac{\partial p_w}{\partial t} - \nabla \cdot \left(\frac{k_{rw}\boldsymbol{K}}{\mu_w} (\nabla p_w + \rho_w g \nabla z) \right) - q_w = 0, \quad (2.49)$$

where $q_i = Q_i / \rho_i$, which represents quantity of flow into or out in unit volume unit time, S^{-1}.
Assuming condition as follows:

$$\lambda_l = \frac{k_{rl}\boldsymbol{K}}{\mu_l} \quad (2.50)$$

The flow potential of l phase can be defined as

$$\Phi_l = p_l + \rho_l g z \quad (2.51)$$

Then the capillary pressure flow potential can be defined as

$$\Phi_c = \Phi_o - \Phi_w = P_c + (\rho_o - \rho_w) g z \quad (2.52)$$

Based on the above definition, permeability scalar K is employed to replace permeability tensor in homogeneous isotropic matrix. Neglecting gravity, the immiscible flow control equation of two-phase slightly compressible fluid can be expressed as

For oil phase:

$$\phi \frac{\partial S_o}{\partial t} + \phi S_o C_o \frac{\partial p_o}{\partial t} - \nabla \cdot (\lambda_o \nabla p_o) - q_o = 0 \tag{2.53}$$

For water phase:

$$\phi \frac{\partial S_w}{\partial t} + \phi S_w C_w \frac{\partial p_w}{\partial t} - \nabla \cdot (\lambda_w \nabla p_w) - q_w = 0 \tag{2.54}$$

Add control equations of oil phase and water phase up and keep the water phase control equation. Combine two auxiliary equations and define composite compressibility as $C_t = S_w C_w + S_o C_o$, $\frac{\partial p_c}{\partial t} \approx 0$. Then above mathematical model can be expressed as two partial differential equations:

$$-\phi C_t \frac{\partial p_w}{\partial t} + \nabla \cdot ((\lambda_o + \lambda_w)\nabla p_w) + \nabla \cdot (\lambda_o \nabla p_c) + (q_o + q_w) = 0 \tag{2.55}$$

$$\phi \frac{\partial S_w}{\partial t} + \phi S_w C_w \frac{\partial p_w}{\partial t} - \nabla \cdot (\lambda_w \nabla p_w) - q_w = 0 \tag{2.56}$$

Equation (2.55) is pressure equation, and Eq. (2.56) is saturation equation. The mathematical model initial conditions are

$$p_i(\boldsymbol{x}, 0) = p_i(\boldsymbol{x}), \ S_i(\boldsymbol{x}, 0) = S_i(\boldsymbol{x}), \ t = 0, \ l = \{w, o\} \tag{2.57}$$

The boundary conditions can be all forms of Dirichlet boundary condition, Neumann boundary condition, and mixed mode.

Dirichlet condition is

$$p_i(\boldsymbol{x}, t) = p_i, \ S_i(\boldsymbol{x}, t) = S_i, \ i = \{w, o\}, \quad \text{on } \Gamma_D \tag{2.58}$$

Neumann condition is (assume the boundary is impermeable)

$$v_i \cdot \overrightarrow{\boldsymbol{n}} = -(\lambda_i \nabla p_i) \cdot \overrightarrow{\boldsymbol{n}} = 0, \ \nabla S_i \cdot \overrightarrow{\boldsymbol{n}} = 0, \ i = \{w, o\}, \quad \text{on } \Gamma_N \tag{2.59}$$

Then the immiscible flow mathematical model of two-phase slightly compressible fluid is developed.

2.3.2 Discrete Fracture Mathematical Model

Based on the equivalence principle of quantity of flow for single fracture, discrete fracture model can be developed, as shown in Fig. 2.14. The model is built upon a parallel plate of single fracture and is fracture aperture. Assume that flow of fluid in fracture follows N-S equation. The flow in parallel plate is laminar flow when fluid velocity is small. The velocity distribution and quantity of flow along the fracture aperture can be obtained. According to Darcy's law, the quantity of flow gives the fractures' equivalent permeability and the equivalent flow velocity distribution. And the value is kept constant along the direction of fracture in aperture. Evidently, the instability of the flow parameters and correlative physical quantities along the direction of the fracture in aperture reduce the dimension of the fracture in aperture and develop the discrete fracture model.

Fractures are simplified into 1-D line element for 2-D problem, and 2-D surface area element for 3-D problem, as shown in Fig. 2.15.

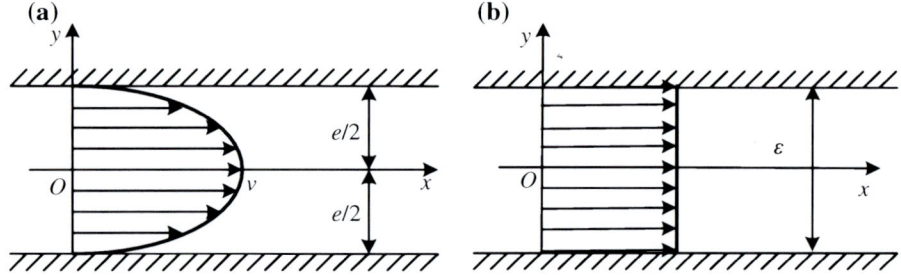

Fig. 2.14 Velocity distribution in single fracture. **a** Realistic velocity, **b** equivalent velocity

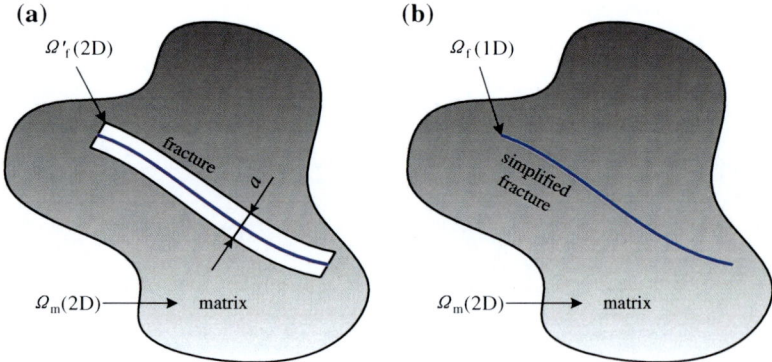

Fig. 2.15 Schematic of discrete fracture model. **a** Single-porosity model, **b** discrete fracture model

Considering the 2-D porous media region in single fracture, the whole area is Ω, and the matrix area is Ω_m, the fracture area is Ω_f in single fracture model and is $a\Omega_f'$ in discrete fracture model. So based on the discrete fracture model, the whole region of reservoir can be expressed as

$$\Omega = \Omega_m + a\Omega_f' \tag{2.60}$$

where a is fracture aperture. In above 2-D discrete fracture model, the 2-D control equation system of the matrix region is

$$-\phi^m C_t \frac{\partial p_w^m}{\partial t} + \nabla \cdot \left(\left(\lambda_o^m + \lambda_w^m\right)\nabla p_w^m\right) + \nabla \cdot \left(\lambda_o^m \nabla p_c^m\right) + \left(q_o^m + q_w^m\right) = 0 \tag{2.61}$$

$$\phi^m \frac{\partial S_w^m}{\partial t} + \phi^m S_w^m C_w \frac{\partial p_w^m}{\partial t} - \nabla \cdot \left(\lambda_w^m \left(\nabla p_w^m\right)\right) - q_w^m = 0 \tag{2.62}$$

The 1-D control equation system of the matrix region is

$$-\phi^f C_t \frac{\partial p_w^f}{\partial t} + \left(\left(\lambda_o^f + \lambda_w^f\right)\frac{\partial p_w^f}{\partial \xi}\right) + \frac{\partial}{\partial \xi}\left(\lambda_o^f \frac{\partial p_c^f}{\partial \xi}\right) + \left(q_w^f + q_o^f\right) = 0 \tag{2.63}$$

$$\phi^f \frac{\partial S_w^f}{\partial t} + \phi^f S_w^f C_w \frac{\partial p_w^f}{\partial t} - \frac{\partial}{\partial \xi}\left(\lambda_w^f \left(\frac{\partial p_w^f}{\partial \xi}\right)\right) - q_w^f = 0, \tag{2.64}$$

where ξ is coordinate system along the direction of fracture in aperture.

In single fracture model, if f is used to represent pressure equation and saturation equation system, the integral form of the whole equation can be written as

$$\int_\Omega f d\Omega = \int_{\Omega_m} f^m d\Omega_m + \int_{\Omega_f} f^f d\Omega_f = 0 \tag{2.65}$$

According to $\Omega = \Omega_m + \varepsilon\Omega_f'$, the integral form of pressure equation and saturation equation in the discrete fracture model can be expressed as

$$\int_\Omega f d\Omega = \int_{\Omega_m} f^m d\Omega_m + \varepsilon \int_{\Omega_f} f^f d\Omega_f' = 0 \tag{2.66}$$

That way, the two-phase discrete fracture model flow equation is developed. In theory, the discrete fracture model can be applied to any fractured porous media that has complex form. Compared to single fracture model, the integration in fractures of the discrete fracture model can simplify the problem to a great extent. And fracture aperture will appear in front of 1-D integral form as a coefficient in order to keep the integral form.

2.3.3   Saturation Discontinuity Treatment at the Interface

After discretized space and time, and performing linearization of the nonlinear term in Eq. (2.66), we can get the discrete equation of system:

$$\int_{\Omega} f d\Omega = A^{m} x^{m} - b^{m} + A^{f} x^{f} - b^{f} = 0, \tag{2.67}$$

where

$$x = [p_{w}, S_{w}]^{T}$$

Neglecting the compressibility of fluid, Karimi-Fard, Firoozabadi, Kim, Deo (Karimi-Fard and Firoozabadi 2003; Kim and Deo 2000) have changed Eq. (2.67) into Eq. (2.68).

$$(A^{m} + A^{f}) x - b^{m} - b^{f} = 0 \tag{2.68}$$

There is an implicit relationship $x^{m} = x^{f} = x$ in Eq. (2.68) and it is only applied to particular circumstances. Hence, the relationship between the matrix system and fracture as well as the corresponding equation in the interface of the matrix and fracture needs to be developed based on a real physical meaning.

There is no change of fluid mass in the interface Γ_{mf} of the matrix and fracture, so the quantity of flow and direction of normal velocity are continuous, i.e.,

$$q_{i}^{f*} = q_{i}^{m*}, \ v_{i}^{m} \cdot \mathbf{n}_{mf} = v_{i}^{m} \cdot \mathbf{n}_{mf}, i = \{w, o\}, \quad \text{on } \Gamma_{mf}, \tag{2.69}$$

where $\mathbf{n}_{mf}$ is the normal vector at the matrix and fracture.

Superposition principle is applied to integrate flow equations in discrete fracture mode and these terms will be eliminated when flow equations are added up. So the quantity of flow at interface of the matrix and fracture can be neglected in flow equations.

The coordinate of arbitrary fixed point z at interface of the matrix and fracture is same and $\Phi_{i}^{m} = \Phi_{i}^{f}$ can be known from Eq. (2.51). The capillary pressure potential is same too as shown in Eq. (2.52), i.e.,

$$\Phi_{c}^{m}(S_{w}^{m}) = \Phi_{c}^{f}(S_{w}^{f}) \tag{2.70}$$

It is equivalent to the continuity of capillary pressure. Figure 2.16 is capillary pressure at the interface of two different phases. Because the capillary pressure at the interface of the matrix and fracture is same, $p_{c}^{f} = p_{c}^{m} = p_{c}^{*}$, and the water saturation deciding capillary pressure is discontinuous at the interface. The physical relationship of S_{w}^{f} and S_{w}^{m} at the interface can be developed using the continuity condition of capillary pressure.

Fig. 2.16 Capillary pressure at the interface between matrix and fracture

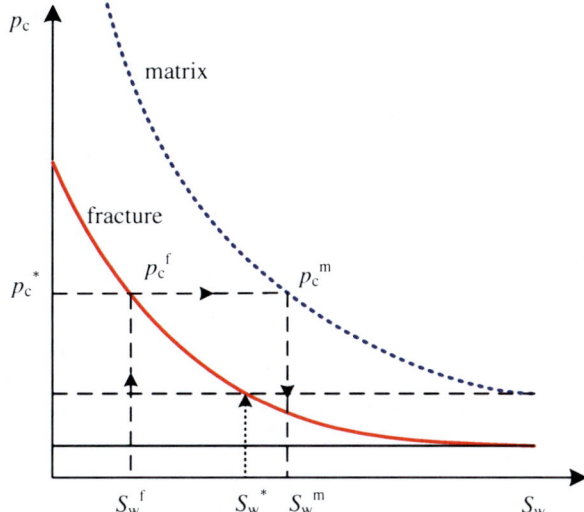

$$S_w^m = \begin{cases} 1, & S_w^f > S_w^* \\ [p_c^m]^{-1} p_c^f (S_w^f), & S_w^f \leq S_w^* \end{cases} \qquad (2.71)$$

Making use of Eq. (2.71) and compound function derivation law, the saturation equation of fractures can be applied to the water saturation S_w^m of the matrix and be expressed as follows:

$$\phi^f \frac{dS_w^f}{dS_w^m} \frac{\partial S_w^m}{\partial t} + \phi^f S_w^f C_w \frac{\partial p_w^f}{\partial t} - \frac{\partial}{\partial \xi}\left(\lambda_w^f \left(\frac{\partial p_w^f}{\partial \xi}\right)\right) - q_w^f = 0 \qquad (2.72)$$

According to the assumption $x^m = x^f = x$ in Eq. (2.68), it can be applied only when the capillary pressure function of the matrix and fracture is same, namely when $dS_w^f / dS_w^m = 1$. Thus the corresponding dS_w^f / dS_w^m should be calculated for different capillary pressure functions of the matrix and fracture. The fluid exchange term can be neglected, because it will be eliminated when added up in control volume element.

2.3.4 Control Volume Numerical Formulation

The control volume method was originally used in computational fluid dynamics. And it is essentially a finite volume computation format based on Delaunay mesh dual element. Hence, the Delaunay dual mesh is needed first when developing the computation format based on control volume method. Then by integrating the pressure equation and the saturation equation at each control volume element, the numerical computation format can be established.

(1) Discrete fracture model control volume mesh generation
Unstructured mesh is applied to accomplish geometrical discrete of discrete fracture model for the distribution of fractures in fractured reservoir is random. For 2-D problem, first, Delaunay triangle mesh is generated. Triangle element will be employed to discrete the matrix and 1-D line element represents the fracture area, as shown in Fig. 2.17a. So-called control volume is a polygonal area. Each area controlled by a node is connected by the center of gravity of adjacent triangle element and midpoint of side that links to the node. And volume element will be divided into three areas by lines connected by the center of gravity and midpoint of side.

As shown in Fig. 2.17b, the adjacent nodes of node a are $\{b_1, b_2, \ldots, b_6\}$, triangles which take node a as the vertex are $\{T_1, T_2, \ldots, T_6\}$, the center of gravity of triangles are $\{G_1, G_2, \ldots, G_6\}$, midpoints of sides which take node a as vertex are $\{M_{ab_1}, M_{ab_2}, \ldots, M_{ab_6}\}$. The control volume element of node a is polygon $G_1M_{ab_1}G_2M_{ab_2}G_3M_{ab_3}G_4M_{ab_4}G_5M_{ab_5}G_6M_{ab_6}$, which can be obtained by connecting the center of gravity of triangles with corresponding midpoint of sides. Control volume element of the other nodes in research region can be obtained in the same way, where ab_1 represents fractures. In standard control volume element, Delaunay triangle is local homogeneous while control volume element might be nonhomogeneous. The major characteristics of polygon control volume are covering the entire calculation region without overlap and keeping calculation accuracy by the cross-distribution of triangle elements and control volume.

The matrix of fractured porous media is homogeneous. Considering saturation variable (S_w, S_o) is constant in each control volume element while flow pressure variable (p_w, p_o, p_c) can be estimated with linear approximation by the value of Delaunay mesh element (triangle or tetrahedron) which comprise control volume element:

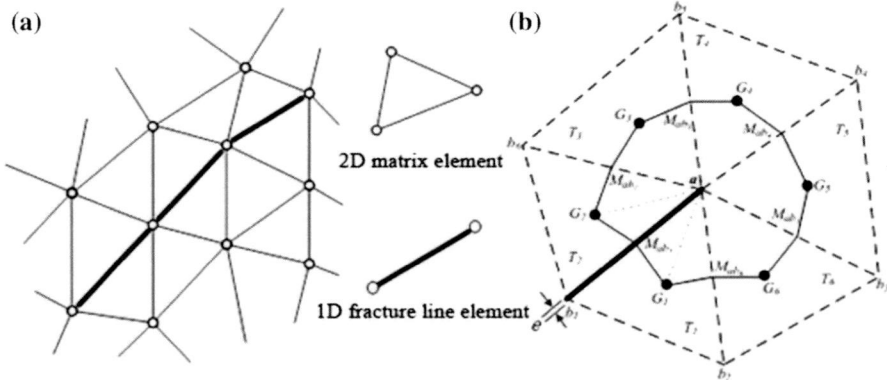

Fig. 2.17 Delaunay triangle mesh and the dual control volume mesh of 2-D discrete fracture model

$$\Psi(\boldsymbol{x}) = \sum_{i=1}^{m} S_i(\boldsymbol{x})\Psi_i, \tag{2.73}$$

where x is coordinate in the dimension of corresponding control volume element; m is the number of vertex; Ψ_i is arbitrary variable of node i at coordinate $\boldsymbol{x}_i$; S_i is shape factor and defined as follows:

For triangle:

$$S_i(\boldsymbol{x}) = \frac{\alpha_i + \beta_i x + \gamma_i y}{2A} \tag{2.74}$$

For tetrahedron:

$$S_i(\boldsymbol{x}) = \frac{\alpha_i + \beta_i x + \gamma_i y + \delta_i z}{6V}, \tag{2.75}$$

where A is area of triangle element, m^2; V is volume of tetrahedron element, m^3; (α_i, β_i, γ_i, δ_i) are constants about geometric coordinates of element nodes.

Assume that constants are used to coding triangle element nodes and anti-clockwise direction is positive direction. Thus α_i, β_i, γ_i in Eq. (2.74) can be expressed as follows, respectively (Wang 2003):

$$
\begin{aligned}
\alpha_i &= \begin{vmatrix} x_j & y_j \\ x_k & y_k \end{vmatrix} = x_j y_k - x_k y_j \\
\beta_i &= -\begin{vmatrix} 1 & x_j \\ 1 & y_k \end{vmatrix} = y_j - y_k \\
\gamma_i &= \begin{vmatrix} 1 & x_j \\ 1 & x_k \end{vmatrix} = -x_j + x_k
\end{aligned}
\tag{2.76}
$$

From Eq. (2.73), the gradient of arbitrary variable in a triangle is

$$\nabla\Psi = \sum_{i=1}^{m} \Psi_i \nabla S_i(\boldsymbol{x}) \tag{2.77}$$

For 3-D tetrahedron, it is similar to the above 2-D triangle.

(2) The establishment of control volume computation format

To establish numerical computation format of mathematical model by control volume method, integral should be done first for pressure Eq. (2.55) and saturation Eq. (2.56) in every control volume element, respectively. In 2-D discrete fracture model, the control volume discrete computation format method has to establish saturation equation which is:

Applying integral to Eq. (2.56) in arbitrary control volume element CV_i, we can obtain

$$\iint_{\Omega} \left(\phi \frac{\partial S_w}{\partial t} + \phi S_w C_w \frac{\partial p_w}{\partial t} \right) dA - \iint_{\Omega} \nabla \cdot (\lambda_w \nabla p_w) dA - \iint_{\Omega} q_w dA = 0$$

$$(2.78)$$

Assuming that porosity only changes at space and transforms surface integral into line integral for the second term of the left side of Eq. (2.78) with Gauss divergence theorem, we obtain

$$\iint_{\Omega} \left(\phi \frac{\partial S_w}{\partial t} + \phi S_w C_w \frac{\partial p_w}{\partial t} \right) dA - \int_{\Gamma} (\lambda_w \nabla p_w) \cdot \mathbf{n} d\Gamma - \iint_{\Omega} q_w dA = 0, \quad (2.79)$$

where Γ is boundary of control volume element CV_i; n is unit outward normal vector on boundary Γ.

The water saturation of the matrix and fractures can be connected with each other by Eq. (2.71), the first term of the left side of Eq. (2.79) can be approximately expressed as

$$\iint_{\Omega} \left(\phi \frac{\partial S_w}{\partial t} + \phi S_w C_w \frac{\partial p_w}{\partial t} \right) dA \approx A_{\phi i} \left(\frac{\partial S_w^m}{\partial t} + S_w C_w \frac{\partial p_w}{\partial t} \right), \quad (2.80)$$

where

$$A_{\phi i} = \sum_{k=1}^{t} \varphi_k A_k \varphi_k^m + \sum_{l=1}^{s} \frac{dS_w^f}{dS_w^m} e_l |L_l| \varphi_l^f, \quad (2.81)$$

where $A_{\phi i}$ is pore volume of CV_i; t is the number of Delaunay triangle element which takes node i as vertex; φ_k is triangle k's area radio to Delaunay triangle k in control volume element CV_f; A_k is the area of Delaunay triangle k; ϕ_k^m is porosity of the matrix system in triangle k; s is the number of fracture in CV_f; ϕ_l^f, e_l, and $|L_l|$ are porosity, aperture, and length of fracture l in control volume element CV_i, respectively.

The first term of the right side of Eq. (2.81) represents pore volume of the matrix in CV_i; the second term represents pore volume of fractures in CV_i. To express the whole equation with the matrix water saturation, the second term will be multiplied by $\frac{dS_w^f}{dS_w^m}$.

The second integral of the left side of Eq. (2.79) can be expressed as

$$\int_{\Gamma} (\lambda_w \nabla \Phi_w) \cdot \mathbf{n} d\Gamma \approx \sum_{k=1}^{t} |s_k| \left[\lambda_w^m \left(S_w^{m,up} \right) \nabla p_w \right]_k \cdot \mathbf{n}_k + \sum_{l=1}^{s} e_l \lambda_w^f \left(S_w^{f,up} \right) \frac{\partial p_w^f}{\partial \xi}, \quad (2.82)$$

where $|s_k|$ is internal boundary of CV_i in triangle k and has unit outward normal vector; ∇p_w is water phase flowing pressure gradient at $|s_k|$, which can be estimated by Eq. (2.47); ξ is local coordinate along the direction of fracture; $\partial p_w^f / \partial \xi$ is flowing potential gradient on fracture l.

The first term of the right side of Eq. (2.82) represents quantity of flow through control volume element CV_i's boundary; the second term represents quantity of flow through every fracture in CV_i. The value of saturation comply with upstream standard, where superscript up represents upstream value. For the flow in fractures can be viewed as 1-D, $\partial p_w^f / \partial \xi$ can be estimated by the following equation:

$$\frac{\mathrm{d}p_w^f}{\mathrm{d}\xi} = \frac{p_j - p_i}{2|L_l|}, \tag{2.83}$$

where p_i, p_j represent pressure of 1-D adjacent element.

The third term of the left side of Eq. (2.79) can be approximately expressed as

$$\iint_\Omega q_w \mathrm{d}A \approx q_{wi} A_i, \tag{2.84}$$

where A_i is the area of 2-D control volume element CV_i and can be calculated by the following equation:

$$q_{wi} A_i = q_{wi}^m \sum_{k=1}^{t} \varphi_k A_k + \sum_{l=1}^{s} e_l |L_l| q_{w,l}^f \tag{2.85}$$

Based on above approximation, the numerical computation format of saturation equation in every control volume element can be written as

$$A_{\varphi i}\left(\frac{\partial S_w^m}{\partial t} + S_w C_w \frac{\partial p_w}{\partial t}\right) - \left[\sum_{k=1}^{t} |s_k| \left[\lambda_w^m\left(S_w^{m,up}\right)\nabla p_w\right]_k \cdot \mathbf{n}_k + \sum_{l=1}^{s} e_l \lambda_w^f\left(S_w^{f,up}\right)\frac{\partial p_w^f}{\partial \xi}\right]$$
$$- q_{wi} A_i = 0$$

$$\tag{2.86}$$

The assumption that flow pressure of corresponding mesh at interface of the matrix and fractures is similar has been done, above step could be applied to pressure equation too. For 2-D matrix system and 1-D fracture system, the numerical computation format of flowing pressure equation can be written as

$$A_{\varphi i} C_t \frac{\partial p_w}{\partial t} - \left[\sum_{k=1}^{t} |s_k| \left[\lambda^m \nabla p_w + \lambda_o^m \nabla p_c\right]_k \cdot \mathbf{n}_k + \sum_{l=1}^{s} \left[\lambda^f \frac{\partial p_w}{\partial \xi} + \lambda_o^f \frac{\partial p_c}{\partial \xi}\right] e_l\right]$$
$$- (q_{wi} + q_{oi}) A_i = 0$$

$$\tag{2.87}$$

where $\lambda = \lambda_w + \lambda_o$ is the total fluidity. Fluidity in equation should comply with upstream standard and capillary pressure gradient in fractures can be estimated with following equations: Eqs. (2.86) and (2.87) are the numerical computation format of pressure equation and saturation equation that are based on control volume method in 2-D research region. The method can be easily extended to 3-D discrete fracture model.

(3) Inhomogeneous matrix discrete fracture model numerical formulation
Control volume computation format of discrete fracture model for homogeneous matrix has been established. There have been some researchers who studied the inhomogeneous problem with the method of control volume. While they mostly had focus on the inhomogeneity of absolutely permeability and the anisotropy of unidirectional flow (Edwards 2002). For permeability, as it changes rapidly at the interface of inhomogeneous media, there will be imprecise velocity field when we use the standard control volume method at the interface (Durlofsky 1994). As shown in Fig. 2.18a, Delaunay triangle is locally homogeneous and control volume element polygon is inhomogeneous in standard control volume method. Some researchers put forward locally homogeneous control volume element to get precise velocity field, as shown in Fig. 2.18b. The fact that Delaunay triangle is inhomogeneous can be known from Fig. 2.18b.

To pressure equation and saturation equation, control volume method is used to process spatial discretization. First, do the integral to equations in a control volume element, which is the dual mesh connected by the center of gravity and midpoint of 2-D Delaunay triangle or 3-D tetrahedron. Figure 2.19a is fractured porous media control volume element schematic for homogeneous matrix, and the two-phase numerical computation format of slightly compressible fluid in discrete fracture model has been established. Figure 2.19b is fractured porous media control volume element schematic for inhomogeneous matrix, and the two-phase numerical computation format of slightly compressible fluid will be established below.

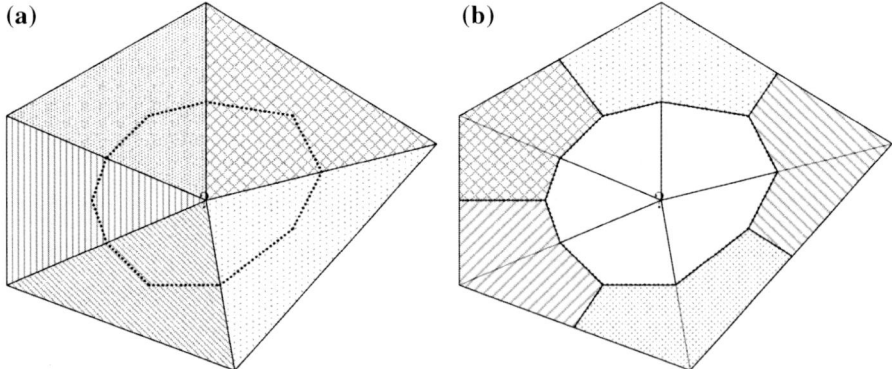

(a)   **(b)**

Fig. 2.18 Inhomogeneous media control volume element. **a** Standard control volume element, **b** local homogeneous control volume element

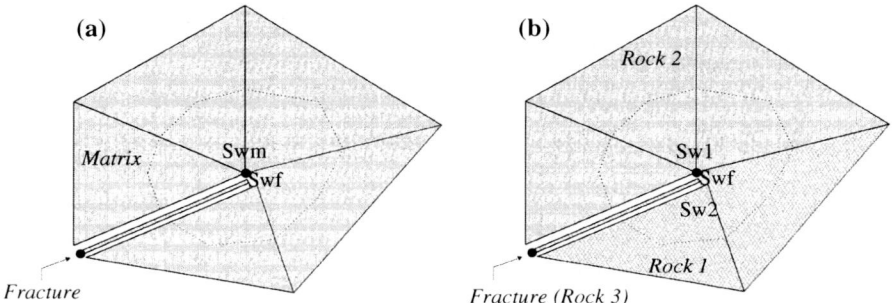

Fig. 2.19 Control volume element with fractures. **a** Homogeneous matrix, **b** Inhomogeneous matrix

Flowing pressure equation and saturation equation both are composed of three terms:

(1) Time derivative term, it is $\phi C_t \frac{\partial p_w}{\partial t}$ in flowing pressure equation and is $\phi \frac{\partial S_w}{\partial t}$ and $\phi S_w C_w \frac{\partial p_w}{\partial t}$ in saturation equation;

(2) Source term q_i, where $i = (o, w)$;

(3) Divergence term $\nabla \cdot F$, flow vector $F = F(S_w)$, which is equal to $(\lambda_o + \lambda_w) \nabla p_w$ and $\lambda_o \nabla p_c$ in flow pressure equation and is equal to $\lambda_w \nabla p_w$ in saturation equation.

As shown in Fig. 2.19b, there exists one fracture and two kinds of matrix rock mass in one control volume element. As same as the interface of matrix and fractures in above section, capillary pressure between different matrix is continuous. Hence, saturations of inhomogeneous matrix rock mass can be connected:

$$\int_A \left(\phi \frac{\partial S_w}{\partial t} + \phi S_w C_w \frac{\partial p_w}{\partial t} \right) dA = \sum_{k=1}^{m} \left(\frac{\partial}{\partial t} S_w^k \phi^k A^k + S_w^k C_w^k \phi^k A^k \frac{\partial p_w}{\partial t} \right), \quad (2.88)$$

where m represents the number of media in control volume element.

Equation (2.88) can be expressed on the basis of reference media S_w^+ which comply with the parameter B in capillary pressure model Eq. (2.89).

$$P_c^k = -B^k \ln S_w^k, \quad (2.89)$$

where superscript $k = 1 \cdots n_m$ is the index of media in control volume element.

The relationship between different media's saturations can be established based on the concept of capillary pressure continuation. For example, as far as fracture media:

$$S_w^m / S_w^f = \exp\left(-B^m / B^f\right)$$

The maximum of B^k in Eq. (2.89) is chosen as reference media in here. Based on Eq. (2.88) and compound function derivation method, we obtain

$$\int_A \left(\phi \frac{\partial S_w}{\partial t} + \phi S_w C_w \frac{\partial p_w}{\partial t} \right) dA = \left(\sum_{k=1}^{m} \frac{dS_w^k}{dS_w^+} \phi^k A^k \right) \frac{\partial}{\partial t} S_w^+ + \left(\sum_{k=1}^{m} S_w^k C_w^k \phi^k A^k \right) \frac{\partial p_w}{\partial t}$$

(2.90)

The integral of source term can be written as

$$\int_A q_w dA = \sum_{k=1}^{m} A^k q_w^k$$

(2.91)

The integral of divergence term can be confirmed in the light of equation below:

$$\int_A \nabla \cdot \mathbf{F} \, dA = \int_{\Gamma_A} \mathbf{F} \cdot \mathbf{n} \, d\Gamma_A \approx \sum_{j=1}^{n_b} [\mathbf{F} \cdot \mathbf{n}]_j,$$

(2.92)

where n_b is the number of boundary element.

Every triangle included in control volume element is local homogeneous. Saturation between different media can be calculated by Eq. (2.71). Combining above three equations, we obtain the numerical computation format of pressure equation and saturation equation.

$$\left(\sum_{k=1}^{m} \frac{dS_w^k}{dS_w^+} \phi^k A^k \right) \frac{\partial}{\partial t} S_w^+ + \left(\sum_{k=1}^{m} S_w^k C_w^k \phi^k A^k \right) \frac{\partial p_w}{\partial t} - \sum_{j=1}^{n_b} [\mathbf{F} \cdot \mathbf{n}]_j - \sum_{k=1}^{m} A^k q_w^k = 0$$

(2.93)

With this, based on control volume method, the discrete fracture model numerical computation format that considers inhomogeneous matrix has been established.

2.3.5   Numerical Examples

As shown in Fig. 2.20, there is a simple 1/4 five-point water injection scheme, the size of porous media model is 1 m × 1 m, initial pressure is $p_i = 10$ MPa, porosity of homogeneous isotropic matrix is $\phi = 0.2$, permeability is $K_m = 1 \times 10^{-3}$ μm². Considering the existence of fractures in porous media with azimuthal angle of $\theta = 0°$, $\theta = 45°$, $\theta = 90°$, and $\theta = 135°$, and the center of fracture and porous media are overlapping, the length of fractures is $L = 60 \sqrt{2}$ cm, fracture aperture is

Fig. 2.20 Schematic of 2-D
reservoir model

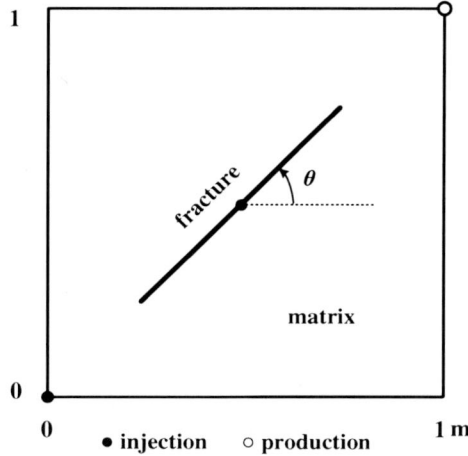

 injection   ○ production

$a = 1$ mm, permeability is $K_f = a^2/12 = 8.33 \times 10^4 \, \mu m^2$. There is a water injection well in the lower left corner and a production well in the top right corner. The injection rate is $q_{in} = 0.01$ PV/day (PV is the acronym of Pore Volume which represents multiple of pore volume) and production rate is $q_{out} = 0.01$ PV/day. Viscosity of water phase is $\mu_w = 1$ mPa s, viscosity of oil phase is $\mu_o = 5$ mPa s, density of water phase is $\rho_w = 1 \, kg/m^3$, density of oil phase is $\rho_o = 0.8 \, kg/m^3$, compressibility of oil phase is $C_o = 10 \times 10^{-4}$ MPa^{-1}, compressibility of water phase is $C_w = 5 \times 10^{-4}$ MPa^{-1}, irreducible water saturation is $S_{wc} = 0$, residual oil saturation is $S_{or} = 0$, normalized saturation is $S_e = (S_w - S_{wc})/(1 - S_{wc} - S_{or})$, relative permeability of water phase for matrix and fractures is $k_{rw} = S_e$, relative permeability of oil phase for matrix and fractures is $k_{ro} = 1 - S_e$, initial water saturation is 0. Ignore the impact of capillary pressure and gravity.

As shown in Fig. 2.21, porous media models which have one fracture with angle of $\theta = 0°$, $\theta = 45°$, $\theta = 90°$, or $\theta = 135°$ (in degrees from horizontal) are described with Delaunay triangle mesh based on three models. There are 874, 878, 875, and 878 control volume elements and 1646, 1654, 1648, and 1654 triangle mesh elements after mesh generation, respectively. For single-porosity model, the fracture aperture that is 1 mm and the scale of research region can differ by three magnitudes, so it is essential to do mesh refinement for real fractures. In single-porosity model I, the number of control volume element node is 9466 and the number of elements is 18,854 for horizontal fracture model whose azimuthal angle is $\theta = 0°$; the number of control volume element node is 9343 and the number of triangle elements is 18,620 for the fracture whose azimuthal angle is $\theta = 45°$; the number of control volume element node is 9294 and the number of triangle elements is 18,510 for vertical fracture whose azimuthal angle is $\theta = 90°$; the number of control volume element node is 9236 and the number of triangle elements is 18,406 for the fracture whose azimuthal angle is $\theta = 135°$. The single-porosity model II that have not been done mesh refinement around fracture are meshed to verify the high

(a)

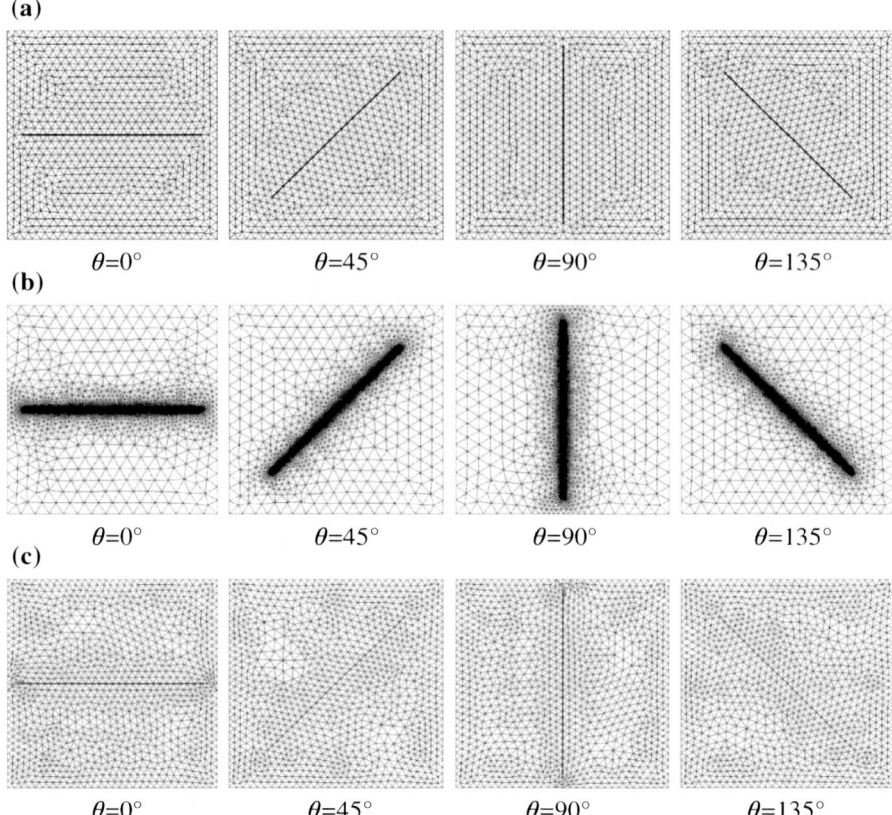

(b)

(c)

Fig. 2.21 Mesh distribution for porous media that have single fracture with different azimuthal angle. **a** Discrete fracture model, **b** single-porosity model I, **c** single-porosity model II

efficiency of discrete fracture model method. 1094, 1088, 1096, and 1092 control volume element nodes and 2068, 2054, 2070, and 2062. Delaunay triangle elements can be obtained after mesh generation. It is observed that the number of control volume element for discrete fracture model and single-porosity model tends to be similar, and both of them are much less than the control volume element number for single-porosity model I.

Water saturation section for the injection volume of 0.5 PV can be obtained based on control volume method, as shown in Fig. 2.22. As can be seen from the figure, fractures have a significant impact on fluid flow, and the computed result of discrete fracture model and single-porosity model is almost same.

Figure 2.23 is schematic of recovery degree for single-porosity media that have different azimuthal angles based on discrete fracture model and single-porosity

(a)

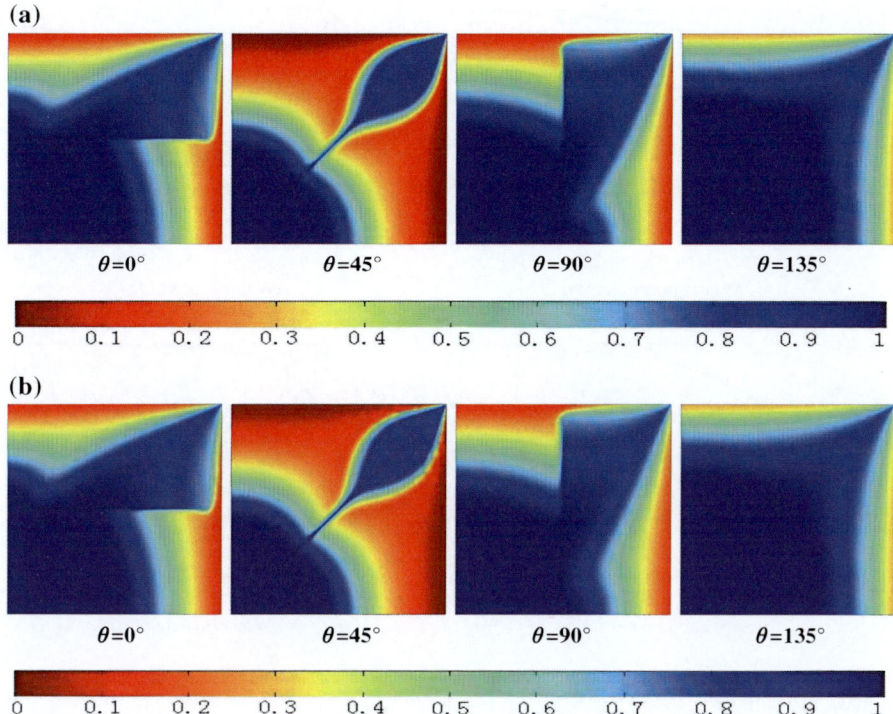

(b)

Fig. 2.22 Schematic of water saturation distribution for two models when the injection volume is 0.5 PV. **a** Discrete fracture model, **b** single-porosity model I

model I when the well has been produced for 300 days. As can be seen from the figure, the computed result of discrete fracture model and single-porosity model I showed a great consistency.

As shown in Figs. 2.22 and 2.23, regarding the single-porosity model I which has local mesh refinement as reference solution, the validity of numerical method can be verified for the computed result of discrete fracture model correlated well with the reference solution.

To verify the efficiency of discrete fracture model, we consider geometric models with different dips and computing times based on discrete fracture model or single-porosity model which has local mesh refinement or not, where CPU clock speed is 2.93 GHz. The corresponding computing time is shown in Table 2.1.

As can be seen from Table 2.1, discrete fracture model has the same calculation accuracy with single-porosity model while the former has less computing time than the latter, which could explain that discrete fracture model have high efficiency. In addition, the latter has a poorer convergence than discrete fracture model for the wide difference between meshes around fracture.

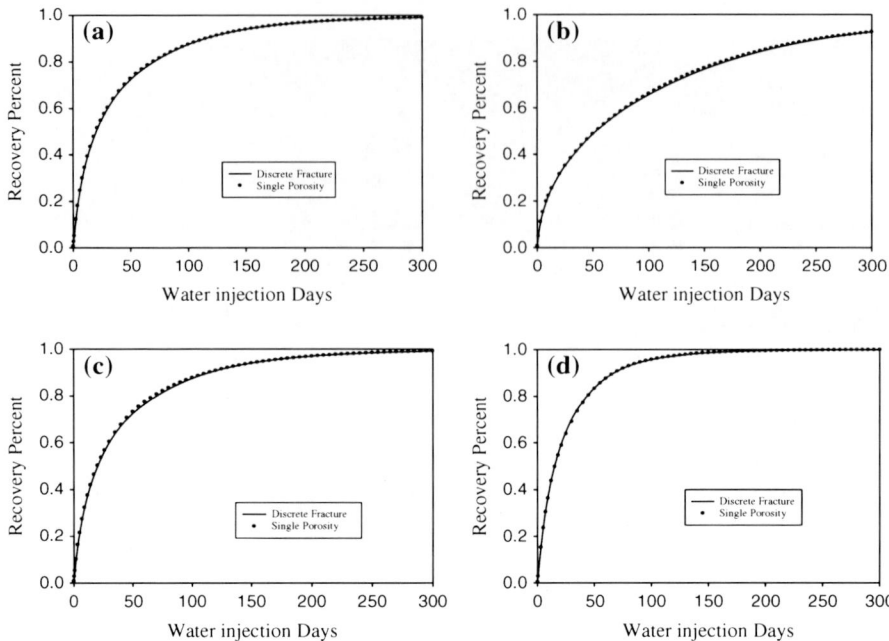

Fig. 2.23 Comparison diagram of recovery degree for single-porosity media based on two models. **a** $\theta = 0°$, **b** $\theta = 45°$, **c** $\theta = 90°$, **d** $\theta = 135°$

Table 2.1 Computing time of different porous media based on different models	Computation times/s	$\theta = 0°$	$\theta = 45°$	$\theta = 90°$	$\theta = 135°$
	Discrete fracture model	47.42	48.81	52.33	54.63
	Single-porosity model I	402.83	397.00	382.06	390.59
	Single-porosity model II	144.93	132.07	138.74	141.22

2.4   Mimetic Finite Difference Numerical Simulation

The existing numerical calculation methods of discrete fracture flow mainly include two categories as below: finite volume method and finite element method. The former needs to simplify and equivalent the processes, which leads to reducing calculation accuracy when processing the mass calculation; the latter has some defects in conservation-type calculation format and computational stability. As a new numerical calculation method, MFD (Mimetic Finite Difference) gets a successful application in the numerical simulation calculation of fluid mechanics, electromagnetic field, and oil reservoir, because of its good local conservation and the applicability of the complex grid. In this section, we have further put this method into use to flow numerical simulation research of discrete fracture model, elaborated the basic principle of Mimetic Finite Difference, set up a corresponding discrete fracture numerical format, and we have solved the two-phase flow problem by the method of IMPES (Implicit Pressure and Explicit Saturation Scheme). In the end, we have proved the validity of this method by an example.

2.4.1   Two-Phase Fluid Flow Mathematical Model

For brevity, we only consider incompressible oil–water two-phase flow problems, and other problems' methods remain the same. Here, we use the classical fractional flow mathematical model, of which the pressure equation is

$$\boldsymbol{v} = -\boldsymbol{K}\lambda \cdot \nabla p + \boldsymbol{K} \cdot (\lambda_{\mathrm{w}} \cdot \rho_{\mathrm{w}} + \lambda_{\mathrm{o}}\rho_{\mathrm{o}})\boldsymbol{G}, \quad \nabla \cdot \boldsymbol{v} = q, \tag{2.94}$$

where $\boldsymbol{v} = \boldsymbol{v}_{\mathrm{w}} + \boldsymbol{v}_{\mathrm{o}}$ stands for total seepage velocity; K is permeability tensor; $\lambda = \lambda_{\mathrm{w}} + \lambda_{\mathrm{o}}$ stands for overall coefficient of fluidity, of which we impose $\lambda = k_{rl}/\mu_l$ ($l = \mathrm{w}, \mathrm{o}$), and define shunt function $f_l = \lambda_l/\lambda$; k_{rl} is relative permeability of l-phase fluid; μ_l is viscosity of l-phase fluid; ρ_l is the density of two-phase fluid; $\boldsymbol{G} = -g\nabla z$ is gravity item, of which g is gravitational acceleration; z is reservoir depth (positive upward); $q = q_{\mathrm{w}} + q_{\mathrm{o}}$ is source or sink term; and the global pressure p is defined as below:

$$p = p_{\mathrm{o}} - \int_{1}^{S_{\mathrm{w}}} f_{\mathrm{w}}(\xi)\frac{\partial p_c}{\partial S_{\mathrm{w}}}(\xi)\mathrm{d}\xi \tag{2.95}$$

where p_c is capillary force, and S_{w} is water phase saturation.
The corresponding water phase saturation equation is

$$\phi\frac{\partial S_{\mathrm{w}}}{\partial t} + \nabla \cdot v_{\mathrm{w}} = q_{\mathrm{w}} \tag{2.96}$$

$$v_{\mathrm{w}} = f_{\mathrm{w}}[v + \boldsymbol{K}\lambda_{\mathrm{o}} \cdot \nabla p_c + \boldsymbol{K}\lambda_{\mathrm{o}} \cdot (\rho_{\mathrm{w}} - \rho_{\mathrm{o}})\boldsymbol{G}] \tag{2.97}$$

where ϕ denotes porosity.

Assume that the flow in the matrix and fracture meet the Darcy's law, therefore, the above equations are applicable to the entire area of the fractured media. In this text, we use IMPES solution to solve Eqs. (2.94) and (2.96) in turns: where we employ IMPES to solve pressure Eq. (2.94), and use finite volume method to get an explicit solution of formula (2.96).

In order to adapt to complex geometry of A discrete fracture model, we adopt unstructured mesh generation technology to discretize the research area, as shown in Fig. 2.24. Due to the small fracture aperture, for this fracture, we employ

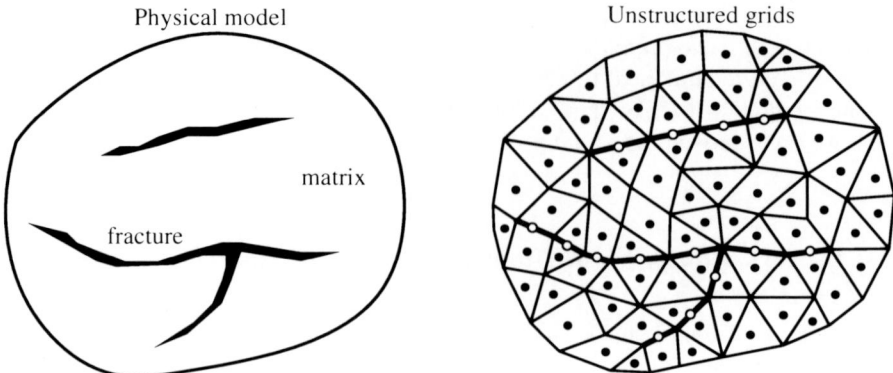

Fig. 2.24 The discrete fracture model and unstructured mesh generation diagram

dimensionality reduction, that is, fracture is simplified as fracture line element in the 2-D problems, and simplified as fracture plane unit in the 3-D problems. By dimension reduction process, the number of grids can be cut down so that computational efficiency is enhanced; however, fracture aperture is only considered in specific numerical calculation.

2.4.2   Solution Strategies for the Pressure Equation

(1) Matrix section

Assume that the research area $\Omega \in \mathbf{R}^d$ is subdivided by a set of nonoverlapping polygon ($d = 2$) or polyhedron ($d = 3$) grids $\Omega_h = \{\Omega_i\}$. As shown in Fig. 2.21, we can take any unit Ω_i to analyze, and Ω_j is adjacent unit. $A_k = \Omega_i \cap \Omega_j$ is interface, $n_k = |A_k|\hat{n}_k$ is the area-weighted normal vector of interface area A_k. $\hat{n}_k$ is the unit outward normal vector (Fig. 2.25).

First of all, on the unit center x_i and boundary surface center x_k, we can, respectively, define unit pressure p_i and boundary surface pressure π_k as follows:

Fig. 2.25 Analysis schematic diagram of finite difference grid cell simulation

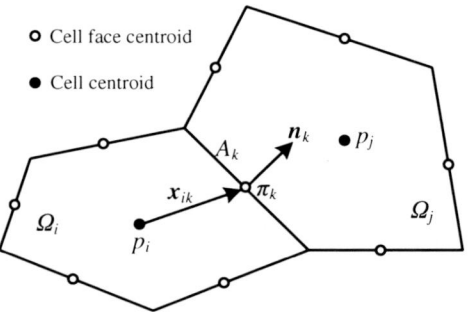

$$p_i = \frac{1}{|\Omega_i|} \int_{\Omega_i} p \, d\Omega, \quad \pi_k = \int_{A_k} p \, dA \tag{2.98}$$

Notice that if the gravity is taken into account, the pressure equations described above should be regarded as flow potential. By Darcy's law, it is easy to know that the normal seepage velocity v_i on boundary surface can be written as the following formula:

$$v_i = T_i \cdot (e_i p_i - \pi_i) \tag{2.99}$$

In this formula, T_i is transmission matrix, $v_i = [v_1, \ldots, v_m]^{\mathrm{T}}$, m is the number of boundary surface of the unit Ω_i, $e_i = [1, \ldots, 1]^{\mathrm{T}}$. The structure of the matrix $e_i = [1, \ldots, 1]^{\mathrm{T}}$ is the key of MFD simulation.

Suppose that pressure on the unit has linear variation, that is, $p = a \cdot x + b$, so by the Darcy law, we can get binding equations:

$$v_k = -\mu^{-1} |A_k| \hat{\boldsymbol{n}}_k \cdot \boldsymbol{K} \cdot \nabla p = \mu^{-1} |A_k| \hat{\boldsymbol{n}}_k \cdot \boldsymbol{K} \cdot \boldsymbol{a} \tag{2.100}$$

Combining Eqs. (2.99) and (2.100), and considering $p_i - \pi_k = \boldsymbol{a} \cdot (x_i - x_k)$, we can get the following equation:

$$v_i = T_i \cdot \begin{bmatrix} x_1 - x_i \\ \vdots \\ x_k - x_i \\ \vdots \\ x_m - x_i \end{bmatrix} \cdot \boldsymbol{a} = \mu^{-1} \begin{bmatrix} |A_1| \hat{\boldsymbol{n}}_1 \\ \vdots \\ |A_k| \hat{\boldsymbol{n}}_k \\ \vdots \\ |A_m| \hat{\boldsymbol{n}}_m \end{bmatrix} \cdot \boldsymbol{K} \cdot \boldsymbol{a} \Rightarrow T_i X = \mu^{-1} N K \tag{2.101}$$

In this formula, $X = [X_1], \ldots, [X_d]$, $N = [N_1], \ldots, [N_d]$, and $N^{\mathrm{T}} X = [Z_{ij}]_{d \times d}$. In this definition, $x^{(i)}$ denotes the ith dimension Cartesian coordinates of x, so we get

$$Z_{ij} = N_i^{\mathrm{T}} X_j = \sum_{k=1}^{m} |A_k| \hat{\boldsymbol{n}}_k^{(i)} (x_k - x_i)^{(j)} \tag{2.102}$$

Notice, $x_k - x_i = \frac{1}{|A_k|} \int_{A_k} (x - x_i) \, dA$, and combine divergence theorem, then we can obtain format as follows:

$$Z_{ij} = \sum_{k=1}^{m} |A_k| \hat{\boldsymbol{e}}_i \cdot \hat{\boldsymbol{n}}_k \frac{1}{|A_k|} \int_{A_k} (x - x_i)^{(j)} \, dA = \sum_{k=1}^{m} \hat{\boldsymbol{e}}_i \cdot \int_{A_k} (x - x_i)^{(j)} \cdot \hat{\boldsymbol{n}}_k \, dA$$

$$= \hat{\boldsymbol{e}}_i \cdot \int_{\Omega_i} \nabla \cdot (x - x_i)^{(j)} \, d\Omega = \hat{\boldsymbol{e}}_i \cdot \hat{\boldsymbol{e}}_j |\Omega_i| = \delta_{ij} |\Omega_i| \tag{2.103}$$

$$\delta_{ij} \begin{cases} 0 & i \neq j \\ 1 & i = j \end{cases}$$

Namely, $N^T X = |\Omega_i| E_d$, where E_d is d order unit matrix, therefore. Through the equations, we can obtain conductance matrix T_i as follows:

$$T_i = \frac{1}{\mu |\Omega_i|} NKN^T + T_2 \tag{2.104}$$

where $T_2 Z = 0$. In order to ensure the existence of inverse matrix, we apply Brezzi–Lipnikov–Simoncini theorem (Thomas et al. 1983) to construct matrix. In this paper, we employ the following form:

$$T_i = \frac{1}{\mu |\Omega_i|} \left[NKN^T + \frac{6}{d} \text{trace}(K) A (E_m - QQ^T) A \right] \tag{2.105}$$

in which, $A = \text{diag}(|A_k|)$, $Q = \text{orth}(AX)$. For the continuity equation in this equation, we can directly integral divergence theorem in the unit Ω_i, and get

$$\sum_{k=1}^{m} v_k^f = \int_{\Omega_i} q_i d\Omega \tag{2.106}$$

Consider that speed continuity conditions on the surface of the cell boundaries, combine Eqs. (2.102) and (2.106), so MFD numerical calculation format can be obtained as follows:

$$\begin{bmatrix} B & -C & D \\ C & 0 & 0 \\ D^T & 0 & 0 \end{bmatrix} \begin{bmatrix} v \\ p \\ \pi \end{bmatrix} = \begin{bmatrix} g \\ q \\ f \end{bmatrix} \tag{2.107}$$

where $v = [v_k]$ is seepage velocity array of unit boundary surface; $p = [p_i]$ is unit center pressure array; $\pi = [\pi_k]$ is pressure array on unit boundary surface center. $g = [g_k]$ is gravity item. $q = [q_i]$ is source sink term of unit Ω_i. $f = [f_i]$ is flow boundary conditions. $f = 0$ represents impermeable barrier. The first line of equation corresponds to the Darcy's law. The second line corresponds to the continuity equation. The third line is continuity conditions for normal speed on the surface of the cell boundaries. The coefficient matrix of above equation is specific as follows:

$$B = \begin{bmatrix} T_1^{-1} & & \\ & \ddots & \\ & & T_{N_e}^{-1} \end{bmatrix}, C = \begin{bmatrix} e_1 & & \\ & \ddots & \\ & & e_{N_e} \end{bmatrix}, D = \begin{bmatrix} I_1 & & \\ & \ddots & \\ & & I_{N_e} \end{bmatrix} \tag{2.108}$$

where N_e is the sum of grid cells: $I_i = E_m$.

Conclusion can be drawn from the above derivation process: MFD method is only based on a single grid cell to construct the numerical format, which is suitable for any complicated grid system. Besides, it has good local conservation property which is similar to the hybrid finite element. However, for complex grid system, the structure of the mixed finite element numerical calculation format has a big difficulty.

(2) Numerical solution of the discrete fracture model

As mentioned earlier, flow of fractures and matrix all meet the Darcy's law. If we consider closed outer boundary, the corresponding equations are as follows:

$$\begin{bmatrix} B_m & -C_m & D_m \\ C_m^T & 0 & 0 \\ D_m^T & 0 & 0 \end{bmatrix} \begin{bmatrix} v_m \\ p_m \\ \pi_m \end{bmatrix} = \begin{bmatrix} g_m \\ q_m \\ 0 \end{bmatrix} \tag{2.109}$$

$$\begin{bmatrix} B_f & -C_f & D_f \\ C_f^T & 0 & 0 \\ D_f^T & 0 & 0 \end{bmatrix} \begin{bmatrix} v_f \\ p_f \\ \pi_f \end{bmatrix} = \begin{bmatrix} g_f \\ q_f \\ 0 \end{bmatrix} \tag{2.110}$$

In these equations, the subscripts m and f, respectively, denote matrix and fracture. Notice that, in this paper, the fracture has been reducing dimensions. Therefore, the space dimension of Eq. (2.110) is low 1-D than Eq. (2.109).

The key of MFD discrete fracture numerical format's structure consists in the coupling of pressure equation of matrix and fracture (Fig. 2.26).

Fig. 2.26 Fracture–matrix coupled flow analysis diagram

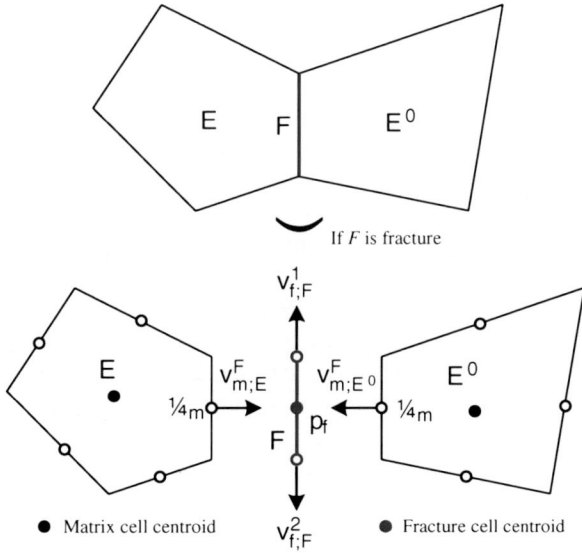

Withal, consider fracture–matrix coupled hybrid flow grid analysis diagram as shown in Fig. 2.22. Fracture grid cell can be treated as the boundary of the matrix grid cell surface, thus fracture unit pressure p_f and boundary surface pressure π_m of adjacent matrix unit are equal. Therefore, we can just reserve π_m in the numerical format. Seepage velocity term is coupling on the fracturing unit in Eqs. (2.109) and (2.110) in accordance with the following conditions.

(1) If F is diversion fractures, the total flow exchange between adjacent rock element and fractures element can be denoted as Q_f^F. For fracture unit, this flow can be used as a source/sink term. So, the equation is as follows:

$$\begin{cases} v_{m,E}^F + v_{m,E'}^F = Q_f^F \\ \sum_i v_{f,F}^i = Q_f^F + q_f^F \end{cases} \tag{2.111}$$

Where $v_{m,E}^F$, $v_{m,E'}^F$, respectively, are exchange to fracture from matrix elements E and E'; q_f^F represents sources/sinks; $\sum_i v_{f,F}^i$ in the second line of above equations corresponds to equation of continuity of fracture element.

(2) If F is flow barrier, it will be processed in accordance with impermeable barrier.

At the moment, Eqs. (2.109) and (2.110) can be coupled together to form the corresponding discrete fracture numerical formats as follows:

$$\begin{bmatrix} B_m & -C_m & D & 0 & 0 \\ C_m^T & 0 & 0 & 0 & 0 \\ D_m^T & 0 & 0 & -C_f^T & 0 \\ 0 & 0 & -C_f & B_f & D_f \\ 0 & 0 & 0 & D_f^T & 0 \end{bmatrix} \begin{bmatrix} v_m \\ p_m \\ \pi_m \\ v_f \\ \pi_f \end{bmatrix} = \begin{bmatrix} g_m \\ q_m \\ -q_f \\ g_f \\ 0 \end{bmatrix} \tag{2.112}$$

2.4.3 The Solution of the Saturation Equation

(1) The calculation format of finite volume method

In this work, we use the IMPES (implicit pressure, explicit saturation) method, which used to be quite popular in the industry. In the IMPES method, the fluid pressure equations (flow equations) are solved implicitly while the saturation field is fixed, yielding the velocities of the fluid phases. These velocities are used to calculate the mass balance of the fluid phases in the transport equations while the pressure field remains fixed. For saturation equation, we apply the finite volume method for solving. We can directly integral the formula (2.96) on element and it can be written as

$$\int_{\Omega_i} \phi \frac{\partial S}{\partial t} d\Omega + \int_{\partial \Omega_i} \{f_w [v + K\lambda_o \cdot \nabla p_c + K\lambda_o \cdot (\rho_w - \rho_o)G]\} \cdot n_i d\Gamma = \int_{\Omega_i} q_w d\Omega$$

$$\tag{2.113}$$

For convenience of writing, we have removed the subscript w of the water saturation S_w. For the time dimension, if we apply θ-rules, the following finite volume numerical discrete format can be obtained:

$$\frac{\phi_i}{\Delta t}(S_i^{n+1} - S_i^n) + \frac{1}{|\Omega_i|}\sum_{k=1}^{m}[\theta F_k(S^{n+1}) + (1 - \phi)F_k(S^n)] = q_w(S_i^n) \qquad (2.114)$$

where

$$F_k(S) = \int_{A_k}[f_w(S)]_k(\boldsymbol{v} \cdot \hat{\boldsymbol{n}}_k + \boldsymbol{K}\lambda_o \cdot \nabla p_c \cdot \hat{\boldsymbol{n}}_k + \boldsymbol{K}\lambda_o \cdot (\rho_w - \rho_o)\boldsymbol{G} \cdot \hat{\boldsymbol{n}}_k)\mathrm{d}A \qquad (2.115)$$

where superscript n stands for time step.

On boundary surface A_k, we have applied following format $[f_w(S)]_k$, which is the upstream windward format

$$[f_w(S)]_k = \begin{cases} f_w(S_i) & \text{if } \boldsymbol{v} \cdot \hat{\boldsymbol{n}}_k \geq 0 \\ f_w(S_j) & \text{if } \boldsymbol{v} \cdot \hat{\boldsymbol{n}}_k < 0 \end{cases} \qquad (2.116)$$

We can solve explicit solution for the saturation equation, namely $\theta = 0$. In order to calculate stability, the time step applies the CFL condition as follows:

$$\Delta t \leq \frac{\phi_i|\Omega_i|}{v_i^{in}\max\{f_w'(S)\}_{0 \leq S \leq 1}} \qquad (2.117)$$

where

$$v_i^{in} = \max(q_i, 0) - \sum_{A_k}\min(v_k, 0) \quad , \quad \frac{\partial f_w}{\partial S} = \frac{\partial f_w}{\partial S^*}\frac{\partial S^*}{\partial S} = \frac{1}{1 - S_{wc} - S_{or}}\frac{\partial f_w}{\partial S^*}$$

In these equations, S^* is the water phase saturation after the normalization; S_{wo} is irreducible water saturation; S_{ro} is residual oil saturation.

(2) The saturation calculation at fractures' intersections

When two or more fractures intersect, the key in the discrete fracture flow simulation is saturation calculation. At present there are mainly two kinds of processing methods: one is the upstream windward format of conductivity calculation based on Delta–Star (Karimi-Fard et al. 2004), which simplifies and equivalently deals with the crossed fracture. Another is upstream wind weighted format (Hoteit and Firoozabadi 2006), which is of high calculation precision, but needs to get real seepage of velocity of every fracture unit at intersections. In this paper, we apply the latter one. As shown in Fig. 2.27, assume that there are N_I fracture elements e_i intersect at I; each fracture element corresponds to distribution function f_{w,e_i}. v_{f,e_i} is the seepage velocity at the intersection. We can define the inflows and outflows on the intersection I as follows:

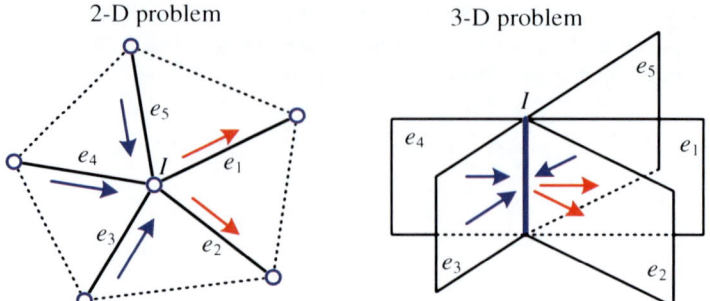

Fig. 2.27 The fracture's intersections saturation calculation diagram

$$
\begin{cases}
v_{f,e_i} \leq 0, & 0 < i \leq N \quad \text{(efflux)} \\
v_{f,e_i} \geq 0, & N < i < N_1 \quad \text{(influx)}
\end{cases}
\tag{2.118}
$$

By the law of conservation of mass, the equation is written as

$$
\sum_{i=N+1}^{N_I} v_{f,e_i} = -\sum_{i=1}^{N} v_{f,e_i}
\tag{2.119}
$$

Further, by the definition of upstream windward format, we can get

$$
\sum_{i=N+1}^{N_I} f_{w,i} v_{f,e_i} = -\sum_{i=1}^{N} f_{w,I} v_{f,e_i} = -f_{w,I} \sum_{i=1}^{N} v_{f,e_i}
\tag{2.120}
$$

Thus, upstream windward weighted distribution function in fracture's intersections I is as follows:

$$
f_{w,I} = -\frac{\sum_{i=N+1}^{N_I} f_{w,i} v_{f,e_i}}{\sum_{i=1}^{N} v_{f,e_i}}.
\tag{2.121}
$$

2.4.4 Numerical Example

First, this section presents two simple numerical examples of discrete fracture model. And through the comparison of the experimental results, the validity of these methods and procedures have been verified. Then, the calculation examples of complex discrete fracture model have further verified the correctness of the method and the robustness of the program.

(1) The simple calculation example of discrete fracture model
Consider the one-well injection and one-well production physical model, as shown in Figs. 2.24 and 2.25, whose size is 1 m × 1 m × 0.025 m and can be treated as

planar flow problem. Figure 2.24 is for a single fracture model and Fig. 2.25 is for two intersecting fractures model. They all are produced by glass (160–180 mesh) sand combined with epoxy resin by compaction and cementation, and then encapsulated by transparent organic glass. Matrix can be regarded as homogeneous isotropic medium, whose porosity is $\phi \approx 0.4$, and permeability $K_m = 10\ \mu m^2$. Fractures are replaced by ultrathin sheet steel when modeling. It will be dissociated after model's cementation, its aperture is about 1 mm, and its permeability is $K_f = a^2/12 = 8.33 \times 10^4\ \mu m^2$. The flow of water injection well is $q_{in} = 0.01$ PV/min. And production well connects to the barometric pressure. The viscosity of water is $\mu_w = 1$ mPa s, the oil viscosity is $\mu_o = 5$ mPa s, the density of water is $\rho_w = 1000\ kg/m^3$, and the density of oil is $\rho_o = 800\ kg/m^3$

The initial value of oil saturation, irreducible water saturation, and residual oil saturation of model is all zero. The water phase relative permeability of matrix and fractures is $k_{rw} = S_w$, and the oil phase relative permeability is $k_{ro} = 1 - S_w$. When we use the method mentioned in this paper to do numerical modeling, we can ignore the influence of capillary force and gravity in the calculation. The corresponding Delaunay triangle mesh subdivision and the results of numerical simulation are shown in Figs. 2.24 and 2.25. By comparison with true flow process in experiment we can see that the results of numerical calculation and experimental results are basically identical. Thus we have verified the correctness of the method and procedure in this paper. It is worth noting that, the rapid flow phenomenon appears on the left border in Fig. 2.28(a), which is due to the poor sealing of the experimental model (Fig. 2.29).

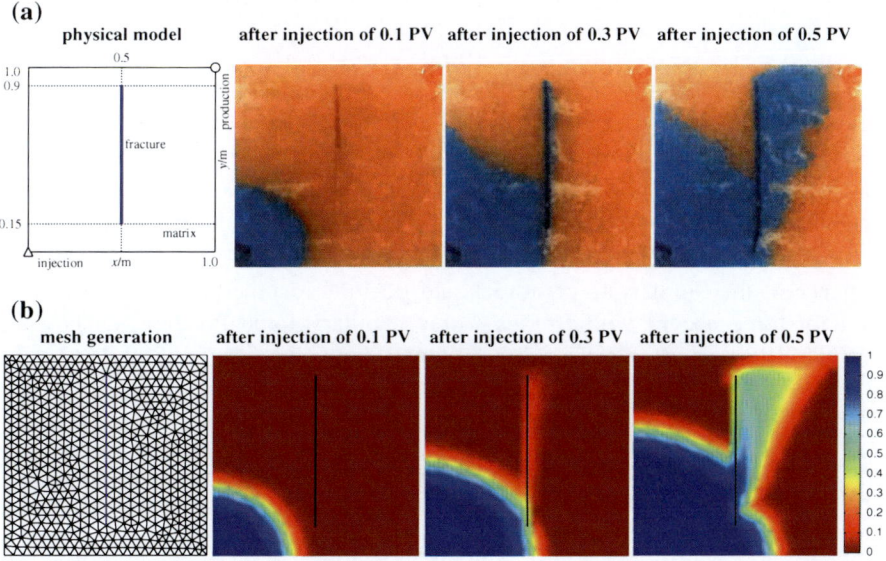

Fig. 2.28 The comparison between results of water saturation and single fracture model. **a** Experimental result, **b** numerical result

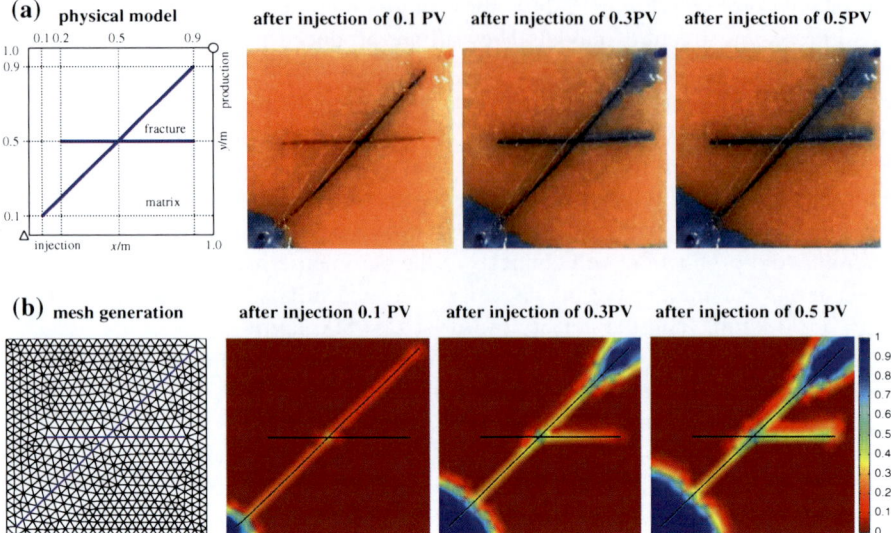

Fig. 2.29 The comparison between results of water saturation and double fracture model. **a** Experimental result, **b** numerical result

(2) Complex discrete fracture model

In this section, we consider a complex fractured model, dimension of 100 m × 50 m (x × y), as shown in Fig. 2.30, the blue lines represent the fractures which generated random based on the geological statistics. The Delaunay triangular gridding is used to discrete the geometrical model (Fig. 2.26 right). Homogeneous isotropic matrix's porosity $\phi = 0.2$, permeability $K_m = 10$ mD (1 mD = 10^{-3} μm^2), fracture aperture $a = 1$ mm, permeability $K_f = a^2/12 = 8.33 \times 10^7$ mD, and physical property parameters of oil and water is in accordance with 5.1 calculation example. The initial reservoir pressure is 10 MPa, initial water saturation is zero, and the speeds of injection wells and production wells are 0.01 PV/day. Water phase relative permeability of matrix and fracture is $K_{rw} = S_w^2$, oil phase relative permeability is $K_{ro} = (1 - S_w)^2$. Assume that model is water-wet reservoir. If we consider the influence of the capillary force in rock and fracture, assume that both types of the capillary force accord with Brooks–Corey capillary force function as shown in formula (2.122). For the matrix, threshold pressure value is $p_d = 1000$ Pa, and λ is 2.0. For fractures, threshold pressure value is $p_d = 1000$, and λ is 1.0.

$$p_c(S_w) = p_d \left(\frac{S_w - S_{wc}}{1 - S_{wc} - S_{or}} \right)^{-\frac{1}{\lambda}}, \quad 0.2 < \lambda < 3.0. \tag{2.122}$$

Figure 2.31 shows the water saturation distribution at different times. The analogous calculation results indicate: the induced water flows into the fracture quickly; and the existence of fractures results in the strong heterogeneity of

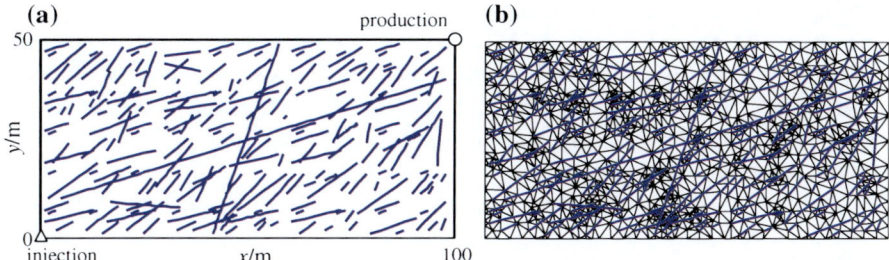

Fig. 2.30 The mode of complex discrete fracture and unstructured grid subdivision. **a** The mode of complex discrete fracture, **b** unstructured grid subdivision

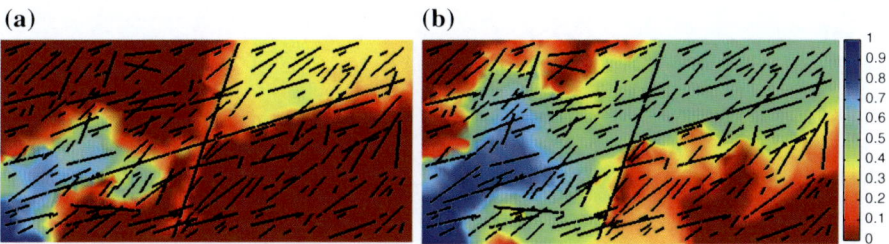

Fig. 2.31 The water saturation distribution at different times. **a** 10 days later, **b** 40 days later

medium; the existence of the capillary force makes the water-flood swept area increased, but the overall effect is still controlled by the macroscopic fracture (macro fracture). Through this calculation example, we further verified the correctness of this suggested method, at the same time we can see that this method still has good applicability for extremely complicated grid system.

2.5 The Embedded Discrete Fracture Numerical Simulation

At present, fractured oil reservoir numerical simulation is mostly based on the double medium model, but this model is only applicable to highly matured fracture in the reservoir. When there are several large fractures that control the direction and the scale of fluid flow, the error in calculation results is bigger. To solve this problem, the discrete fracture model was set up, and with the wide use of artificial fracturing technology in unconventional reservoirs, its corresponding flow simulation technique has a rapid development. However, the existing numerical discrete fracture models are all based on matched grid, that is, we treat the fractures as internal boundary and constrain face for grid subdivision. Due to the complexity of the fracture's geometrical morphology, we need to adopt the unstructured grid technique, whose subdivision process is very complicated and tedious. Especially when the distance or the angle between the fractures is very small, the mesh generation often is of poor quality, which leads to deviation calculation, as shown in Fig. 2.32a. However, the embedded discrete fracture model does not need to consider the internal fracture morphology when partitioning grid, where matrix system separately generates grids, fracture part generates grids according to the intersection of fracture and matrix grids, as shown in Fig. 2.32b, that greatly reduces the complexity of meshing, so that it can improve the calculation efficiency.

To this end, Lee and Moinfar et al. (2012) put forward embedded discrete fracture model. This model will directly embed fracture network into the matrix structured grid system, which has avoided the complex unstructured grid subdivision process. Although we need to calculate geometry information between the

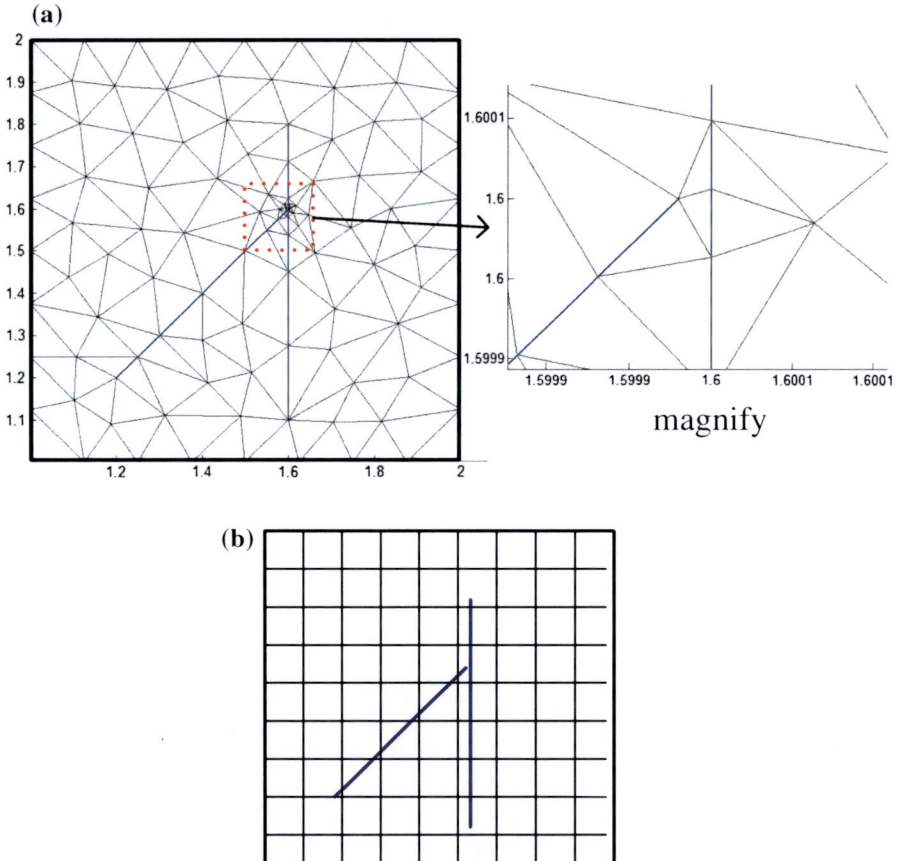

Fig. 2.32 The discrete fracture model and the embedded discrete fracture model mesh generation contrast. **a** Matched unstructured grid, **b** Non-matched structured grid

fracture and grid, computation complexity is significantly reduced to improve computational efficiency, relative to the complex unstructured grid subdivision process.

However, the existing embedded discrete fracture models all adopt finite difference method to solve, so it cannot process accurately the permeability situation of full tensor, and applies only to structured grid. To this, it is necessary to set up a new embedded discrete fracture numerical format, based on the simulation of finite difference method, to apply to numerical simulation of complex fractured reservoir.

Discrete fractures numerical model regards fractures as internal constraint face to generate mesh. Due to the complexity of the fracture's geometrical morphology, we need to adopt the unstructured grid technique, whose subdivision process is very complicated and tedious. Especially when the distance or the angle between the fractures is very small, the mesh generation often is of poor quality, which leads to deviation calculation, as shown in Fig. 2.32a. However, the embedded discrete fracture model do not need to consider the internal fracture morphology when partitioning grid, where matrix system separately generates grids, fracture part generates grids according to the intersection of fracture and matrix grids, as shown in Fig. 2.32b, that greatly reduces the complexity of meshing, so that it can improve the calculation efficiency.

2.5.1   The Mathematical Model of Embedded Discrete Fracture Model

For the convenience of study, illustrate the basic ideas and principals of embedded numerical simulation, based on the 2-D single-phase flow. Assume that the fluid flow process is of constant temperature, regardless of the matrix and the fluid compressibility; matrix system and fluid flow in fracture system meet Darcy's law; ignore the influence of gravity and capillary pressure.

Matrix system mathematical model:

$$v_m = -\frac{K_m}{\mu} \cdot \nabla p_m \tag{2.123}$$

$$\nabla \cdot v_m = q_m + \frac{q_{mf}}{V_m} \delta_{mf} \tag{2.124}$$

Fracture system mathematical model:

$$\frac{K_f}{\mu} \frac{\partial^2 p_f}{\partial \xi^2} = q_f + \frac{q_{mf} + q_{ff}\delta_{ff}}{V_f} \tag{2.125}$$

where

$$
\delta_{mf} = \begin{cases} 1 & \text{if there are fractures embedding on the bedrock grid} \\ 0 & \text{if there is no fracture embedding on the bedrock grid} \end{cases}
$$

$$
\delta_{ff} = \begin{cases} 1, & \text{if one fracture element intersect with another} \\ 0, & \text{if one fracture element doesn't intersect with any other} \end{cases}
$$

where v_m is rock seepage velocity; K_m is rock permeability tensor; K_f is fracture permeability (scalar); μ is fluid viscosity; p_m and p_f, respectively, are basement and fracture of the pressure (or streaming potential); V_f and V_m, respectively, are fracture element and the volume of rock unit; q_m and q_f, respectively, represent basement and fracture source sink term; ξ is local coordinate system along the fracture direction; q_{mf} denotes quantity flow between basement and fracture; q_{ff} denotes quantity flow between intersecting fracture elements;

(1) Flow between matrix and fracture element
The fracture aperture is very small compared with mesh scale, and the fracture permeability is greater than the matrix permeability, so we can think that the pressure is of succession on both sides of the fractures. The expression of quantity flow calculation between matrix and fracture element can be written as

$$
q_{mf} = -T_{mf}(p_m - p_f) \tag{2.126}
$$

where

$$
T_{mf} = \frac{k_{mf}A_{mf}}{\mu\widehat{d}}, \quad \widehat{d} = \frac{\int x_{mf}\,dS}{S}, \quad K_{mf} = \frac{1}{1/K_f + 1/K_m}
$$

where $\widehat{d}$ represents equivalent distance between the matrix grid and fracture section, that is, the average of vertical distance between all the points in the matrix grids and fracture section; S is the volume of matrix element; K_m denotes permeability (scalar), in the direction perpendicular to fracture; A_{mf} denotes contact area of fracture section and bed rock; x_{mf} is perpendicular distance from all points in the matrix grids to fracture.

(2) Flow between fracture elements
Calculate quantity flow between fractures referring to Karimi-Fard's using transfer coefficient method in calculating intersecting fracture section:

$$
q_{ff} = T_{ff}(p_{fi} - p_{fj}) \tag{2.127}
$$

Fig. 2.33 Fracture section–
intersection diagram

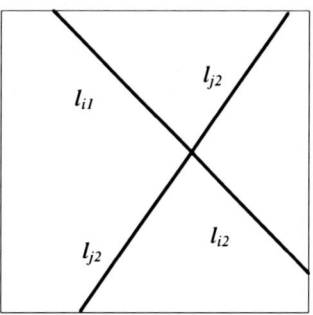

where

$$T_{\text{ff}} = \frac{T_{\text{f}i}T_{\text{f}j}}{T_{\text{f}i} + T_{\text{f}j}}, \quad T_{\text{f}i} = \frac{k_{\text{f}i}a_{\text{f}i}}{\mu \widehat{d}_i}, \quad T_{\text{f}j} = \frac{K_{\text{f}j}a_{\text{f}j}}{\mu \widehat{d}_j}$$

$$\widehat{d}_i = \frac{l_{i1}}{l_{i1} + l_{i2}} \cdot \frac{1}{2} l_{i1} + \frac{l_{i2}}{l_{i1} + l_{i2}} \cdot \frac{1}{2} l_{i2}$$

$$\widehat{d}_j = \frac{l_{j1}}{l_{j1} + l_{j2}} \cdot \frac{1}{2} l_{j1} + \frac{l_{j2}}{l_{j1} + l_{j2}} \cdot \frac{1}{2} l_{j2}$$

where a_{f} denotes fracture aperture, and l denotes the length of fracture (Fig. 2.33).

2.5.2 Numerical Solution of Mathematical Model

(1) Finite difference solution for matrix
Matrix is subdivided by a set of nonoverlapping polygon mesh shown as in
Fig. 2.34, we analyze any unit Ω_i, Ω_j is the adjacent cell, the interface $A_k = \Omega_i \cap \Omega_j$,
$\boldsymbol{n}_k = |A_k|\hat{\boldsymbol{n}}_k$ is vector by the area-weighted method of the interface A_k, $\hat{\boldsymbol{n}}_k$ is the unit

Fig. 2.34 Simulation of finite
difference grid cell analysis
diagram

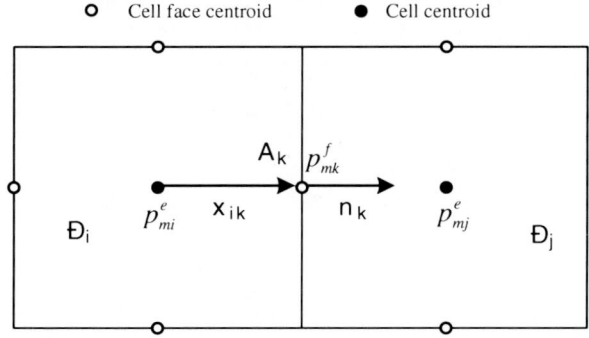

outward normal vector. In the central point $\boldsymbol{x}_i$ of unit Ω_i and central point $\boldsymbol{x}_k$ of boundary surface, it is defined, respectively, unit pressure p_{mi}^e and the pressure p_{mk}^f on the boundary surface, as follows:

$$p_{mi}^e = \frac{1}{|\Omega_i|}\int_{\Omega_i} p_m \, d\Omega, \quad p_{mk}^f = \frac{1}{|A_k|}\int_{A_k} p_m \, dA \tag{2.128}$$

It is easy to know by Eq. (2.123) that on the surface of the matrix boundary, the normal seepage velocity and pressure gradient have the relationship as follows:

$$\boldsymbol{v}_m^f = \boldsymbol{T}_{mi} \cdot \left(\boldsymbol{e}_i p_{mi}^e - \boldsymbol{p}_m^f\right) \tag{2.129}$$

where $\boldsymbol{T}_{mi}$ is conductance matrix of matrix grid, $\boldsymbol{v}_m^f = [v_{m1}, v_{m2}, \cdots, v_{mn}]^T$, n is number of boundary surface of the unit Ω_i, $\boldsymbol{e}_i = [1, \cdots, 1]^T$. Therefore, the key of the finite difference simulation is gaining the matrix $\boldsymbol{T}_{mi}$, hereon, assume that the pressure is linearly varying, $p_m = \boldsymbol{a}_m \cdot \boldsymbol{x} + b_m$. Evidently, based on Eq. (2.123), we can get equations as follows:

$$\boldsymbol{v}_{mk}^f = -\mu^{-1}|A_k|\widehat{\boldsymbol{n}}_k \cdot \boldsymbol{K}_m \cdot \nabla p_m = -\mu^{-1}|A_k|\widehat{\boldsymbol{n}}_k \cdot \boldsymbol{K}_m \cdot \boldsymbol{a}_m \tag{2.130}$$

Meanwhile, $p_{mi}^e - p_{mk}^f = \boldsymbol{a}_m \cdot (\boldsymbol{x}_i - \boldsymbol{x}_k)$, and combine Eqs. (2.129) and (2.130), then we can obtain formula as follows:

$$\boldsymbol{v}_m^f = \boldsymbol{T}_{mi} \cdot \begin{bmatrix} \boldsymbol{x}_1 - \boldsymbol{x}_i \\ \vdots \\ \boldsymbol{x}_k - \boldsymbol{x}_i \\ \vdots \\ \boldsymbol{x}_n - \boldsymbol{x}_i \end{bmatrix} \cdot \boldsymbol{a}_m = \mu^{-1} \begin{bmatrix} |A_1|\widehat{\boldsymbol{n}}_1 \\ \vdots \\ |A_k|\widehat{\boldsymbol{n}}_k \\ \vdots \\ |A_n|\widehat{\boldsymbol{n}}_n \end{bmatrix} \cdot \boldsymbol{K}_m \cdot \boldsymbol{a}_m \quad \Rightarrow \quad \boldsymbol{T}_{mi}\boldsymbol{X} = \mu^{-1}\boldsymbol{N}\boldsymbol{K}_m$$

$$\tag{2.131}$$

where $\boldsymbol{X} = [\boldsymbol{X}_1|, \cdots, |\boldsymbol{X}_n]$, $\boldsymbol{N} = [\boldsymbol{N}_1|, \cdots, |\boldsymbol{N}_n]$, and $\boldsymbol{N}^T\boldsymbol{X} = [Z_{ij}]_{n \times n} = |\Omega_i|\,\boldsymbol{E}_d$, where $\boldsymbol{E}_d$ is d order unit matrix.

Therefore, conduction matrix can be obtained by Eq. (2.131), as shown in the following formula:

$$\boldsymbol{T}_{mi} = \frac{1}{\mu|\Omega_i|}\boldsymbol{N}\boldsymbol{K}_m\boldsymbol{N}^T + \boldsymbol{T}_2 \tag{2.132}$$

$$\boldsymbol{T}_{mi} = \frac{1}{\mu|\Omega_i|}\left[\boldsymbol{N}\boldsymbol{K}_m\boldsymbol{N}^T + \frac{6}{d}\mathrm{trace}(\boldsymbol{K}_m)\boldsymbol{A}\left(\boldsymbol{E}_m - \boldsymbol{Q}\boldsymbol{Q}^T\right)\boldsymbol{A}\right] \tag{2.133}$$

where

$$
A = \begin{bmatrix} |A_1| & & & & \\ & \ddots & & & \\ & & |A_k| & & \\ & & & \ddots & \\ & & & & |A_n| \end{bmatrix} , \quad Q = \mathrm{orth}(AX) \tag{2.134}
$$

For this equation, integral directly and use the divergence theorem on the matrix grid cell:

$$
\sum_{k=1}^{n} v_{mk}^{f} = \int_{\Omega_i} q_{mi} d\Omega + q_{mf} \delta_{mf} \tag{2.135}
$$

Consider the velocity continuity conditions on the cell boundaries surface, combine equation, and obtain simulation of the finite difference numerical formats of matrix section:

$$
\begin{bmatrix} B_m & -C_m & D_m \\ C_m^{T} & 0 & 0 \\ D_m^{T} & 0 & 0 \end{bmatrix} \begin{bmatrix} v_m \\ p_m \\ \pi_m \end{bmatrix} = \begin{bmatrix} 0 \\ f_m + Q_{mf} \\ 0 \end{bmatrix} \tag{2.136}
$$

where $v_m = \left[v_{mk}^{f}\right]$; $p_m = \left[p_{mi}^{e}\right]$; $\pi_m = \left[p_{mk}^{f}\right]$; $f_m = \left[f_{mi}\right]$, of which $f_{mi} = \int_{\Omega_i} q_{mi} d\Omega$; $Q_{mf} = \left[q_{mfi} \delta_{mfi}\right]$, in order to convenient writing, this item should be on the right side of the equation, and shift it to the left in the final calculation format.

Obviously in this equation, the first line corresponds to the Darcy's law, the second line corresponds to the law of conservation of mass, and the third line represents the normal velocity continuity conditions on the cell boundaries surface. The specific expression of coefficient matrix of above equation is as follows:

$$
B_m = \begin{pmatrix} T_{m1}^{-1} & & \\ & \ddots & \\ & & T_{mN_e}^{-1} \end{pmatrix} , \quad C_m = \begin{pmatrix} e_1 & & \\ & \ddots & \\ & & e_{N_e} \end{pmatrix} , \quad D_m = \begin{pmatrix} I_1 & & \\ & \ddots & \\ & & I_{N_e} \end{pmatrix} \tag{2.137}
$$

where the subscript N_e is the total number of grid cells; $I_i = E_n$.

For this Eq. (2.137), the coefficient matrix of Eq. (2.136) only is related to the geometry information and reservoir parameters of grid cell, however, has no requirement for the grid geometry, so it is easy to solve, and applicable to any complex grid in principle.

(2) Finite Difference Solution for the fracture parts
For 1-D fracture system, we employ implicit difference, and equation multiplies grid cell volume V_f at both ends, so its difference equation is shown as (2.138)

$$T_{\xi_{i+\frac{1}{2}}}(p_{fi+1} - p_{fi}) - T_{\xi_{i-\frac{1}{2}}}(p_{fi} - p_{fi-1}) = f_{fi} + q_{mfi} + q_{ffi}\delta_{ffi} \qquad (2.138)$$

where

$$T_{\xi_{i+\frac{1}{2}}} = \frac{K_f}{\mu} \frac{d_{fi}}{0.5(\Delta\xi_{i\pm1} + \Delta\xi_i)}; \quad f_{fi} = V_{fi}q_{fi}.$$

(3) The embedded discrete fracture model format
Note that we take the situation with two fractures for example in this section, the calculation format for other situation with more fractures is analogous.

$$
\begin{bmatrix}
B_m & -C_m & D_m & 0 & 0 \\
C_m^T & T_{mf1} + T_{mf2} & 0 & -T_{mf1} & -T_{mf2} \\
D_m^T & 0 & 0 & 0 & 0 \\
0 & T_{mf1} & 0 & T_{f1} - T_{mf1} - T_{ff} & T_{ff} \\
0 & T_{mf2} & 0 & T_{ff} & T_{f2} - T_{mf2} - T_{ff}
\end{bmatrix}
\begin{bmatrix}
v_m \\
p_m \\
\pi_m \\
p_{f1} \\
p_{f2}
\end{bmatrix}
$$

$$
=
\begin{bmatrix}
0 \\
f_m \\
0 \\
f_{f1} \\
f_{f2}
\end{bmatrix}
$$

$$(2.139)$$

where $T_{mfi} = [T_{mfi}]$ represents transmissibility matrix between the i-th fracture and matrix; $T_{ff} = [T_{ff}]$ represents transmissibility matrix between fractures; and, respectively, represent finite difference conductivity coefficient matrix of i-th fracture.

2.5.3   Numerical Examples

(1) Fractured medium single-phase flow experimental verification
Consider one-injection and one-production physical model as shown in Fig. 2.35, whose size is 7 cm × 17 cm × 1 cm, and can be treated as situation of plane flow. This model is made up of quartz sand (80–100 mesh) and epoxy resin through the cementation and compaction, then it is encapsulated by the transparent organic glass. Matrix can be regarded as homogeneous isotropic media, its porosity is $\phi = 0.3$, and its permeability is $K_m = 10\,\mu m^2$ while the model is manufactured, fractures are replaced by stalloys. When the model is cemented, we will dissociate the stalloys. In the model, the opening is about 2 mm, fracture permeability is $K_f = 6.67 \times 10^2\,\mu m^2$. Viscosity of water is $\mu_w = 1\,mPa\,s$, and density of water is $\rho_o = 1000\,kg/m^3$. The model injects and produces stable discharge with constant pressure difference.

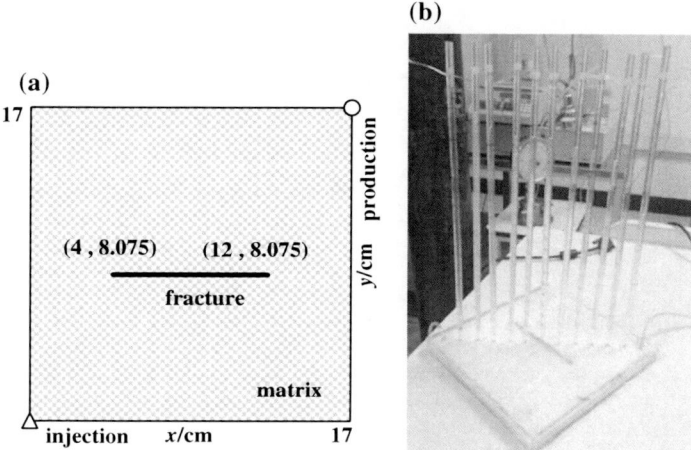

Fig. 2.35 Physical model and experimental model diagrams. **a** Physical model, **b** experimental model

To saturated water of this model, at initial moment, we have calculated its fluid pressure by measuring liquid column height in the glass tube at each point under steady flow state. In this section, we apply the discrete fracture model and the embedded discrete fracture model to simulate the above physical experiment, while gravity influence is ignored. The corresponding numerical simulation results are shown in Fig. 2.36. Figure 2.37 shows pressure curve measured by these two methods in the straight line between the injection–production two points, and it shows comparison results of pressure value measured experiments. From Fig. 2.36, the results of numerical calculation and results of experiment are basically identical, thus the correctness of the method and procedure in this paper is verified. It is worth noting that in this model, the profiles of organic glass and quartz sand glue joint so

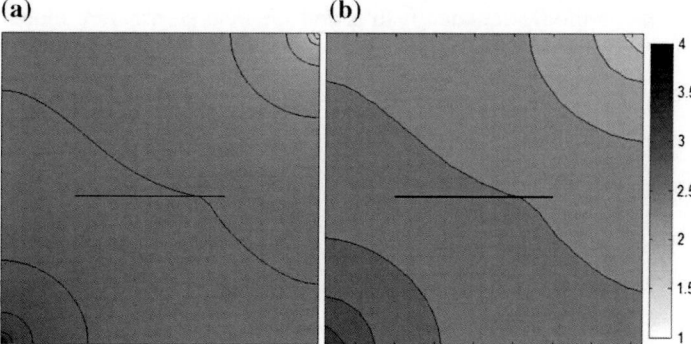

Fig. 2.36 Pressure field distribution by these two methods (KPa). **a** Discrete fracture model, **b** the embedded discrete fracture model (MFD)

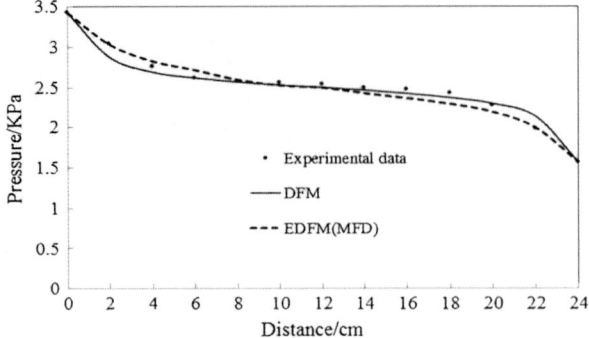

Fig. 2.37 Comparison of pressure distribution on the injection–production diagonal

that model is difficult to achieve completely sealed and quartz sand is hard to be completely homogeneous filling in the model. Therefore, the results will have certain error.

(2) Irregular quadrilateral fractured reservoir

As shown in Fig. 2.38a, it is an irregular quadrilateral fractured reservoir geometry model, matrix permeability is full tenser format. Figure 2.38b, c, respectively, are the embedded triangular mesh which treats the fracture as inner boundary and matched triangular mesh which do not consider fracture subdivision. The basic parameters of the model are as follows: the fracture permeability is $K_f = 1 \times 10 \, \mu m^2$, fracture aperture is $a = 1$ mm, fluid viscosity is 1 mPa s, and rock permeability is $K_m = \begin{bmatrix} 3 & 1 \\ 1 & 2 \end{bmatrix} \times 10^{-3} \, \mu m^2$.

Based on the above two kinds of grid system, we, respectively, use the discrete fracture model (Fig. 2.39a) and the embedded discrete fracture model combined with simulation of finite difference (Fig. 2.39b) to do single-phase flow numerical simulation of the fractured reservoir. Figure 2.13 indicates two methods on the straight line of two-point source and sink to obtain pressure curve. In the figure, we can see that the method calculated results and the discrete fracture model reference solution are basically identical. The error norm is 2.0 %, and the local maximum

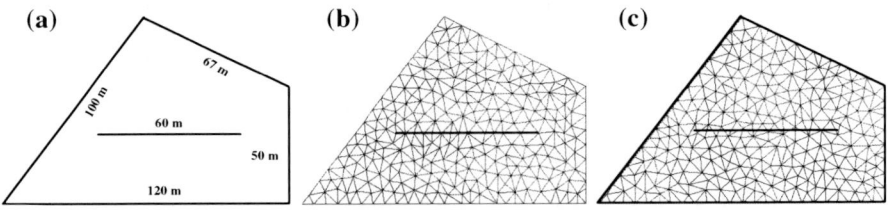

Fig. 2.38 Geometry model of fractured reservoir and result of meshing. **a** Geometry model, **b** matching triangular mesh system, **c** embedded triangular mesh system

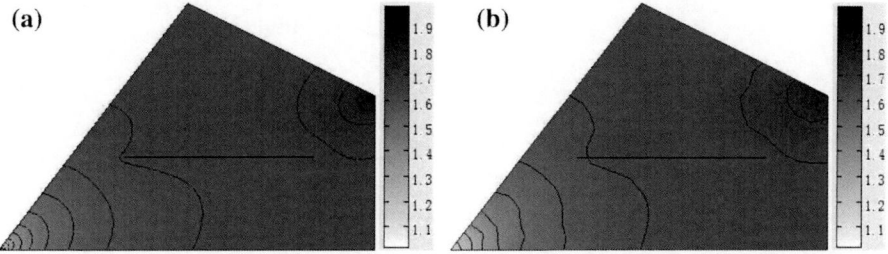

Fig. 2.39 Pressure field distributions calculated by these two methods (MPa). **a** The discrete fracture model, **b** the embedded discrete fracture model (MFD)

error norm is 5.0 %. It can be seen that the method mentioned by this paper also applies to the triangle grid system. Therefore, this method will be applicable for all kinds of complicated boundary shape fractured reservoir flow simulation when combining with a triangular mesh or hybrid mesh (Fig. 2.40).

(3) Complex fractured reservoir
According to an actual fractured reservoir, fracture statistical information data include the fracture density, length, aperture, and direction; we have to generate the corresponding actual complex fractured reservoir model, as shown in Fig. 2.14, where fracture penetrates the ground in the vertical direction. Some parameters in the model are shown as follows: the fracture permeability is $K_f = 1 \times 10\,\mu m^2$, fracture aperture is $a = 1$ mm, fluid viscosity is 1 mPa s, and rock permeability is

$$K_m = \begin{pmatrix} 3 & 1 & 0 \\ 1 & 2 & 0 \\ 0 & 0 & 1 \end{pmatrix} \times 10^{-3}\,\mu m^2 \text{ (Fig. 2.41)}$$

We, respectively, use the discrete fracture model (Fig. 2.42a) and the embedded discrete fracture model (Fig. 2.42b), combined with closed constant pressure boundary condition to carry out single-phase flow numerical simulation to the complex fractured reservoir. Figures 2.43 and 2.44, respectively, show pressure

Fig. 2.40 Comparison of pressure distribution on the injection–production diagonal

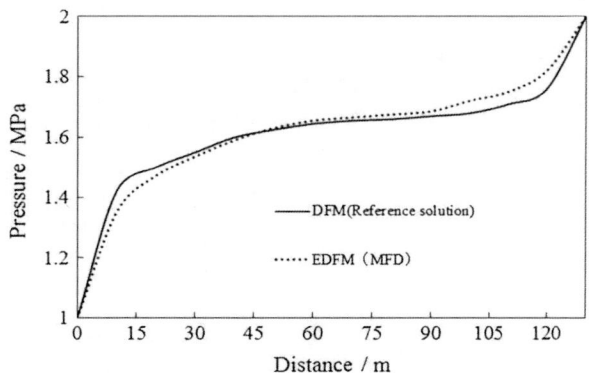

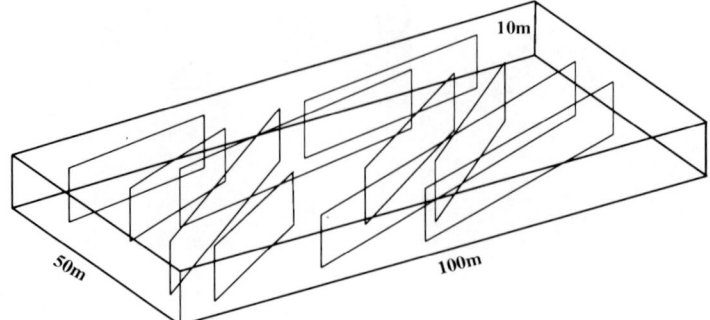

Fig. 2.41 Complex fractured reservoir model

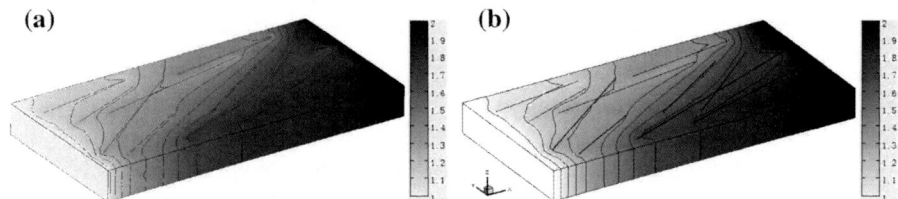

Fig. 2.42 Pressure field distributions by the discrete fracture model and MFD (MPa). **a** The discrete fracture model, **b** the embedded discrete fracture model (MFD)

curve on the two straight lines $y = 26.25$ m and $x = 48.75$ m, which are measured by these two methods. And both results are basically identical.

(4) Calculation example of two-phase flow fractured reservoirs
Based on the embedded discrete fracture single-phase flow model and its solution, and combining the saturation equation by limited volume method in Sect. 2.4.3, the embedded discrete fracture model can be extended to two-phase flow simulation. It is worth noting that this model adapts upstream weighting method to calculate

Fig. 2.43 Comparison of pressure field distribution on the line $y = 26.2$ m

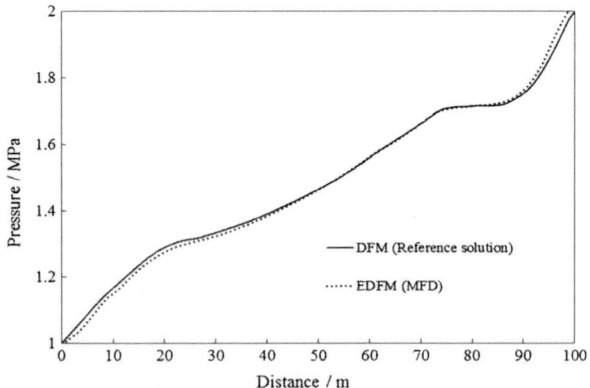

coefficient of fluidity of two-phase flow channeling item in between fracture and matrix.

Three-dimensional fractured reservoir geometry model is shown in Fig. 2.45, whose size is 40 m × 100 m × 100 m ($x \times y \times z$). In the reservoir, there are six large fractures, and to the matrix permeability, we should consider scalar form and full tensor form. Parameters of the model are shown in Table 2.2.

We used the embedded discrete fracture model for Water flooding Displacement numerical simulation of the fractured reservoir, and we have, respectively, considered two forms of matrix permeability, scalar form and full tenser form. Figure 2.46 shows the water saturation distribution in the two forms of matrix permeability when exchange of injection water is 0.5 PV. Figure 2.47 shows the injection–production relation curve under different conditions.

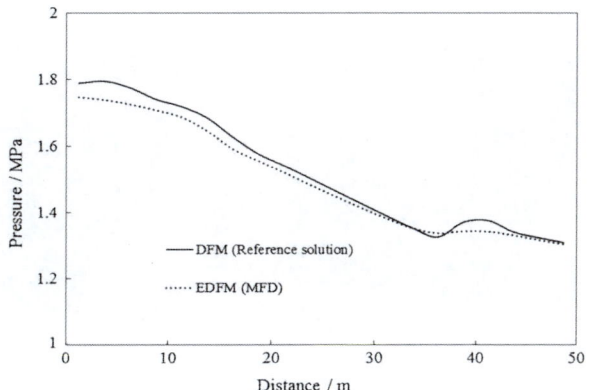

Fig. 2.44 Comparison of pressure field distribution on the line $x = 48.75$ m

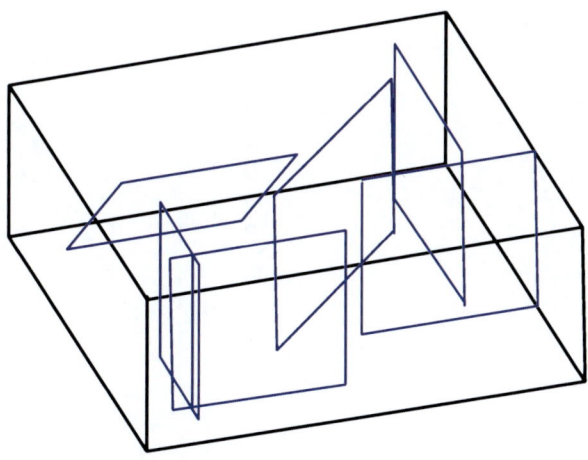

Fig. 2.45 Sketch of geometrical of the 3-D fractured porous medium

Table 2.2 The parameters of physics models

Name of properties	Value
Matrix properties	$\phi_{\mathrm{m}} = 0.4, K_{\mathrm{m}} = 1 \times 10^{-15}\,\mathrm{m}^2, \boldsymbol{K}_{\mathrm{m}} = \begin{bmatrix} 1 & 0.5 & 0.8 \\ 0.8 & 1 & 0.5 \\ 0.5 & 0.5 & 1 \end{bmatrix} \times 10^{-15}\,\mathrm{m}^2$
Fracture properties	$\phi_{\mathrm{f}} = 1.0, K_{\mathrm{f}} = 8.33 \times 10^{-8}\,\mathrm{m}^2, a_{\mathrm{f}} = 1 \times 10^{-3}\,\mathrm{m}$
Fluid properties	$\mu_{\mathrm{w}} = \mu_{\mathrm{o}} = 1.0\,\mathrm{mPa\,s}, \rho_{\mathrm{w}} = \rho_{\mathrm{o}} = 1000\,\mathrm{kg/m}^3$
Residual saturations in matrix and fractures	$S_{\mathrm{wc}} = 0.0, S_{\mathrm{or}} = 0.0$
Relative permeabilities in matrix and fractures	$k_{\mathrm{rw}} = S_{\mathrm{e}}, k_{\mathrm{ro}} = 1 - S_{\mathrm{e}}, S_{\mathrm{e}} = (1 - S_{\mathrm{w}})/(1 - S_{\mathrm{wc}} - S_{\mathrm{or}})$
Capillary pressure	Neglected
Water injection and oil production rates	0.01 PV/d

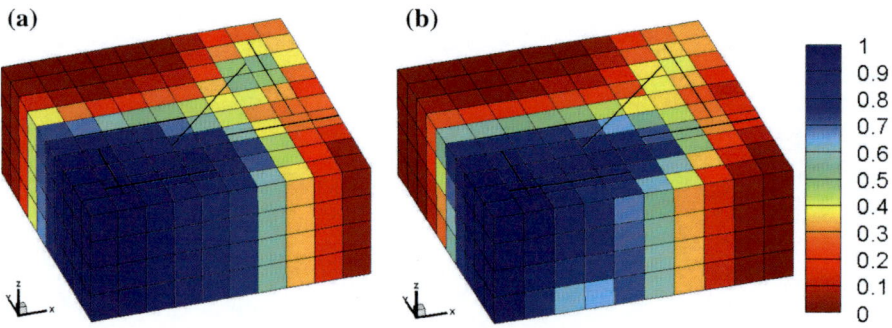

(a)   **(b)**

Fig. 2.46 Water saturation profiles after 0.5 PV water injection with different rock permeability.
a Scalar permeability field, **b** tenser permeability field

Fig. 2.47 Cumulative oil production curve

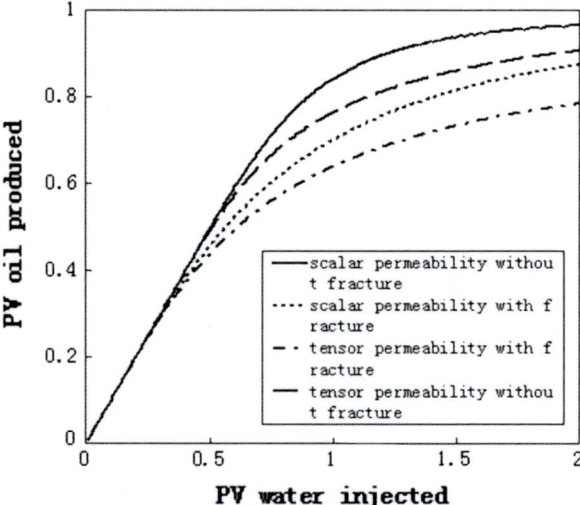

scalar permeability without fracture
scalar permeability with fracture
tensor permeability with fracture
tensor permeability without fracture

PV oil produced

PV water injected

2.6 Summary and Remarks

(1) Discrete fracture model gives an explicit representation to each fracture on the medium, and has good advantages of high calculation precision and good accuracy. But the large amount of calculation is its disadvantage. With the rapid development of computer technology, based on this model, detailed flow simulation will be possible. At the same time, this model, as a tool, can obtain related parameters of double medium and the equivalent medium model. So, it has broad application prospects. Based on the concept of equivalent single fracture, we have set up the discrete fracture model in this section, and elaborated the basic principle of the model. And a variety of numerical solution of the model is given, including the finite element method, finite volume method, and simulation finite difference method. The correctness of the model and algorithm is verified by calculation examples.

(2) The finite volume method needs to be simplified and equivalent process in fracture's intersections makes calculation accuracy to reduce during large-scale computation; finite element method has certain defects, in the aspects of conservation-type format structure and calculation stability; simulation finite difference method is only based on a single grid node and surface information when structuring numerical computational formulation, and it is applicable to any complex grid system in theory. It also has a good local conservation, and can be applied to the discrete fracture flow simulation research, which indicates that it has a broad application prospect.

(3) The embedded discrete fracture model do not need to consider the reservoir fracture morphology when partitioning grid. It only needs to do simple grid subdivision on the matrix system. Therefore, it can effectively avoid the situation of poor quality of the grid which is caused by too much small distance or angle between the fractures. This model needs to compute fluid channeling information between the fracture element and the matrix grid, however, it treats fracture as the inner boundary and the constraint to do unstructured grid subdivision relative to the discrete fracture model, so that its complexity of grid is reduced greatly, which improves efficiently the computational efficiency. From simulation finite difference method to the embedded discrete fracture model, the latter overcomes the limitations of the former. The limitations of the former are as follows: based on the finite difference method, it cannot effectively deal with the permeability of full tensor, and it is not applicable to complicated boundary shape fractured reservoir.

References

Bastian P, Chen Z, Ewing RE, Helmig R, Jakobs H, Reichenberger V (2000) Numerical simulation of multiphase flow in fractured porous media. In: Numerical treatment of multiphase flows in porous media. Springer, pp 50–68

Brezzi F, Lipnikov K, Simoncini V (2005) A family of mimetic finite difference methods on polygonal and polyhedral meshes. Math Model Methods Appl Sci 15:1533–1551

Dalen V (1979) Simplified finite-element models for reservoir flow problems. Soc Pet Eng J 19:333–343

Durlofsky LJ (1993) A triangle based mixed finite element—finite volume technique for modeling two phase flow through porous media. J Comput Phys 105:252–266

Durlofsky LJ (1994) Accuracy of mixed and control volume finite element approximations to Darcy velocity and related quantities. Water Resour Res 30:965–973

Edwards MG (2002) Unstructured, control-volume distributed, full-tensor finite-volume schemes with flow based grids. Comput Geosci 6:433–452

Feng J, Cheng L, Chang Y, Feng J, Fu N, Luo C (2009) Percolation characteristics of fractured anisotropic reservoir. J China Univ Pet (Edition Nat Sci) 33:78–82

Geiger S, Roberts S, Matthäi SK, Zoppou C, Burri A (2004) Combining finite element and finite volume methods for efficient multiphase flow simulations in highly heterogeneous and structurally complex geologic media. Geofluids 4:284–299

Granet S, Fabrie P, Lemmonier P, Quitard M (1998) A single phase flow simulation of fractured reservoir using a discrete representation of fractures. In: 6th European conference on the mathematics of oil recovery (ECMOR VI), Peebles, Scotland, UK

Helmig R, Huber R (1998) Comparison of Galerkin-type discretization techniques for two-phase flow in heterogeneous porous media. Adv Water Resour 21:697–711

Hoteit H, Firoozabadi A (2006) Compositional modeling by the combined discontinuous Galerkin and mixed methods. SPE J 11:19–34

Huang Z, Yao J, Lv X, Li Y (2010) Influence of fluid exchanging between rock matrix and fractures on seepage of fractured porous medium. J China Univ Pet (Edition Nat Sci) 34:93–97

Huang Z, Yao J, Wang Y, Lv X (2011) Numerical simulation on water flooding development of fractured reservoirs in a discrete-fracture model. Chin J Comput Phys 28:41–49

Huang Z, Yan X, Yao J (2014) A two-phase flow simulation of discrete-fractured media using mimetic finite difference method. Commun Comput Phys 16:799–816

Karimi-Fard M, Firoozabadi A (2003) Numerical simulation of water injection in fractured media using the discrete fractured model and the Galerkin method. SPE Reserv Eval Eng 6(2): 117–126

Karimi-Fard M, Durlofsky LJ, Aziz K (2003) An efficient discrete fracture model applicable for general purpose reservoir simulators. In: SPE reservoir simulation symposium. Society of Petroleum Engineers

Karimi-Fard M, Durlofsky LJ, Aziz K (2004) An efficient discrete-fracture model applicable for general-purpose reservoir simulators. SPE J 9:227–236

Kim JG, Deo MD (1999) Comparison of the performance of a discrete fracture multiphase model with those using conventional methods. In: SPE symposium on reservoir simulation, pp 359–371

Kim JG, Deo MD (2000) Finite element, discrete-fracture model for multiphase flow in porous media. AIChE J 46(6):1120–1130

Lange A, Basquet R, Bourbiaux B (2004) Hydraulic characterization of faults and fractures using a dual medium discrete fracture network simulator. In: Abu Dhabi international conference and exhibition. Society of Petroleum Engineers

Lee SH, Lough MF, Jensen CL (2001) Hierarchical modeling of flow in naturally fractured formations with multiple length scales. Water Resour Res 37:443–455

Li L, Lee SH (2008) Efficient field-scale simulation of black oil in a naturally fractured reservoir through discrete fracture networks and homogenized media. SPE Reserv Eval Eng 11(4): 750–758

Lie K, Krogstad S, Ligaarden IS, Natvig JR, Nilsen HM, Skaflestad B (2012) Open-source MATLAB implementation of consistent discretization on complex grids. Comput Geosci 16:297–322

Lipnikov K, Manzini G, Shashkov M (2014) Mimetic finite difference method. J Comput Phys 257:1163–1227

Lv X (2010) Numerical simulation of the discrete fracture network model based on control volume method. China University of Petroleum (East China)

Lv X, Yao J, Huang Z, Zhao J (2012) study on discrete fracture model two-phase flow simulation based on finite volume method. J Southwest Pet Univ Technol Ed 34:123–130

Matthäi SK, Belayneh M (2004) Fluid flow partitioning between fractures and a permeable rock matrix. Geophys Res, Lett 31

Moinfar A, Varavei A, Sepehrnoori K, et al. (2012) Development of a novel and computationally-efficient discrete-fracture model to study IOR processes in naturally fractured reservoirs. In: SPE improved oil recovery symposium. Society of Petroleum Engineers

Monteagudo JEP, Firoozabadi A (2007) Control-volume model for simulation of water injection in fractured media: incorporating matrix heterogeneity and reservoir wettability effects. SPE J 12:355–366

Noorishad J, Mehran M (1982) An upstream finite element method for solution of transient transport equation in fractured porous media. Water Resour Res 18:588–596

Panfili P, Cominelli A, Scotti A (2013) Using embedded discrete fracture models (EDFMs) to simulate realistic fluid flow problems. In: Second EAGE workshop on naturally fractured reservoirs

Raviart P-A, Thomas J-M (1977) A mixed finite element method for 2-nd order elliptic problems. In: Mathematical aspects of finite element methods. Springer, pp 292–315

Reddy JN (1993) An introduction to the finite element method. McGraw-Hill, New York

Reichenberger V, Jakobs H, Bastian P, Helmig R (2006) A mixed-dimensional finite volume method for two-phase flow in fractured porous media. Adv Water Resour 29:1020–1036

Rutqvist J, Wu Y-S, Tsang C-F, Bodvarsson G (2002) A modeling approach for analysis of coupled multiphase fluid flow, heat transfer, and deformation in fractured porous rock. Int J Rock Mech Min Sci 39:429–442

Sandve TH, Berre I, Nordbotten JM (2012) An efficient multi-point flux approximation method for Discrete Fracture-Matrix simulations. J Comput Phys 231:3784–3800

Slough KJ, Sudicky EA, Forsyth PA (1999) Numerical simulation of multiphase flow and phase partitioning in discretely fractured geologic media. J Contam Hydrol 40:107–136

Thomas LK, Dixon TN, Pierson RG (1983) Fractured reservoir simulation. Soc Pet Eng J 23(1):42–54

Van Golf-Racht, T.D., 1982. Fundamentals of fractured reservoir engineering. Elsevier

Wang X (2003) Finite element method. Tsinghua University Press, Beijing

Yan X, Huang Z, Yao J, Huang T (2014) The embedded discrete fracture model based on mimetic finite difference method. Sci Sin Tech 44:1333–1342

Yao J, Wang Z, Zhang Y, Huang Z (2010) Numerical simulation method of discrete fracture network for naturally fractured reservoirs. Acta Pet Sin 31:284–288

Yuan S, Song X, Ran Q (2004) Fractured reservoir development technology. Petroleum Industry Press, Beijing

Zhang N, Yao J, Huang Z, Wang Y (2013) Local conservative finite element analysis of two-phase porous flow. Chin J Comput Phys 30:667–674

Zhang Q, Wu A (2010) Three-dimensional arbitrary fracture network seepage model and its solution. Chin J Rock Mech Eng 29:720–730

Zhang Y (2005) Rock hydraulics and engineering. China Water Power Press, Beijing

Zhou F, Shi A, Wang X (2014) An efficient finite difference model for multiphase flow in fractured reservoirs. Pet Explor Dev 41(2):262–266

Zhou Z (2007) Theory on dynamics of fluids in fractured medium. China Higher Education Press, Beijing

Zhou Z, Wang J (2004) Dynamics of fluids in fractured media. China WaterPower Press, Beijing

Chapter 3
Discrete Fracture-Vug Network Model

Abstract In this chapter, a novel conceptual model named discrete fracture-vuggy network (DFVN) model, which is assumed to be composed of free flow region (macrovug system) and porous medium region (macrofracture system and porous rock matrix system), has been developed to demonstrate the flow characterization: coupling of free flow and porous media flow. Based on DFVN, the macromathematical model of two-phase coupling flow, including coupling interface conditions, is developed by upscaling micro Navier–Stokes equation through Volume Average Method. Darcy's law is utilized to describe the flow behavior in the porous medium, while in the free flow region, Navier–Stokes equation is applied. Besides, normal stress and mass continuity conditions, as well as Beavers–Joseph–Saffman boundary condition are added to coupling the two different flow subdomains. The whole mathematical model is solved by upwind Petrov–Galerkin finite element method and the coupling is implemented via alternative solution method. Finally, several numerical cases are given to validate the effectivity and accuracy of the coupling model.

Keywords DFVN · Beavers–Joseph–Saffman condition · Coupling fluid–porous flow · Volume average method · Finite element method

Vugs and macrofractures are developed in earth crust under the effect of karst and tectonic movement. Their spatial scales are much larger than intergranular pore and also have an important influence on the physical properties of rocks and fluids flow in the stratum, which is defined as fractured-vuggy medium in the exploration and development of petroleum. Figure 3.1 illustrates the outcrop of the typical fractured-vuggy medium. This kind of medium, such as carbonate reservoir, underground karst aquifer, etc., exist extensively in nature and has a close relationship with human beings. A fracture is the smallest geologic structure, almost existing in all the stratums. In book of "Carbonate Reservoir Characterization," Lucia defined vugs as a sort of caves whose pore spaces are larger than intergranular pore in porous medium which are formed because of dissolution and decomposition occurring in carbonate and sulfate (Lucia 2007). Vugs can become

© Petroleum Industry Press and Springer-Verlag Berlin Heidelberg 2017
J. Yao and Z.-Q. Huang, *Fractured Vuggy Carbonate Reservoir Simulation*,
Springer Geophysics, DOI 10.1007/978-3-662-55032-8_3

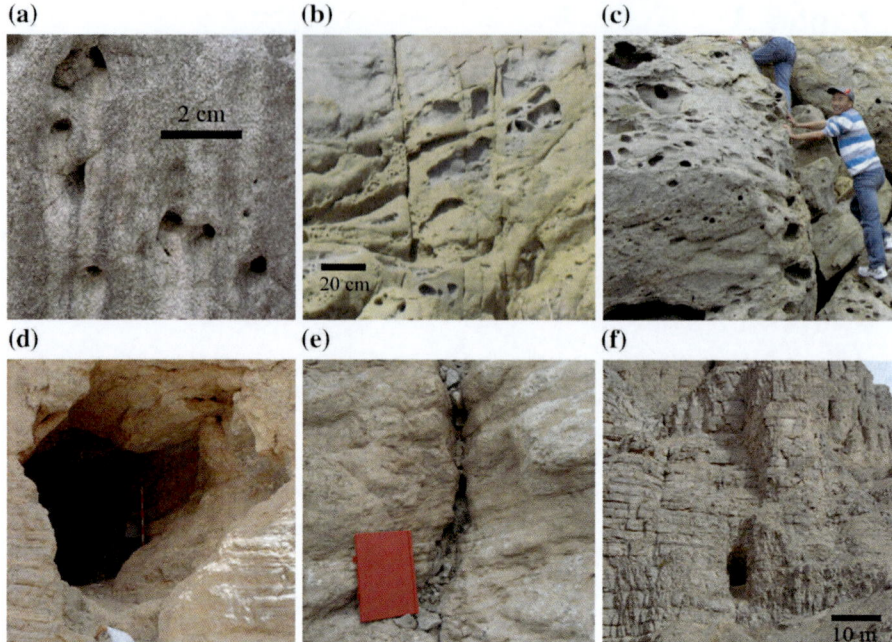

Fig. 3.1 Typical fractured-vuggy medium field outcrop

more complicated fractured-vuggy network by connecting with fractures in stratum as in Fig. 3.1b, f.

Fractured-vuggy medium has complicated internal structure, including not only bed rocks and fractures but also multi-scale spatial vugs, whose volumes range from millimeter scale to meter scale (Erzeybek and Akin 2008). Meanwhile a large number of logging, core, and outcrop materials indicate that due to tectonic movement in the late period, fractures and vugs are filled severely not only by physical filling process such as with sand, mud, etc., but also by chemical filling process with silicon, calcite, etc., which aggravate the heterogeneity of the stratums as Fig. 3.1e shows.

3.1 Mathematical Models: The State of the Art

The complication and multi-scale in fractured-vuggy medium structure make it a challenge task to describe and simulate the fluid flow behavior accurately (Popov et al. 2009b). Because of the presence of vugs which are connected via fracture networks at vuggy multiple scales, the main challenge is the coexistence of seepage mechanics and free flow in the state of laminar or/and turbulent flow in such reservoirs (Zheng et al. 2009). Similar to the research on fluid flow in fracture medium, the mathematical model used in the present research on fluid flow in fractured-vuggy porosity can be divided into two classes: conventional continuous porosity model and discrete porosity model. The former can be subdivided into triple porosity model, equivalent porosity model, and its evolutionary model.

3.1.1   Conventional Continuous Media Model

(1) Triple porosity model
In China, the concept of triple porosity model was initially proposed by Wu and Ge (1983) while in foreign countries, Abdassah and Ershagi were the first who proposed it (1986). All of their researches are about natural fractured reservoirs. They found that the pressure character curves in some fractural reservoirs vary from the previous research results and cognition. Even the double-porosity fluid flow theory could not make a perfect explanation, so they proposed the concept of triple porosity. In this research, they divide rock matrix into two classes: one has a good connectivity with fracture systems and the other class has a bad one. There are two possible reasons contributing these two classes pore systems. One reason is due to the heterogeneity of primary pores and connectivity and the other one is that there are isolated vugs in some rock matrixes. As Fig. 3.2 shows it is considered that the comprehensive permeability of rock matrix with vugs are better than that of without vugs. Thus rock system can be considered as two pore systems, which compose triple porosity system together with fracture system by interporosity flow function.

Liu et al. made a research on the characteristic of fluid flow in fractured lithophysa rock with triple porosity model (Liu et al. 2003). Lithophysa structure is a common innate orbicular structure. There are multilayer concentric spheres—some with cavities as a result of gas escape and volume shrinkage during the process of solidification. Most of the lithophysa are cavities and this type of rock structure is a typical fractured-vuggy medium. In their research, fluid only flows in the fracture

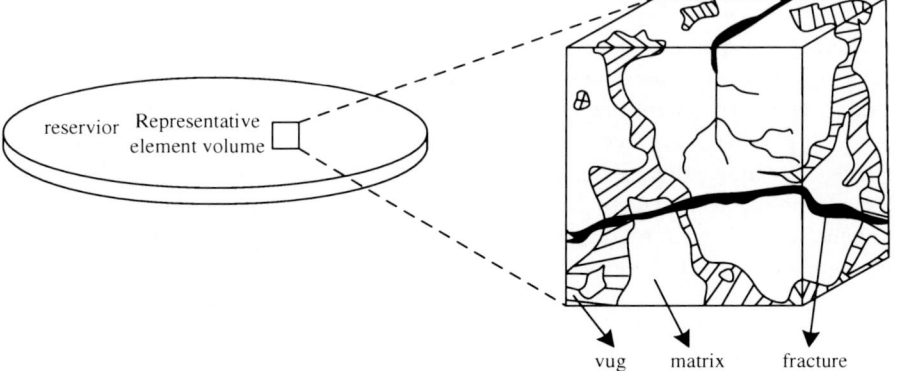

Fig. 3.2 Schematic of fractured-vuggy medium triple pores system

network system while rock matrix and vugs act as the primary reservoir space. Fluid flows into fracture system and flow between systems can be described by quasi-steady state interporosity flow function. Under the condition of variable well-bore storage and aiming at triple porosity reservoirs consisting of rock matrix systems, fracture systems and vugs systems, Jun Yao et al. made a research on well test interpretation model where there is a connectivity between vugs and wellbores (Chang et al. 2004; Yao et al. 2004; Yang et al. 2005). In their research, wellbore is assumed to be only connected with vug systems and the liquid from fractures and rock matrix are neglected. In other words, both rock matrix and fractures are only considered as source and quasi-steady interporosity flow happens among matrix, fractures, and vugs.

Afterwards Velazquez et al. (2005), Wu et al. (2004, 2006, 2007), Yao and Zisheng (2007) and Zhang et al. (2008) made further development for triple porosity model based on the characteristic of fractured-vuggy porosity. They divided this type of porosity into three parallel continuous systems, namely high permeability fracture system, low permeability rock system, and vug system. Relationships can be built up among these three systems by Quasi-steady state function. Based on this function the character curves reflecting pressure variation of natural fracture reservoir is studied and well test curves template was produced which enhanced our understanding for the characteristic of fluid flow in fractured-vuggy reservoir. Although triple porosity model describes the phenomenon of preferential flow in fractured-vuggy media to some degree and considers the exchange of matter among fracture systems, rock matrix systems and vug systems, which conforms to actual models, the characteristics such as anisotropy, discontinuity, multi-scale, etc., of fractured-vuggy media cannot be displayed as we have assumed that rock matrix and vug systems divided by fracture systems have the same sizes and shapes that are overly simplified. Meanwhile the coefficient of matter exchange is difficult to determine and the triple continuous assumption will be appropriate under the condition that fractures and vugs have a good development. What's more important is that this model does not show the multi-scale coupling characteristics of fluid flow in fractured-vuggy reservoir. Therefore in many cases the calculated results have a huge deviation from practice.

(2) Equivalent porosity model

Varying from the triple continuous assumption in triple media model, equivalent media model considers the whole fractured-vuggy media as a continuum where physical parameters in every coordinate are in a local equilibrium state under the effects of adequate fluids exchange among fracture rock matrix and vugs. This model focuses on macroflow characteristic showing in the entire media. Some parameters like the solute concentration and permeability, etc., are averaged equivalently in the whole media, which therefore is considered as anisotropy media with permeability tensor and physical structure of single fracture and vug will be neglected. This media will be thought as common porous media, whose flow characteristic and numerical simulation can be explained with present fluid flow theory. It is critical to gain equivalent permeability and reservoir parameters in media.

There are two methods to determine permeability in porous media. They are experimental measurement and theoretical calculation. The former one mainly used logging, well testing, tracer, and core to determine the permeability. However, there are many problems (Arbogast et al. 2004) existing in this method because of the spatial multi-scale of fractures and vugs in fractured-vuggy media, which cannot reflect the real permeability. Thus theoretical calculation has been paid more attention gradually and has become a primary method to predict the permeability in complex media. The first step in theoretical calculation is to build corresponding model to describe the flow condition in media. The secondary step is to obtain the equivalent permeability by using equivalence principle.

Equivalent media model has many advantages. This model is simple and its theory is mature and it is easy to calculate. This model is appropriate for fluid flow in large-range areas where there is an intensive fractures and vugs distribution while it will generate huge errors if using the theory of continuous media making an analysis when the fractures and vugs are highly discrete. It is a controversial topic about whether we can utilize equivalent media model to study fractured-vuggy media model. The key point is to judge whether the Representative Element Volume (REV) exists. Actually effective REV does not exist because the influence from heterogeneity and multi-scale in most of fractured-vuggy media exist.

In conclusion, conventional continuous media model is not appropriate for most of fractured-vuggy media. Therefore it is realized that it is not feasible to study this type of media based on conventional fluid flow. We must establish a new set of appropriate study method and mathematical description for this special media. At present there is no mature study foundation in this area and it is still at the stage of exploration.

3.1.2 Discrete Media Model

After experiencing a long-term geological process, rock system will generate discrete surface with different types, sizes and mechanical properties, including joints, fractures, and faults. Therefore, all the rock systems with developed fractures belong to discrete media. In 1971, Louis and Wittke have come up with line element network model (Louis and Wittke 1971) which is similar with cyclic current method in circuit analysis. The mathematical model in this method is developed based on node flux conservation, loop pressure conservation, and bar-type element water pressure difference conservation, which is the earliest discrete media model. This model was developed into discrete fracture network model afterwards by Wilson and Witherspoon, where permeability in rock matrix (Wilson and Witherspoon 1974) were not considered. Then Noorishad et al. (1982) and Baca et al. (1984) established discrete media model with the consideration of rock matrix permeability, called discrete fracture model. During the past decade, with the continuous development of fractured reservoir, tight sandstone reservoir and low permeability reservoir, discrete fracture model is gradually causing scholars' attention abroad and becoming the research focus. In recent years in Yao et al. pursued professional research work (Yao et al. 2010c; Huang et al. 2011a, b) in areas of oil–water flow simulation in fracture reservoir based on discrete fracture model. The results showed that this model highlights the hydraulic power characteristics of single fracture and can depict fluid flow characteristics in fracture rock matrix accurately.

Based on the research methods and results on fractured reservoir and given the characteristics of fractured-vuggy medium, Yao et al. (2010a) proposed Discrete Fractured-Vuggy Network Model (DFVN). This model combines vug system to the former discrete fracture model, which made it appropriate to the research on fractured-vuggy media. This is an effective extension and expansion, which reflects the structure characteristics of fractured-vuggy media and fully illustrates the multi-scale coupling flow feature.

At present DFVN model is limited in single phase flow research (Yao et al. 2010b; Huang et al. 2011b), whose fluid flow area conformed to Darcy flow. The area of free flow is described by Navier–Stokes equations and two-phase region can be coupled by Beavers–Joseph–Saffman boundary conditions. The result shows that DFVN model can describe the flow conditions in fractures-vugs accurately and depict the single flow characteristics in fracture-vugs explicitly.

3.1.3   Discrete Fractured-Vuggy Network Model(DFVN)

Fractured-vuggy media varies from conventional clasolite reservoir and common fracture media in form, distribution, attitude, etc., due to the effect of structure fracture, corrosion, diagenesis, and epigenesis.

The main structure characteristics are listed as follows:

(1) Reservoir spaces are various, including pores, fractures, and vugs. Pores are void space whose spatial scales are similar and no less than 2 mm in three directions. There are many formation factors and types of pores, whose porosity and permeability are poor. Similar to pores, vugs are also void space whose spatial scales are close and less than 2 mm, but they can be recognized completely in cores and they are a variety of important reservoir space. Fractures are void space whose one direction scale is small and the other two are very large (usually less than 1:10). Fractures are the tiniest geological structure. They exist in reservoir widely and they can link to vugs effectively to form primary reservoir and fluid flow space.

(2) The scale of reservoir space range from several micrometers to scores of meters, having obvious scale characteristics and heterogeneity, Table 3.1 shows the spatial scale classification of fractures and vugs. It also shows the geological cause of formation.

Table 3.1 Classification of fractures and vugs

Form	Classification	Diameter/aperture (μm)	Geologic process
Cavity	Large cavity	$>5 \times 10^5$	Denudation
	Medium cavity	5×10^5 to 1×10^4	
	Small cavity	1×10^4 to 2×10^3	
Fracture	Tectonic corroded fracture	Vary in size	Tectonic denudation
	Tectonic fracture	<1	Tectogenesis
	Interlayer fracture	10–200	Sedimentation
	Pressolved fracture	Several microns	Sedimentation and lithogenesis

Vugs and fractures are filled seriously. A large amount of logging, cores, and outcrops materials show that there exists mechanical–physical filling (by sand and mud) and chemical filling (by siliceous and calcite) in fractures and vugs, which aggravate the heterogeneity of reservoir due to the tectonic movement in later stage.

In terms of above structure characteristics, fractured-vuggy medium is actually a huge discrete fractured-vuggy network space where both free flow and porous flow exist, as Fig. 3.3a shows. This model is such a complex coupling flow system that the present theories are not suitable for this type of media. Thus we propose discrete fractured-vuggy network model, aiming to describe the real flow behavior in fractured-vuggy media.

As Fig. 3.3b shows, in DFVN model, fractured-vuggy medium is divided into rock system (including matrix, microfracture, and microvugs), fracture system and vug system, among which fracture and vug are imbedded into rock, forming a network. Vugs are considered as free flow region while rock and fractures are thought as free flow region. Fluid flow region can be regarded as typical discrete fracture model. Thus DFVN model is an effective extension and expansion of discrete fracture. As Fig. 3.4 shows, DFVN model can be divided into two different flow regions: free flow region and porous flow region. How to establish corresponding coupling flow mathematic model is the main purpose of our study. Obviously, DFVN model is concerned with two critical scientific problems: the coupling of porous flow and free flow and the establishment of discrete fracture mathematic model in porous flow region.

3.2   The Coupling Theory of Porous Flow–Free Flow

The coupling of porous flow and free flow is a universal phenomenon in nature. For example, the evaporation of water in soil with air flows, proton exchange membrane fuel cell technology, well-reservoir coupling in oil production and blood–organ interactions, etc.

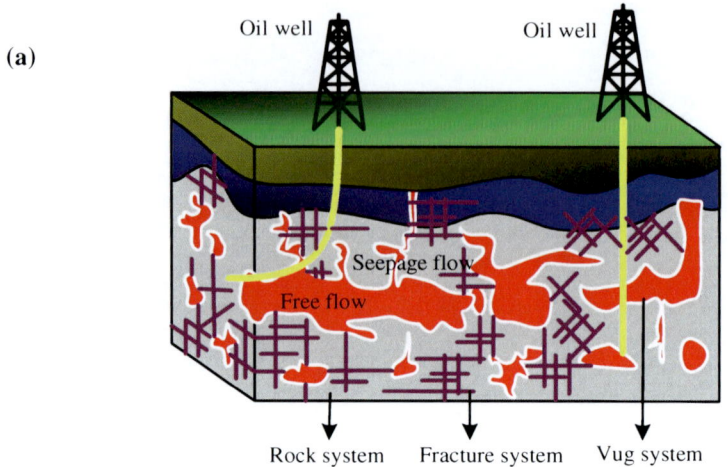

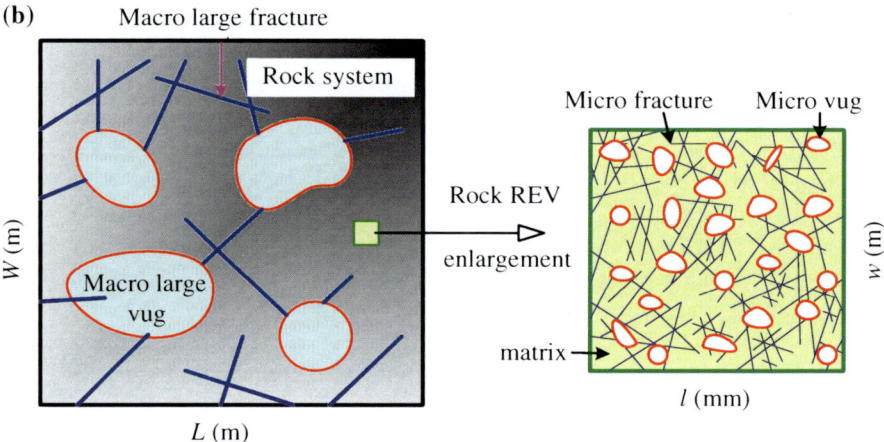

Fig. 3.3 Scheme of natural fractured-vuggy medium

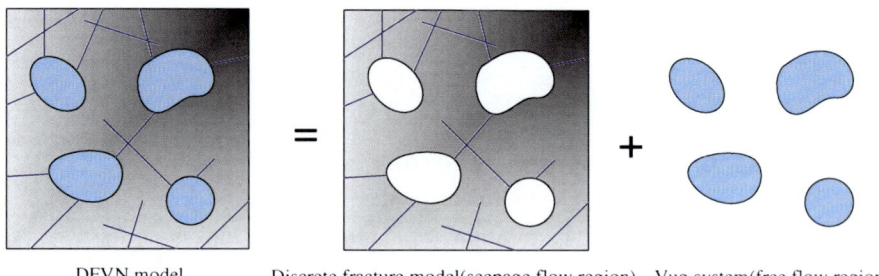

Fig. 3.4 Decomposition chart of DFVN model

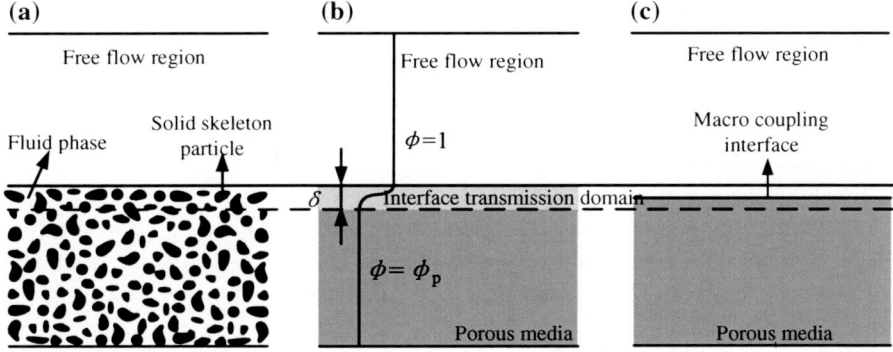

Fig. 3.5 Comparison of interface description in different coupling methods

As Fig. 3.5a shows, the fluid flow is always conformed to Navier–Stokes equation in microscope pore scale not only in free pathway but also in pores. Theoretically, the particular description on fluid flow in coupling flow area can be obtained if geometry structure information is known. However, it is nearly impossible to explicitly describe the complex geometry structure in porous media apart from some particularly simple conditions (such as straight capillary model). To overcome this difficulty, we usually transformed it into much larger scale to describe, namely macroscale representative elementary volume (REV) scale. Darcy law is used to describe macroscale flow in REV scale. The control differential equations in two flow regions vary in physical significance and differential order, which causes some difficulties to the coupling of two kinds of flows. At present there are two methods to deal with this coupling physical problem: single-domain approach (SDA) and two-domain approach (TDA).

3.2.1 Coupling Flow Method and Its Surface Condition

(1) Single-Domain Approach
In this method, the whole coupling flow region is considered as continuous system by introducing the definition of interface transition region. Among this method, physical attribute of porous media changes continuously in spatial distribution, such as porosity and permeability. The whole region can be described by a set of unified flow equations. Assuming fluid and heterogeneous porous media are incompressible, the unified flow equations can be written as follows:

$$\nabla \cdot v = 0 \qquad (3.1)$$

$$\phi^{-1}\frac{\partial(\rho v)}{\partial t} + \phi^{-2}\nabla \cdot (\rho v v) = -\nabla p + \rho \mathbf{g} + \mu_e \nabla^2 v - \underbrace{\mu \mathbf{K}^{-1} \cdot v}_{\text{Darcy term}} \qquad (3.2)$$

where ρ, v, μ and p are density, velocity vector, viscosity, and pressure of the fluid, respectively; $\mathbf{g}$ is acceleration of gravity; ϕ is porosity; μ_e is effective viscosity; $\mathbf{K}$ is permeability tensor of porous media.

In free flow region, $\phi = 1$, $\mu_e = \mu$, $\mathbf{K} \rightarrow \infty$. Thus the Darcy term in Eq. (3.2) tends to be zero and this equation is simplified into Navier–Stokes equation. In porous media fluid flow region, $\phi = \phi_p$. μ_e is relevant to structure characteristic of porous media, generally $\mu_e = \mu/\phi$. Although all the terms that have relationships with velocity are kept, only Darcy term plays the primary role. The advantage of this method is that the whole region is described by a unified set of equations. Thus the requirement for continuity is satisfied automatically in interface. There is no need to introduce extra interface conditions. In SDA, it is critical to determine the parameters in interface transition region while present research method could not be able to determine these parameters and their change. This set of unified equations is only suitable for single phase flow. There is no unified equation now and it is very difficult to study for two phases flow.

(2) Two-Domain Approach, TDA

TDA will establish flow mathematic model in two different flow regions respectively, which is different from SDA. Then this two flow models are coupled by introducing specific interface conditions, as Fig. 3.5c shows. In free flow region, fluid flow can be described by classical Navier–Stokes equations, while in porous media, flow model can be described by Darcy equations or its modified equations. Meanwhile appropriate interface conditions are required to couple this two flow regions. There are many scholars doing some work on this problem and they have gained a series of research results during the past half century.

Although Rhodes and Rouleau had made a discussion (Rhodes and Rouleau 1966) on this coupling flow problem in 1966, systematical experiment and theoretic research were finished initially by Beavers and Joseph (1967). Based on the experimental results and theoretical analysis, Beavers and Joseph proposed a semiempirical formula to couple Stokes equations and Darcy equations. This formula is the famous BJ velocity slip condition as follows:

$$\frac{du}{dy}\Big|_{y=0^+} = \frac{\alpha}{\sqrt{K}}(u_B - Q) \qquad (3.3)$$

where u is velocity component in x axis in free flow region; 0^+ is the underlying boundary in free flow region, as shown in Fig. 3.6a; u_B is slip velocity on boundary; Q is seepage velocity in x axis in porous media; α is nondimensional slip coefficient, which is used to represent porous structure characteristic in interface region; K is the permeability of porous media.

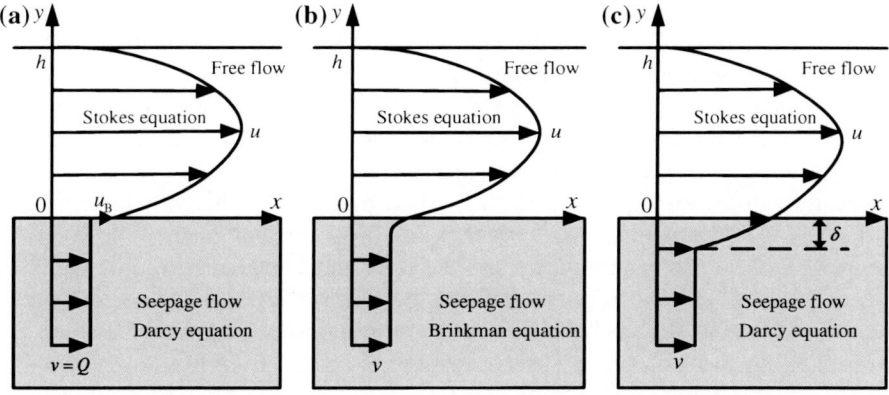

Fig. 3.6 Comparison of different coupling flow modes (h is the height of free flow)

Afterwards, Beavers et al. (1970) additionally did a series of experiment to validate the accuracy of interface conditions further. To validate the universality of the interface conditions, Beavers et al. did experiment with gas (Beavers et al. 1974) and the experiment results validated the accuracy and universality of Eq. (3.3) again. Saffman proved the effectiveness of BJ condition from the perspective of physics and mathematics meanwhile Saffman pointed that BJ condition was only suitable for models similar to Beavers–Joseph experiment equipment and he also proposed a more universal interface condition (Saffman and Saffman 1971), namely Beavers–Joseph–Saffman (BJS) conditions:

$$u_{\mathrm{B}} = \frac{\sqrt{K}}{\alpha}\frac{\partial u}{\partial n} + O(K) \tag{3.4}$$

where n is normal direction of the interface.

Compared with Eq. (3.3), we can see that $Q = Q(K)$. Generally, seepage velocity Q is far less than slip velocity u_{B}. Thus Q can be neglected. Slip velocity can be neglected if slip velocity is smaller than the largest seepage velocity in porous media, namely nonslip interface conditions. Dagan (1979) proposed the same conclusion. Taylor (1971) and Richardson (1971) proved the correctness of BJ theoretically further. Afterwards Jonse re-clarified the mathematical and physical significance of BJ condition (Jones 1973). He thought the essence of BJ condition is linking shear stress with slip velocity and establishing general boundary condition:

$$\boldsymbol{u} \cdot \lambda = -\frac{\sqrt{\lambda \cdot \boldsymbol{K} \cdot \lambda}}{\mu\alpha}(\boldsymbol{n} \cdot \boldsymbol{\tau} \cdot \lambda) \tag{3.5}$$

where λ is a unit tangent vector of the interface. N is unit normal vector. τ is viscous stress of fluid in free flow region.

Seepage velocity can be neglected in Eq. (3.5).

Recently, Jager and Mikelic (1999, 2009) deduced the same interface condition based on homogenization theory. The interface condition of both Eqs. (3.4) and (3.5) can be considered as the modification and extension of BJ condition.

Compared with nonslip wall condition, the total flux in free flow region will increase after introducing velocity slip condition. Toward this, Beavers and Joseph defined a parameter Φ to represent this variation. They draw the result that is basically agree with experiment data by adjusting slip coefficient whose range is $0.1 \leq \alpha \leq 4.0$. Meanwhile, they think that the slip coefficient α is not relevant to the property of fluid and is only relevant to the structure characteristic of porous media. Then Beavers (1970), Richdson (1971), and Goyeau (2003) made a research on how to determine slip coefficient α. And the results show that α has a close relationship with local geometry structure of the interface. But until now there is no specific expression for it. Recently, Zhang and Prosperetti (2009) showed that as to the same topic there is difference in α between pressure-driven flow and viscous stress-driven flow and α is relevant to Reynolds number. Later Liu and Prosperetti (2010) proposed a new boundary condition to express this characteristic as follows:

$$\frac{du}{dy}\Big|_{y=0+} = \frac{\alpha}{\sqrt{K}}(u_B - \gamma Q) \tag{3.6}$$

where γ is a dimensionless scalar. Liu and Prosperetti predicted that this coefficient has relationship with the volume percentage of void space in interface region. Obviously, this new boundary condition comes from intuition and experience. The problem that how to determine α and γ is present hot spot and difficult point, which will be elaborated in Eq. (3.3).

In TDA, Darcy equation is used in the flow analysis of seepage region while the flow in porous media can be described by Brinkman equation. In 1971, Taylor (1971) had noticed that BJ condition can be deduced from Brinkman equations though he did not use the term of Brinkman equation. After that, the work of Neale and Nader (1974) showed that the results from Brinkman equations and the results that gained by combining Darcy equations with BJ conditions are consistent in seepage region. Furthermore, they obtained $\alpha = \sqrt{\mu_e/\mu}$. Ochoa-Tapia and Whitaker (1995a, b) proposed stress jump condition and velocity continuity condition based on nonlocal average Stokes equation. Brinkman equations will be adopted in seepage region and Stokes equations will be utilized in free flow region, as Fig. 3.6b shows. The stress jump condition can be listed as follows:

$$u_e \frac{\partial v}{\partial y} - \mu \frac{\partial u}{\partial y} = \frac{\beta}{\sqrt{K}} u, \quad y = 0 \tag{3.7}$$

where v is seepage velocity; u is the velocity in free flow region; β is dimensionless coefficient and $\mu_e/\mu = 1/\phi$. The analytical solution of above model can almost matches to Beavers–Joseph experimental data by adjusting coefficient β. In fact stress jump condition is a representative of the rapid porous media structure variation in interface region, which is similar to BJ condition. Chandesris and Jamer

(2006, 2009) and Jamet and Chandesris (2009) proposed a general two steps scale updated method and combined this method with asymptotic expansion method to deduce and establish the corresponding coupling interface condition of single phase laminar flow and turbulent flow.

However, Nield pointed that this coupling model was not suitable to two phases and multiphases flow because it will be very difficult to consider the effect from interfacial tension in interface region. Meanwhile the mutability of porous media will disturb the continuity of Marangoni effect, which brings some difficulties. Thus Nield thinks that the best approach is to introduce a tiny interface transition region and to utilize BJ condition to solve this coupling flow problem.

Apart from the above two primary coupling models in TDA, Bars and Worster (2006) proposed a new coupling model. A viscous transition region was defined in seepage flow–free flow porous media. In this region, Stokes equations are still suitable while outside this region fluid flow meets Darcy law, as Fig. 3.6c shows. The equation can be listed as follows:

$$u(x, -\delta_+) = v(x, -\delta_-), \quad \delta = c\sqrt{K} \tag{3.8}$$

where c is dimensionless scalar coefficient, $c = O(1)$.

In the case of porosity changing continuously within the interfacial region, the analytical solution based on interface condition Eq. (3.8) matches better with Beavers–Joseph experimental data than on BJ condition. However, it is very difficult to apply this interface condition primarily in practical engineering because defining and describing this coupling interface of practical problem accurately is impossible.

According to the above analysis, we have known the coupling flow characteristic of single phase laminar and turbulent flow very well. However, two phases and multiphases have not been studied. There are two reasons contributing to current situation: one is that the research on seepage flow–free flow for two phases and multiphases is very difficult; another one is that engineering single phase flow played a primary role in previous practical. Two phases and multiphases attracted much more attention with the expansion of human activities range and the deepening of research.

Mosthaf et al. (2011) considered air free flow as a single phase and two components flow model when he studied the effectiveness on climate from water evaporation. However, two phases and two components seepage model will be adopted in soil. In this model these two flows can be coupled by applying normal velocity continuity and normal stress continuity and BJS in interface. Obviously, they simply extend the result of single phase flow into the established model but lack of enough support theoretically. Even though, this result still promotes the study on coupling two phases flow. It is still a challenging task to build corresponding two phases coupling flow model. As to this problem, two phases coupling flow mathematic model and its interface condition will be established theoretically by applying volume average method and two scale upscaling method based on TDA, which will be introduced later in Sect. 3.3.

3.2.2 Two Phases Flow of Discrete Fracture-Vug Network

In micro-porous scale, fluid flow can be described by Navier–Stokes equation and interfacial tension equation in both free flow region and porous medium region. Thus there is no difference in the control differential equations of fluid flow in porous scale. However, porous medium flow usually needs to be described and analyzed in a much larger scale by macroscopic Darcy equation or its modified formula (Bear 1972). In this case, scale difference exists in control differential equations between two kinds of flow regions. How to remove the scale difference is the critical problem in establishing the coupling mathematic model of porous fluid flow.

This section solves this problem by twice scale updating based on volume average method. Then the mathematic model of two-phase coupling of porous fluid flow is built. First, make a scale update from the perspective of flow equation in microscale based on volume average method for both the two flow regions directly. A set of universal volume average equations can be obtained, which are suitable in the whole coupling flow regions without introducing any scale limit. Then by introducing particular scale limiting condition, simplify the above volume average equations in free flow region and seepage flow region respectively. In free flow region, it can be simplified into typical two fluids model while in seepage flow region, it can be simplified into two-phase flow Darcy model. To remove the error during the process of simplification, establish corresponding coupling interface condition by introducing surface excess function and the complete mathematic model for two-phase coupling of porous fluid flow is formed. The following part will elaborate the basic concepts and corresponding principles first in order to lay foundations for the later theoretical derivation.

1. The basic principles of volume average method
Considering two-phase flow system as Fig. 3.7 shows, we can assume an average volume V which will not vary with time. This volume V includes solid matrix

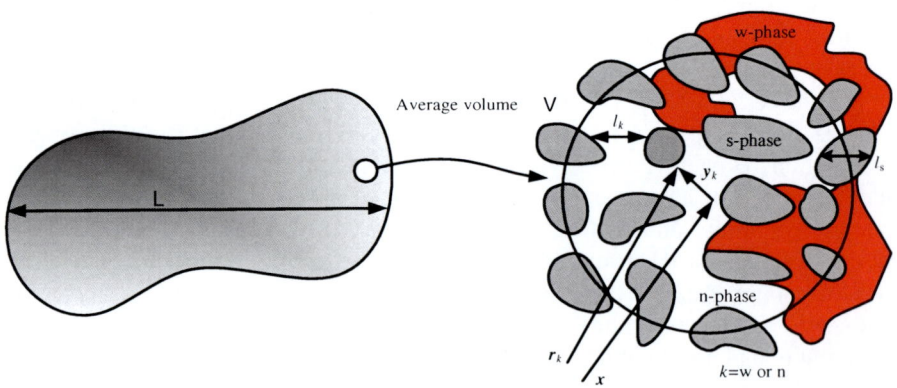

Fig. 3.7 Representative element volume in study region

(s-phase), wetting phase (w-phase), and non-wetting phase (n-phase). Volume average method is mainly based on the following five basic definitions and principles.

(1) Superficial average

$$\langle \psi_k \rangle |_x = \frac{1}{V} \int_{V_k} \psi_k |_{x+y_k} dV \tag{3.9}$$

where ψ_k is one certain physical property of k phase (k = w, n). This property may be scalar, vector or high order tensor, such as pressure, velocity, and stress. V is the average volume of substance region. V_k is the volume percentage of k phase.

When applying volume average method, another average value is required to represent macroscopic physical quantity, namely the intrinsic average.

(2) Intrinsic average

$$\langle \psi_k \rangle^k |_x = \frac{1}{V_k} \int_{V_k} \psi_k |_{x+y_k} dV \tag{3.10}$$

Obviously, $\langle \psi_k \rangle = \varepsilon_k \langle \psi_k \rangle^k$ and $\varepsilon_k = V_k/V$ is the volume percentage of k phase substance.

(3) Space decomposition of physical quantity
Generally in an average volume, the real physical property ψ_k at a random spot is not equal to volume average value $\langle \psi_k \rangle^k$. Its deviation is decided by the inner structure of porous media and flow regime. ψ_k can be written as follows:

$$\psi_k = \langle \psi_k \rangle^k + \tilde{\psi}_k \tag{3.11}$$

where we can see $\langle \psi_k \rangle^k$ only changes in spatial scale L while deviation value $\tilde{\psi}_k$ changes in spatial l_k as Fig. 3.7 shows.

(4) Spatial average principal
Gradient and divergence calculations are generally required in volume average method. In terms of this, Slattery (1967) and Whitaker (1999) created corresponding spatial average principal as follows:

$$\langle \nabla \psi_k \rangle |_x = \nabla \langle \psi_k \rangle |_x + \frac{1}{V} \int_{A_k} \boldsymbol{n}_k \psi_k |_{x+y_k} dA \tag{3.12}$$

If ψ_k is a vector, then

$$\langle \nabla \psi_k \rangle |_x = \nabla \langle \psi_k \rangle |_x + \frac{1}{V} \int_{A_k} n_k \cdot \boldsymbol{\psi}_k |_{x+y_k} dA \tag{3.13}$$

where A_k is the interface of k phase and i phase. i stands for one certain substance in V ($i = $ w, n, s, $i \neq k$). Its outward normal unit vector n_k points to i phase from k phase.

The above space average principal is also suitable for high order tensor.

(5) Mass transfer equation

Assume that $V_k(t)$ is a spatial region which can change with time, whose interface is $A_k(t)$. Then

$$\left\langle \frac{\partial \psi_k}{\partial t} \right\rangle \big|_x = \frac{\partial \langle \psi_k \rangle |_x}{\partial t} - \frac{1}{V} \int_{A_k} \psi_k |_{x + y_k} \boldsymbol{w}_k \cdot \boldsymbol{n}_k \mathrm{d}A \qquad (3.14)$$

where w_k is the velocity of two phases interface.

When $\psi_k = 1$, Eqs. (3.13) and (3.14) can be simplified as follows:

$$\nabla \varepsilon_k = -\frac{1}{V} \int_{A_k} \boldsymbol{n}_k \mathrm{d}A \qquad (3.15)$$

$$\frac{\partial \varepsilon_k}{\partial t} = \frac{1}{V} \int_{A_k} \boldsymbol{w}_k \cdot \boldsymbol{n}_k \mathrm{d}A \qquad (3.16)$$

2. Microscopic scale two-phase flow description

As Fig. 3.7 shows, two-phase flow system on a microscopic scale consists of closed regions of every single phase fluid. Assuming that the flow region of every single phase fluid is considered as continuous media in microporous scale, two-phase flow can be described by the following differential equations.

$$\frac{\partial \rho_k}{\partial t} + \nabla \cdot (\rho_k \boldsymbol{v}_k) = 0 \qquad (3.17)$$

$$\frac{\partial (\rho_k \boldsymbol{v}_k)}{\partial t} + \nabla \cdot (\rho_k \boldsymbol{v}_k \boldsymbol{v}_k) = -\nabla p_k + \mu_k \nabla^2 \boldsymbol{v}_k + \rho_k \boldsymbol{g} \qquad (3.18)$$

$$\boldsymbol{v}_k = 0, \quad \text{at } A_{ks} \qquad (3.19)$$

$$(-p_k \boldsymbol{I} + \tau_k) \cdot \boldsymbol{n} = (-p_m \boldsymbol{I} + \tau_m) \cdot \boldsymbol{n} + 2\sigma\kappa\boldsymbol{n}, \quad \text{at } A_{km} \qquad (3.20)$$

where $m \in \{$w, n$\}$ and $m \neq k$; σ and κ are respectively the curvature of interface tension and A_{km}; $\boldsymbol{I}$ is the unit matrix.

Obviously, if microscopic porous media structure in flow region can be described accurately, flow state can be obtained by the above basic differential equations and boundary conditions. However, it is nearly an impossible task to obtain its inner porous structure with the present techniques. Even if we can get accurate inner structure of pores, there are still many difficulties in solving the above partial differential equation sets, such as the stability of numerical calculation, the method of describing two phases interface and huge calculation. Thus we need to study this problem in much larger scale, namely so-called upscaling. At present, many upscaling methods have been proposed such as volume average method, averaging theory, etc. At this moment we will adopt volume average method.

3. Upscaling based on volume average method
Continuous equation upscaling
   As to Eq. (3.17), apply superficial average operator at both ends of the equation first, and we can get

$$\left\langle \frac{\partial \rho_k}{\partial t} \right\rangle + \langle \nabla \cdot (\rho_k \boldsymbol{v}_k) \rangle = \langle 0 \rangle \tag{3.21}$$

Applying space average principle (3.13) and mass transfer Eq. (3.14), we can get

$$\frac{\partial \langle \rho_k \rangle}{\partial t} + \nabla \cdot \langle \rho_k \boldsymbol{v}_k \rangle = \frac{1}{V} \int_{A_{km}} \rho_k (\boldsymbol{v}_k - \boldsymbol{w}_k) \cdot \boldsymbol{n}_{km} \mathrm{d}A = 0 \tag{3.22}$$

where the final term at the left of this equation is mass transfer representative term of two phases interface. If we do not consider substance diffusion and physical–chemical reaction, then $\boldsymbol{v}_k = \boldsymbol{w}_k$. In this condition we just only consider non-mixed two phases flow. Thus the final term is constantly equal to zero. Equation (3.22) can be written as:

$$\frac{\partial \langle \rho_k \rangle}{\partial t} + \nabla \cdot \langle \rho_k \boldsymbol{v}_k \rangle = 0 \tag{3.23}$$

To simplify the problem, assume that the fluid and porous media skeleton are all impressible. Then Eq. (3.23) can be simplified into

$$\frac{\partial \varepsilon_k}{\partial t} + \nabla \cdot \langle \boldsymbol{v}_k \rangle = 0 \tag{3.24}$$

The above velocity is superficial average value. However in two phases flow, we often use intrinsic average value. Thus Eq. (3.24) can be written as Eq. (3.25)

$$\frac{\partial \varepsilon_k}{\partial t} + \nabla \cdot \left(\varepsilon_k \langle \boldsymbol{v}_k \rangle^k \right) = 0 \tag{3.25}$$

During the above deduction, there is no scale limitation condition introduced. Thus Eq. (3.25) is suitable for the whole flow region.

(1) Momentum equation upscaling
As to momentum Eq. (3.18), similar analysis can be used to deduce equations. At first apply superficial average operator at both end of equation, we can get

$$\left\langle \frac{\partial \langle \rho_k \boldsymbol{v}_k \rangle}{\partial t} \right\rangle + \langle \nabla \cdot \langle \rho_k \boldsymbol{v}_k \boldsymbol{v}_k \rangle \rangle = -\langle \nabla p_k \rangle + \mu_k \langle \nabla^2 \boldsymbol{v}_k \rangle + \varepsilon_k \rho_k \mathbf{g} \qquad (3.26)$$

Applying mass transfer principle and space average principle respectively and combining $\boldsymbol{v}_k = \boldsymbol{w}_k$, we can get

$$\rho_k \frac{\partial \langle \boldsymbol{v}_k \rangle}{\partial t} + \rho_k \nabla \cdot \langle \boldsymbol{v}_k \boldsymbol{v}_k \rangle = -\langle \nabla p_k \rangle + \mu_k \langle \nabla^2 \boldsymbol{v}_k \rangle + \varepsilon_k \rho_k \mathbf{g} \qquad (3.27)$$

Applying space average principle on the first term at right end of the above equation, then

$$\langle \nabla p_k \rangle = \nabla \langle p_k \rangle + \frac{1}{V} \int_{A_{km}} \boldsymbol{n}_{km} p_k \big|_{x + y_k} \mathrm{d}A \qquad (3.28)$$

Transfer Eq. (3.28) into intrinsic average form as follows:

$$\langle \nabla p_k \rangle = \varepsilon_k \nabla \langle p_k \rangle^k + \langle p_k \rangle^k \nabla \varepsilon_k + \frac{1}{V} \int_{A_{km}} \boldsymbol{n}_{km} p_k \big|_{x + y_k} \mathrm{d}A \qquad (3.29)$$

At this moment applying Eq. (3.15), the above equation can be written as Eq. (3.30)

$$\langle \nabla p_k \rangle = \varepsilon_k \nabla \langle p_k \rangle^k + \frac{1}{V} \int_{A_{km}} \boldsymbol{n}_{km} \left(p_k \big|_{x + y_k} - \langle p_k \rangle^k \big|_x \right) \mathrm{d}A \qquad (3.30)$$

Similarly as to Eq. (3.27), apply space average principal on the second term at the right end of equation, and we can get

$$\mu_k \langle \nabla \cdot \nabla \boldsymbol{v}_k \rangle = \mu_k \nabla \cdot \langle \nabla \boldsymbol{v}_k \rangle + \frac{1}{V} \int_{A_{km}} \boldsymbol{n}_{km} \cdot \nabla \boldsymbol{v}_k \big|_{x + y_k} \mathrm{d}A \qquad (3.31)$$

As to Eq. (3.31), apply space average principal on the first term at the right end of equation, and we can get

$$\mu_k \langle \nabla \cdot \nabla \boldsymbol{v}_k \rangle = \mu_k \nabla \cdot \langle \nabla \boldsymbol{v}_k \rangle + \frac{1}{V} \int_{A_{km}} \boldsymbol{n}_{km} \cdot \nabla \boldsymbol{v}_k \big|_{x + y_k} \mathrm{d}A + \frac{1}{V} \nabla$$
$$\cdot \left(\int_{A_{km}} \boldsymbol{n}_{km} \boldsymbol{v}_k \big|_{x + y_k} \mathrm{d}A \right) \qquad (3.32)$$

As to Eq. (3.32) apply intrinsic average and Eq. (3.15) on the first term at the right end of the equation, then we can get

$$\mu_k \langle \nabla \cdot \nabla \boldsymbol{v}_k \rangle = \mu_k \varepsilon_k \nabla^2 \langle \boldsymbol{v}_k \rangle^k + \mu_k \left(\nabla^2 \varepsilon_k \langle \boldsymbol{v}_k \rangle^k + \nabla \varepsilon_k \cdot \nabla \langle \boldsymbol{v}_k \rangle^k \right)$$

$$+ \frac{1}{V} \int_{A_{km}} \boldsymbol{n}_{km} \cdot \left(\nabla \boldsymbol{v}_k |_{x+y_k} - \nabla \langle \boldsymbol{v}_k \rangle^k |_x \right) dA \qquad (3.33)$$

$$+ \frac{1}{V} \nabla \cdot \left(\int_{A_{km}} \boldsymbol{n}_{km} \boldsymbol{v}_k |_{x+y_k} dA \right)$$

Substitute Eqs. (3.30) and (3.33) into Eq. (3.27), then

$$\rho_k \varepsilon_k^{-1} \left(\frac{\partial \langle \boldsymbol{v}_k \rangle}{\partial t} + \nabla \cdot \langle \boldsymbol{v}_k \boldsymbol{v}_k \rangle \right) = - \nabla \langle p_k \rangle^k + \rho_k \boldsymbol{g} + \mu_k \nabla^2 \langle \boldsymbol{v}_k \rangle^k$$

$$+ \mu_k \varepsilon_k^{-1} \left(\nabla^2 \varepsilon_k \langle \boldsymbol{v}_k \rangle^k + \nabla \varepsilon_k \cdot \nabla \langle \boldsymbol{v}_k \rangle^k \right) - \mu_k \boldsymbol{F}_k$$

$$(3.34)$$

$\boldsymbol{F}_k$ can be defined as follows:

$$\mu_k \boldsymbol{F}_k = - \frac{1}{V_k} \int_{A_k} \boldsymbol{n}_k \cdot \left[-\boldsymbol{I} \left(p_k |_{x+y_k} - \langle p_k \rangle^k |_x \right) + \mu_k \left(\nabla \boldsymbol{v}_k |_{x+y_k} - \nabla \langle v_k \rangle^k |_x \right) \right] dA$$

$$- \frac{1}{V_k} \nabla \cdot \left(\int_{A_{km}} \mu_k \boldsymbol{v}_k \boldsymbol{n}_k dA \right)$$

$$(3.35)$$

As to Eq. (3.34), applying space average principal on the second term at the left of equation, we can get

$$\nabla \cdot \langle \boldsymbol{v}_k \boldsymbol{v}_k \rangle = \nabla \cdot \left(\varepsilon_k \langle \boldsymbol{v}_k \rangle^k \langle \boldsymbol{v}_k \rangle^k \right) + \nabla \cdot \left(\boldsymbol{\tau}_k^t + \langle \boldsymbol{v}_k \boldsymbol{v}_k \rangle_{\text{NL}} \right) \qquad (3.36)$$

where

$$\boldsymbol{\tau}_k^t = \langle \tilde{\boldsymbol{v}}_k \tilde{\boldsymbol{v}}_k \rangle \qquad (3.37)$$

$$\langle \boldsymbol{v}_k \boldsymbol{v}_k \rangle_{\text{NL}} = \left\langle \langle \boldsymbol{v}_k \rangle^k \langle \boldsymbol{v}_k \rangle^k \right\rangle + \left\langle \tilde{\boldsymbol{v}}_k \langle \boldsymbol{v}_k \rangle^k \right\rangle + \left\langle \langle \boldsymbol{v}_k \rangle^k \tilde{\boldsymbol{v}}_k \right\rangle - \varepsilon_k \langle \boldsymbol{v}_k \rangle^k \langle \boldsymbol{v}_k \rangle^k \qquad (3.38)$$

Equation (3.37) is called local turbulent term. Equation (3.38) is nonlocal volume average deviation term.

As to Eqs. (3.37) and (3.38), there are no better methods to represent and calculate. But it must be noticed that these two terms can be neglected because they are very small or they tend to be zero when they are in free flow region and seepage flow region without the effect from interface. Thus add the effect to $\mu_k \mathbf{F}_k$ and consider the effect of Eqs. (3.37) and (3.38) by adjusting corresponding parameters.

To simplify the problem, we omit these two terms and put Eq. (3.36) into Eq. (3.34), then we can get Eq. (3.39)

$$
\rho_k \varepsilon_k^{-1} \left[\frac{\partial \left(\varepsilon_k \langle \mathbf{v}_k \rangle^k \right)}{\partial t} + \nabla \cdot \left(\varepsilon_k \langle \mathbf{v}_k \rangle^k \langle \mathbf{v}_k \rangle^k \right) \right] = - \nabla \langle p_k \rangle^k + \rho_k \mathbf{g} + \mu_k \nabla^2 \langle \mathbf{v}_k \rangle^k
$$
$$
+ \mu_k \varepsilon_k^{-1} \left(\nabla^2 \varepsilon_k \langle \mathbf{v}_k \rangle^k + \nabla \varepsilon_k \cdot \nabla \langle \mathbf{v}_k \rangle^k \right)
$$
$$
- \mu_k \mathbf{F}_k
$$

$$(3.39)$$

During above derivation process, we did not introduce any scale limitation or constraint. Thus Eq. (3.39) is suitable in the whole study region.

2. A description on REV scale two phases flow

Equations (3.25) and (3.39) are respectively continuous equation and momentum equation after the first upscaling. In REV scale after introducing some special scale limited condition or constraint, the above two equations can be further simplified. To apply conventional Darcy law, we assume that flow in porous media is under the condition that Reynold number is small. That is to say $Re \ll 1$. In this case, the effect of inertia item in porous media can be neglected. It is not necessary for free flow region. There is a clear interface transition region as Eq. (3.8) shows. There will be a sharp change in the structure attribute of porous media through interface region. Due to this region, two flow region can have different flow state in REV scale. As to porous media fluid flow, we can introduce the following typical constraint conditions:

$$
\frac{r^2}{LL_d} \ll 1, \quad l \ll r \ll L, \quad L \ll L_p
$$

$$(3.40)$$

where L_d is characteristic length relevant to the derivative of volume average physical parameters. L_p is the characteristic scale relevant to inertia term. r is the characteristic length of average volume.

After introducing the above scale constraint, as to Eq. (3.39), there are following order analysis estimated by magnitudes.

$$
\frac{1}{V_k} \int_{A_k} \mathbf{n}_k \cdot \mu_k \left(\nabla \mathbf{v}_k |_{x + y_k} - \nabla \langle \mathbf{v}_k \rangle^k |_x \right) dA = O\left(\frac{\mu_k \mathbf{v}_k}{l_k^2} \right)
$$

$$(3.41)$$

$$
\frac{1}{V_k} \nabla \cdot \left(\int_{A_{km}} \mu_k \mathbf{v}_k \mathbf{n}_k dA \right) = O\left(\frac{\mu_k \mathbf{v}_k}{L l_k} \right)
$$

$$(3.42)$$

$$\mu_k \nabla^2 \langle v_k \rangle^k = O\left(\frac{\mu_k v_k}{L^2}\right) \tag{3.43}$$

$$\mu_k \varepsilon_k^{-1} \nabla^2 \varepsilon_k \langle v_k \rangle^k = O\left(\frac{\mu_k v_k}{L^2}\right) \tag{3.44}$$

$$\mu_k \varepsilon_k^{-1} \nabla \varepsilon_k \cdot \nabla \langle v_k \rangle^k = O\left(\frac{\mu_k v_k}{L^2}\right) \tag{3.45}$$

According to Eq. (3.40), the estimated magnitude order of Eqs. (3.42)–(3.45) is much smaller than Eq. (3.41). Thus as to the fluid flow in porous media, Eqs. (3.42)–(3.45) can be neglected in Eq. (3.39). At this moment, Eqs. (3.25) and (3.39) can be simplified respectively as follows:

$$\frac{\partial \varepsilon_{k\omega}}{\partial t} + \nabla \cdot \left(\varepsilon_{k\omega} \langle v_k \rangle_\omega^k\right) = 0 \tag{3.46}$$

$$0 = -\nabla \langle p_k \rangle_\omega^k + \rho_k g - \mu_k \mathbf{K}_{k\omega}^{-1} \cdot \left(\varepsilon_{k\omega} \langle v_k \rangle_\omega^k\right) \tag{3.47}$$

where subscript ω represents porous media region as Fig. 3.8 shows. The final term in Eq. (3.47) is equal to $\mu_k F_k$ and $K_{rw} = k_{rk} K_\omega$ is the effective permeability of k phase. K_ω is the absolute permeability tensor in porous media. k_{rk} is the relative permeability of k phase fluid.

The process of detailed derivation can be found in Whitaker (1986).

In free flow region Eqs. (3.25) and (3.39) can be written in REV scale as follows:

$$\frac{\partial \varepsilon_{k\eta}}{\partial t} + \nabla \cdot \left(\varepsilon_{k\eta} \langle v_k \rangle_\eta^k\right) = 0 \tag{3.48}$$

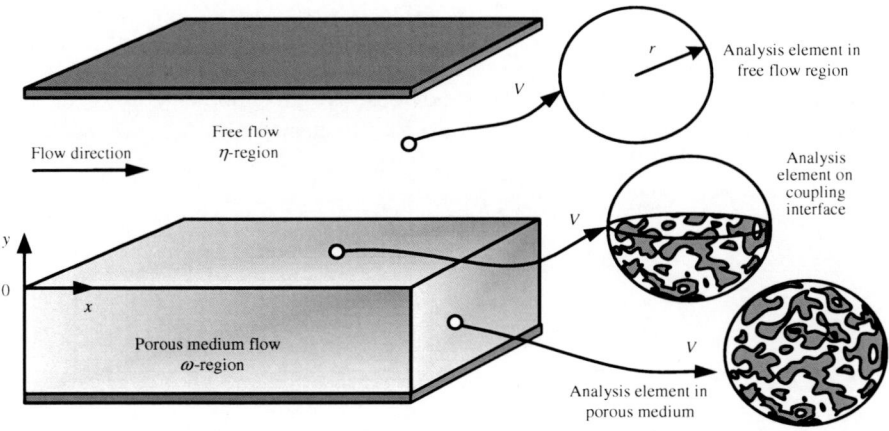

Fig. 3.8 Scheme of coupling flow model

$$\rho_k \frac{\partial \left(\varepsilon_{k\eta} \langle \boldsymbol{v}_k \rangle_\eta^k \right)}{\partial t} + \rho_k \nabla \cdot \left(\varepsilon_{k\eta} \langle \boldsymbol{v}_k \rangle_\eta^k \langle \boldsymbol{v}_k \rangle_\eta^k \right) = - \varepsilon_{k\eta} \nabla \langle p_k \rangle_\eta^k + \varepsilon_{k\eta} \rho_k \boldsymbol{g} \qquad (3.49)$$
$$+ \varepsilon_{k\eta} \mu_k \nabla^2 \langle \boldsymbol{v}_k \rangle_\eta^k - \varepsilon_{k\eta} \mu_k \boldsymbol{F}_{k\eta}$$

where subscript η represents free flow region (Fig. 3.8). We introduce two scale constraints $r^2 \ll LL_d$ and $l_k \ll L.\mu_k \boldsymbol{F}_{k\eta}$ in Eq. (3.49) is momentum term of interface, generally written as follows:

$$\mu_k \boldsymbol{F}_{k\eta} = \mathbf{K}_{k\eta}^{-1} \cdot \left(\langle \boldsymbol{v}_m \rangle_\eta^m - \langle \boldsymbol{v}_k \rangle_\eta^k \right) \qquad (3.50)$$

Equations (3.48) and (3.49) is typical macroscopic average two-fluid model, which is the most popular model at present in area of two-phase flow.

Similarly Eq. (3.20) can be updated by above upscaling method. At first apply Eq. (3.11) in (3.20) and we can get

$$-\tilde{p}_k \boldsymbol{n}_{km} = - \tilde{p}_m \boldsymbol{n}_{km} + \left(\langle p_k \rangle^k - \langle p_m \rangle^m \right) + 2\sigma\kappa \boldsymbol{n}_{km}$$
$$- \left[\mu_k \left(\nabla \tilde{\boldsymbol{v}}_k + \nabla \tilde{\boldsymbol{v}}_k^T \right) \cdot \boldsymbol{n}_{km} - \mu_m \left(\nabla \tilde{\boldsymbol{v}}_m + \nabla \tilde{\boldsymbol{v}}_m^T \right) \cdot \boldsymbol{n}_{km} \right] \qquad (3.51)$$

where the following two equations are needed.

$$\begin{cases} \mu_k \left[\nabla \langle \boldsymbol{v}_k \rangle^k + \nabla \left(\langle \boldsymbol{v}_k \rangle^k \right)^T \right] \ll \mu_k \left(\nabla \tilde{\boldsymbol{v}}_k + \nabla \tilde{\boldsymbol{v}}_k^T \right) \\ \mu_m \left[\nabla \langle \boldsymbol{v}_m \rangle^m + \nabla \left(\langle \boldsymbol{v}_m \rangle^m \right)^T \right] \ll \mu_m \left(\nabla \tilde{\boldsymbol{v}}_m + \nabla \tilde{\boldsymbol{v}}_m^T \right) \end{cases} \qquad (3.52)$$

Similar to the analysis on orders estimated by magnitudes in Eqs. (3.41)–(3.45), consider the scale constraint of Eq. (3.40), we can get

$$\nabla \tilde{\boldsymbol{v}}_k = \mathrm{O}\left(\langle \boldsymbol{v}_k \rangle^k / l_k \right), \quad \nabla \langle \boldsymbol{v}_k \rangle^k = \mathrm{O}\left(\langle \boldsymbol{v}_k \rangle^k / L \right), \quad l_k \ll L \qquad (3.53)$$

Thus the accuracy of Eq. (3.52) is obvious. At this moment, make an area average on every normal component of Eq. (3.51) at interface A_{km}, we can get

$$-\left(\langle p_k \rangle^k - \langle p_m \rangle^m \right) = 2\sigma \langle \kappa \rangle_{km} - \langle \tilde{p}_k - \tilde{p}_m \rangle$$
$$- \left\langle \mu_k \boldsymbol{n}_{km} \cdot \left(\nabla \tilde{\boldsymbol{v}}_k + \nabla \tilde{\boldsymbol{v}}_k^T \right) \cdot \boldsymbol{n}_{km} - \mu_m \boldsymbol{n}_{km} \cdot \left(\nabla \tilde{\boldsymbol{v}}_m + \nabla \tilde{\boldsymbol{v}}_m^T \right) \cdot \boldsymbol{n}_{km} \right\rangle$$
$$(3.54)$$

where:

$$\langle \kappa \rangle_{km} = \frac{1}{A_{km}} \int_{A_{km}} \kappa \, dA \tag{3.55}$$

Apply Eqs. (3.40) and (3.9) and combine corresponding analysis method of orders estimated by magnitudes, we can get

$$-\left(\langle p_k \rangle^k - \langle p_m \rangle^m \right) = 2\sigma \langle k \rangle_{km} + O\left(\frac{\mu_k \langle v_k \rangle^k}{l_k}\right) \tag{3.56}$$

At this moment, if we introduce the following scale constraint:

$$\frac{\mu_k \langle v_k \rangle^k}{2\sigma \langle k \rangle_{km} l_k} \ll 1 \tag{3.57}$$

We can get macroscopic expression Eq. (3.58) on two-phase flow in REV scale.

$$\langle p_k \rangle^k - \langle p_m \rangle^m = 2\sigma \bar{\kappa} \tag{3.58}$$

where $\bar{\kappa}$ is average curvature of k-m interface in average volume V.

The detailed deduction can be found in (Whitaker 1986). The left side in Eq. (3.57) can be defined as capillary number.

Obviously Eqs. (3.41)–(3.44) is not appropriate in interface transition region. Deviation from original problem will appear if Eqs. (3.41)–(3.44) are used in interface region. How to remove the derivation is the current research focus. We will eliminate the above error by introducing jump boundary conditions in coupling interface. In practical engineering interface, transition can be simplified into a simple interface region because it is extremely small relative to the whole region, when corresponding boundary condition needs to be introduced. In the following section we will introduce a surface excess function to build coupling interface condition. This process is the second process of upscaling.

4. Two-phase flow interface condition on the coupling porous fluid flow
As Fig. 3.9 shows, take a volume element V out of the coupling interface region randomly. This outer surface of this element consists of three sections: upper surface A^+ which is connected with free flow region, the lower surface A^- which is connected to porous media fluid flow and lateral surface, whose unit outer normal vector is n_s. To simplify our study, we assume A^+ and A^- are planes and they are trapped by curve C. δ is the thickness of interface region and δC is lateral superficial area. l and Δ are respectively characteristic sizes in porous media flow region and free flow region.

(1) The establishment of velocity jump condition
To deduce the velocity condition in interface, make an integration for continuous equation Eq. (3.25) in interface element V. detail as follows:

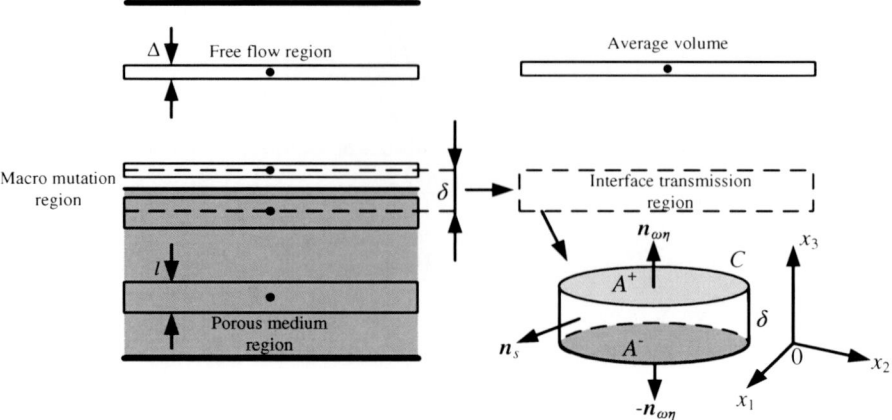

Fig. 3.9 Analysis model for coupling interface condition

$$\int_V \frac{\partial \varepsilon_k}{\partial t} dV + \int_V \nabla \cdot \left(\varepsilon_k \langle v_k \rangle^k \right) dV = 0 \tag{3.59}$$

Applying divergence principal we can get Eq. (3.60)

$$\int_{A^-} n_{\omega\eta} \cdot \varepsilon_{k\omega} \langle v_k \rangle_{\omega}^k dA - \int_{A^+} n_{\omega\eta} \cdot \varepsilon_{k\eta} \langle v_k \rangle_{\eta}^k dA = \int_A \frac{\partial}{\partial t} \left(\int_0^\delta \varepsilon_k dx_3 \right) dA$$
$$+ \oint_C \int_0^\delta n_s \cdot \varepsilon_k \langle v_k \rangle^k dx_3 dC \tag{3.60}$$

Obviously the first term on the right of Eq. (3.60) can be written as follows:

$$\int_A \frac{\partial}{\partial t} \left(\int_0^\delta \varepsilon_k dx_3 \right) dA = \int_A \bar{H}_k \delta \frac{\partial}{\partial t} \left(\varepsilon_{k\omega} - \varepsilon_{k\eta} \right) dA \tag{3.61}$$

where $\bar{H}_k$ is a dimensionless scalar parameter which can be adjusted and $\bar{H}_k = O(1)$.

As to the second term on the right of Eq. (3.60), we introduced the average flux function of interface region in unit volume $\delta \langle v_k \rangle_s^k = \int_0^\delta \varepsilon_k \langle v_k \rangle^k dx_3$ and combined divergence principal, Eq. (3.62) hence obtained as: can be gained.

$$\int_A n_{\omega\eta} \cdot \left(\varepsilon_{k\omega} \langle v_k \rangle_{\omega}^k - \varepsilon_{k\eta} \langle v_k \rangle_{\eta}^k \right) dA = \int_A \left[\bar{H}_k \delta \frac{\partial}{\partial t} \left(\varepsilon_{k\omega} - \varepsilon_{k\eta} \right) + \nabla_s \cdot \left(\delta \langle v_k \rangle_s^k \right) \right] dA \tag{3.62}$$

where $\nabla_s = (\boldsymbol{I} - \boldsymbol{n}_{\omega\eta}\boldsymbol{n}_{\omega\eta}) \cdot \nabla = \partial/\partial x_j, \ j = 1, 2$. Consider the randomicity of interface A, Eq. (3.62) can be simplified into the following normal velocity condition.

$$\boldsymbol{n}_{\omega\eta} \cdot \left(\varepsilon_{k\omega}\langle \boldsymbol{v}_k\rangle^k_\omega - \varepsilon_{k\eta}\langle \boldsymbol{v}_k\rangle^k_\eta\right) = \bar{H}_k\delta\frac{\partial}{\partial t}\left(\varepsilon_{k\omega} - \varepsilon_{k\eta}\right) + \nabla_s \cdot \left(\delta\langle \boldsymbol{v}_k\rangle^k_s\right) \tag{3.63}$$

In practical engineering, δ can be neglected because it is much smaller than characteristic length L. Thus Eq. (3.63) can be simplified into

$$\boldsymbol{n}_{\omega\eta} \cdot \left(\varepsilon_{k\eta}\langle \boldsymbol{v}_k\rangle^k_\eta - \varepsilon_{k\omega}\langle \boldsymbol{v}_k\rangle^k_\omega\right) = H_k\sqrt{K_n}\frac{\partial}{\partial t}\left(\varepsilon_{k\eta} - \varepsilon_{k\omega}\right) \tag{3.64}$$

where H_k is a dimensionless scalar parameter which can be adjusted and $H_k = O(1)$; $\delta = O(\sqrt{K_n})$, $K_n = \boldsymbol{n}_{\omega\eta} \cdot \boldsymbol{K}_\omega \cdot \boldsymbol{n}_{\omega\eta}$ is a component in normal direction of porous media permeability tensor.

Equation (3.64) is called normal velocity jump condition. As to tangential velocity condition, according to the physical significance and characteristic of model, Eq. (3.65) can be written directly.

$$\boldsymbol{\lambda}_{\omega\eta} \cdot \left(\varepsilon_{k\eta}\langle \boldsymbol{v}_k\rangle^k_\eta - \varepsilon_{k\omega}\langle \boldsymbol{v}_k\rangle^k_\omega\right) = u_{ks} \tag{3.65}$$

where $\boldsymbol{\lambda}_{\omega\eta}$ is unit tangent vector and u_{ks} is relative slip velocity of k phase fluid in porous fluid flow interface.

(2) The establishment of stress jump condition
Similarly we can make a derivation and analysis on Eq. (3.39). At first, make an analysis and simplification on the two terms which include $\nabla\varepsilon_k$ and $\nabla^2\varepsilon_k$. Obviously there are in the same orders with the following estimated magnitudes in these two terms:

$$\mu_k\varepsilon_k^{-1}\nabla^2\varepsilon_k\langle \boldsymbol{v}_k\rangle^k = O\left(\frac{\mu_k\langle \boldsymbol{v}_k\rangle^k}{L_{\text{NL}}^2}\right), \quad \mu_k\varepsilon_k^{-1}\nabla\varepsilon_k \cdot \nabla\langle \boldsymbol{v}_k\rangle^k = O\left(\frac{\mu_k\langle \boldsymbol{v}_k\rangle^k}{L_{\text{NL}}^2}\right) \tag{3.66}$$

where L_{NL} is the characteristic length related with nonlocal volume average parameters. Obviously these two terms have the same magnitudes estimate, thus we can rewrite these two terms as:

$$\mu_k\varepsilon_k^{-1}\left(\nabla^2\varepsilon_k\langle \boldsymbol{v}_k\rangle^k + \nabla_{\varepsilon_k} \cdot \nabla\langle \boldsymbol{v}_k\rangle^k\right) = (1 + \xi)\mu_k\varepsilon_k^{-1}\nabla^2\varepsilon_k\langle \boldsymbol{v}_k\rangle^k \tag{3.67}$$

where ξ is a dimensionless scalar parameter and $\xi = O(1)$.

Substitute Eq. (3.67) into Eq. (3.39) and repeat the derivation process of velocity jump condition. Make an integral and neglect the inertia term. Then we can get

$$
\rho_k \left(\int_{A^+} \boldsymbol{n}_{\omega\eta} \cdot \langle v_k \rangle_\eta^k \langle v_k \rangle_\eta^k \mathrm{d}A - \int_{A^-} \boldsymbol{n}_{\omega\eta} \cdot \langle v_k \rangle_\omega^k \langle v_k \rangle_\omega^k \mathrm{d}A + \int_A \nabla_s \cdot \left(\delta U_s^k \right) \mathrm{d}A \right)
$$

$$
+ \rho_k \int_A \int_0^\delta \varepsilon_k^{-1} \nabla \varepsilon_k \cdot \langle v_k \rangle^k \langle v_k \rangle^k \mathrm{d}x_3 \mathrm{d}A = - \int_{A^+} \boldsymbol{n}_{\omega\eta} \langle p_k \rangle_\eta^k \mathrm{d}A + \int_{A^-} \boldsymbol{n}_{\omega\eta} \langle p_k \rangle_\omega^k \mathrm{d}A
$$

$$
- \int_A \nabla_s \left(\delta \langle p_k \rangle_s^k \right) \mathrm{d}A + \int_A \delta \varepsilon_{ks} \rho_k g \mathrm{d}A + \int_A \nabla_s \cdot \left(\delta \tau_s^k \right) \mathrm{d}A + \int_{A^+} \boldsymbol{n}_{\omega\eta} \cdot \mu_k \nabla \langle v_k \rangle_\eta^k \mathrm{d}A
$$

$$
- \int_{A^-} \boldsymbol{n}_{\omega\eta} \cdot \mu_k \nabla \langle v_k \rangle_\omega^k \mathrm{d}A + \int_A \int_0^\delta (1 + \xi) \mu_k \varepsilon_k^{-1} \nabla^2 \varepsilon_k \langle v_k \rangle^k \mathrm{d}x_3 \mathrm{d}A - \int_A \int_0^\delta \mu_k \boldsymbol{F}_k \mathrm{d}x_3 \mathrm{d}A
$$

$$
\tag{3.68}
$$

where surface excess is introduced. And surface excess is defined as follows

$$
\delta U_s^k = \int_0^\delta \langle v_k \rangle^k \langle v_k \rangle^k \mathrm{d}x_3, \quad \delta \langle p_k \rangle_s^k = \int_0^\delta \langle p_k \rangle^k \mathrm{d}x_3
$$

$$
\delta \varepsilon_{ks} = \int_0^\delta \varepsilon_k \mathrm{d}x_3, \quad \delta \tau_s^k = \int_0^\delta \mu_k \nabla \langle v_k \rangle^k \mathrm{d}x_3 \tag{3.69}
$$

ε_k is included in the final term at the left end of Eq. (3.68) and $\varepsilon_k = \delta S_k$ in two-phase flow. S_k is the saturation of k phase. $\varepsilon_k^{-1} \nabla \varepsilon_k$ can be written as follows:

$$
\varepsilon_k^{-1} \nabla \varepsilon_k = \phi^{-1} \nabla \phi + S_k^{-1} \nabla S_k \tag{3.70}
$$

where $\nabla \phi \sim \boldsymbol{n}_{\omega\eta} |\nabla \phi|$ and $\nabla S_k \sim \boldsymbol{B}_k \cdot \boldsymbol{n}_{\omega\eta} |\nabla S_k|$ is obvious. $\boldsymbol{B}_k$ is a dimensionless two orders tensor coefficient and $\boldsymbol{B}_k = O(1)$. Then we can get the following approximate expression

$$
\int_0^\delta \varepsilon_k^{-1} \nabla \varepsilon_k \cdot \langle v_k \rangle^k \langle v_k \rangle^k \mathrm{d}x_3 \sim \varepsilon_k^{-1} \delta |\nabla \varepsilon_k| (\boldsymbol{B}_k + \boldsymbol{I}) \cdot \boldsymbol{n}_{\omega\eta} \cdot \left(\langle v_k \rangle^k \langle v_k \rangle^k \right) \tag{3.71}
$$

Thus the final term at the left end of Eq. (3.68) can be written as follow:

$$
\int_0^\delta \varepsilon_k^{-1} \nabla \varepsilon_k \cdot \langle v_k \rangle^k \langle v_k \rangle^k \mathrm{d}x_3 = (\boldsymbol{B}_K + \chi \boldsymbol{I}) \cdot \boldsymbol{n}_{\omega\eta} \cdot \left(\langle v_k \rangle_\eta^k \langle v_k \rangle_\eta^k - \langle v_k \rangle_\omega^k \langle v_k \rangle_\omega^k \right) \tag{3.72}
$$

where χ is a dimensionless scalar parameter and $\chi = O(1)$.

Considering Reynold number $\mathrm{Re} \ll 1$ in porous flow region, inertia term and viscous force term can be neglected. Equation (3.69) will approach to zero when $\delta \to 0$. Thus Eq. (3.68) can be simplified as Eq. (3.73):

$$
\int_A \mathbf{E}_k \cdot \boldsymbol{n}_{\omega\eta} \cdot \rho_k \langle \boldsymbol{v}_k \rangle_\eta^k \langle \boldsymbol{v}_k \rangle_\eta^k \mathrm{d}A = \int_A \boldsymbol{n}_{\omega\eta} \cdot \left[\boldsymbol{I} \langle p_k \rangle_\omega^k - \langle p_k \rangle_\eta^k + \mu_k \nabla \langle \boldsymbol{v}_k \rangle_\eta^k \right] \mathrm{d}A
$$
$$
+ \int_A \int_0^\delta (1+\xi)\mu_k \varepsilon_k^{-1} \nabla^2 \varepsilon_k \langle \boldsymbol{v}_k \rangle^k \mathrm{d}x_3 \mathrm{d}A - \int_A \int_0^\delta \mu_k \mathbf{F}_k \mathrm{d}x_3 \mathrm{d}A
$$

$$(3.73)$$

where $\mathbf{E}_k = (1+\chi)\boldsymbol{I} + \boldsymbol{B}_k$, a dimensionless two-order tensor coefficient.

The last two terms at the right of Eq. (3.73) cannot be neglected and need to be simplified. According to references (Ochoa-Tapia and Whitaker 1995a, b) the last term at the right of Eq. (3.73) is called excess bulk stress and can be written as follows:

$$
\int_0^\delta \mu_k \mathbf{F}_k \mathrm{d}x_3 = \delta \mu_k \boldsymbol{D}_k \cdot \left[K_{k\omega}^{-1} \cdot \left(\varepsilon_{k\omega} \langle \boldsymbol{v}_k \rangle_\omega^k \right) \right] - \delta \boldsymbol{D}_k \cdot \left[K_{k\eta}^{-1} \cdot \left(\langle \boldsymbol{v}_m \rangle_\eta^m - \langle \boldsymbol{v}_k \rangle_\eta^k \right) \right]
$$

$$(3.74)$$

where $\boldsymbol{D}_k$ is a dimensionless two orders tensor coefficient and $\boldsymbol{D}_k = O(1)$.

Then we will analyze the second term at the right of Eq. (3.73) which involves $\varepsilon_k^{-1} \langle \boldsymbol{v}_k \rangle^k$. In single phase flow, this term can be seen as constant and $\varepsilon_k = \phi$. In two-phase flow, this term can be written as:

$$
\varepsilon_k^{-1} \langle \boldsymbol{v}_k \rangle^k = \left(\phi^{-1} \langle \boldsymbol{v}_k \rangle^k \right) S_k^{-1} k_{rk}
$$

$$(3.75)$$

where $\varepsilon_k^{-1} \langle \boldsymbol{v}_k \rangle^k$ is also constant. Then we can get:

$$
\int_A \int_0^\delta (1+\xi)\mu_k \nabla^2 \varepsilon_k \varepsilon_k^{-1} \langle \boldsymbol{v}_k \rangle^k \mathrm{d}x_3
$$
$$
= \int_A (1+\xi)\mu_k \delta^{-1} \left(\varepsilon_{k\omega} - \varepsilon_{k\eta} \right) \mathbf{A}_k \cdot \left(\varepsilon_{k\eta}^{-1} \langle \boldsymbol{v}_k \rangle_\eta^k + \varepsilon_{k\omega}^{-1} \langle \boldsymbol{v}_k \rangle_\omega^k \right) \mathrm{d}A
$$

$$(3.76)$$

where $\mathbf{A}_k$ is a dimensionless two orders tensor coefficient and $\mathbf{A}_k = O(1)$.

Substituting Eqs. (3.76) and (3.74) into Eq. (3.73) and considering the randomicity of A, we can get the following general stress jump condition:

$$
\boldsymbol{n}_{\omega\eta} \cdot \left[\boldsymbol{I} \left(\langle p_k \rangle_\omega^k - \langle p_k \rangle_\eta^k \right) + \mu_k \nabla \langle \boldsymbol{v}_k \rangle_\eta^k \right] = \mu_k \left(\mathbf{A}_{k\eta} \cdot \langle \boldsymbol{v}_k \rangle_\eta^k - \mathbf{A}_{k\omega} \cdot \varepsilon_{k\omega} \langle \boldsymbol{v}_k \rangle_\omega^k \right)
$$
$$
+ \mathbf{E}_k \cdot \boldsymbol{n}_{\omega\eta} \cdot \rho_k \langle \boldsymbol{v}_k \rangle_\eta^k \langle \boldsymbol{v}_k \rangle_\eta^k
$$
$$
- \boldsymbol{G}_k \cdot \left(\langle \boldsymbol{v}_m \rangle_\eta^m - \langle \boldsymbol{v}_k \rangle_\eta^k \right)
$$

$$(3.77)$$

$$\begin{cases} \mathbf{A}_{k\eta} = (1+\xi)\delta^{-1}\varepsilon_{k\eta}^{-1}\left(\varepsilon_{k\eta} - \varepsilon_{k\omega}\right)\mathbf{A}_k \\ \mathbf{A}_{k\omega} = (1+\xi)\delta^{-1}\varepsilon_{k\omega}^{-2}\left(\varepsilon_{k\eta} - \varepsilon_{k\omega}\right)\mathbf{A}_k - \delta\mathbf{D}_k \cdot \mathbf{K}_{k\omega}^{-1} \\ \mathbf{G}_k = \delta\mathbf{D}_k \cdot \mathbf{K}_{k\eta}^{-1} \end{cases} \tag{3.78}$$

The second term at the right of Eq. (3.77) represents the effect of inertia term in interface in free flow region and it can be neglected when flow velocity is low. This inertia term is relevant to both flow state and the porous structure in interface region. The last term at the right end of Eq. (3.77) means friction force term, which is relevant to volume percentage and relative velocity closely.

(3) The two-phase flow mathematic model which couples seepage flow and free flow

The two-phase flow mathematic model which couples seepage flow and free flow and its corresponding coupling condition has been built theoretically based on volume average method. This model is suitable for REV scale. To simplify our study, change the above basic mathematic model into typical form as follows.

① seepage flow region in porous media

$$\frac{\partial(\phi\rho_k S_k)}{\partial t} + \nabla \cdot (\rho_k \mathbf{v}_k) = 0 \tag{3.79}$$

$$\mathbf{v}_k = -\frac{k_{rk}}{\mu_k}\mathbf{K} \cdot (\nabla p_k - \rho_k \mathbf{g}) \tag{3.80}$$

$$S_w + S_n = 1, \quad p_c(S_w) = p_n - p_w \tag{3.81}$$

where $\mathbf{v}_k$ is the seepage velocity of k phase fluid and p_c is capillary pressure which is the function of saturation of wetting phase.

② free flow region

$$\frac{\partial(\rho_k C_k)}{\partial t} + \nabla \cdot (\rho_k C_k \mathbf{u}_k) = 0 \tag{3.82}$$

$$\frac{\partial(\rho_k C_k \mathbf{u}_k)}{\partial t} + \nabla \cdot (\rho_k C_k \mathbf{u}_k \mathbf{u}_k) = -C_k \nabla p_k + C_k \rho_k \mathbf{g} + C_k \mu_k \nabla^2 \mathbf{u}_k - b_k(\mathbf{u}_w - \mathbf{u}_n) \tag{3.83}$$

$$C_w + C_n = 1, \quad p_n - p_w = 2\sigma\bar{\kappa} \tag{3.84}$$

where C_k is the volume percentage of k phase in free flow region; $\mathbf{u}_k$ is the average velocity of k phase fluid in V; $p_n - p_w$ is the pressure difference between two-phase fluids and it is relevant to flow state and interface attribute. If interfacial tension is neglected, then $p_n = p_w$. $\bar{\kappa}$ is average curvature of two-phase interface. b_k is a coefficient.

According to Eq. (3.83), the kinetic equation of two-phase fluid can be linked by the last term which is called frictional resistance term where b_k can be written as Eq. (3.85)

$$b_w = -b_n = \frac{1}{8} c_d \rho_c a_i |\boldsymbol{u}_w - \boldsymbol{u}_n| \qquad (3.85)$$

where ρ_c is the density of continuous fluid in V. a_i is the area of two-phase interface in V. c_d is an empirical coefficient which is relevant to Reynold number and two-phase flow state.

In the study of Mat and Ilegbusi (2002) a simpler and more convenient empirical equation is proposed as follows:

$$b_w = -b_n = \bar{\rho} c_d C_w C_n \qquad (3.86)$$

where $\bar{\rho} = \sum C_k \rho_k$ is the density of fluid mixture and $c_d = 20$ is an empirical constant.

Equation (3.86) does not include the coefficient a_i. This equation is suitable for two-phase laminar flow and low velocity flow. a_i is very important in more complex flow when Eq. (3.85) is required.

b_k is relevant to the flow state of two-phase flow but there are not a general method to determine this coefficient. Especially under the critical condition at which the flow state transfers, we still do not have effective calculated method and theory mainly because there are too many flow states of two phases and influencing factors, and at the same time, their relationship are complex as shown in Fig. 3.10. Even though the above two models are the main applied model in the study of two-phase flow, especially in practical engineering. In addition, Eq. (3.83) will have a form similar to Darcy's law if the inertial term is neglected, which will guide in the following research.

③ the velocity condition in coupling interface

$$\boldsymbol{n} \cdot (C_k \boldsymbol{u}_k - \boldsymbol{v}_k) = H_k \sqrt{K_n} \frac{\partial}{\partial t} (C_k - \phi S_k) \qquad (3.87)$$

$$\boldsymbol{\lambda} \cdot (C_k \boldsymbol{u}_k - \boldsymbol{v}_k) = u_{ks} \qquad (3.88)$$

The above two equations are normal velocity and tangent velocity conditions respectively.

As to Eq. (3.87), H_k is similar to α in BJ condition, both of which are the representatives of structures in porous media. If the thickness of interfacial area $\delta \ll (\boldsymbol{n} \cdot \boldsymbol{u}_k) t_c$ (t_c is characteristic scale of time), then normal velocity condition Eq. (3.87) can be simplified as:

$$\boldsymbol{n} \cdot (C_k \boldsymbol{u}_k - \boldsymbol{v}_k) = 0 \qquad (3.89)$$

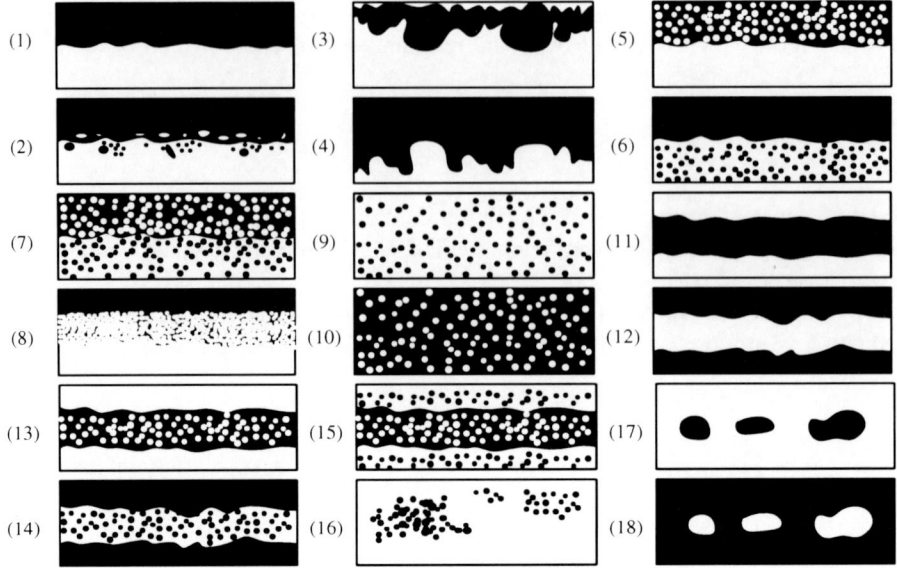

Fig. 3.10 Different flow state of two-phase flow

④ the stress condition in the coupling interface

$$p_k^\omega - \boldsymbol{n} \cdot \left(I p_k^\eta - \mu_k \nabla \boldsymbol{u}_k\right) \cdot \boldsymbol{n} = \mu_k \left(\mathbf{A}_{k\eta} \cdot \boldsymbol{u}_k - \mathbf{A}_{k\omega} \cdot \boldsymbol{v}_k\right) \cdot \boldsymbol{n}$$
$$+ \mathbf{E}_k \cdot \boldsymbol{n}_{\omega\eta} \cdot \rho_k \boldsymbol{u}_k \boldsymbol{u}_k \cdot \boldsymbol{n} - \mathbf{G}_k \cdot (\boldsymbol{u}_m - \boldsymbol{u}_k) \cdot \boldsymbol{n}$$
$$(3.90)$$

$$\boldsymbol{n} \cdot \mu_k \nabla \boldsymbol{u}_k \cdot \boldsymbol{\lambda} = \mu_k \left(\mathbf{A}_{k\eta} \cdot \boldsymbol{u}_k - \mathbf{A}_{k\omega} \cdot \boldsymbol{v}_k\right) \cdot \boldsymbol{\lambda}$$
$$+ \mathbf{E}_k \cdot \boldsymbol{n}_{\omega\eta} \cdot \rho_k \boldsymbol{u}_k \boldsymbol{u}_k \cdot \boldsymbol{\lambda} - \mathbf{G}_k \cdot (\boldsymbol{u}_m - \boldsymbol{u}_k) \cdot \boldsymbol{\lambda}$$
$$(3.91)$$

The above two equations are normal stress and tangent stress conditions respectively in coupling interface, in which $\mathbf{G}_k = \delta \boldsymbol{D}_k b_k$.

In practical study especially in numerical simulation, time step length is generally much larger than time scale t_c.

If assuming thermodynamic equilibrium is reasonable in coupling interface, we can neglect nonlinear inertia term. Meanwhile velocity in free flow region is over three magnitudes larger than flow velocity in porous media. Thus velocity at the right of Eqs. (3.90) and (3.91) can be neglected. Then we can get:

$$p_k^\omega - \boldsymbol{n} \cdot \left(I p_k^\eta - \mu_k \nabla \boldsymbol{u}_k\right) \cdot \boldsymbol{n} = \mu_k \mathbf{A}_{k\eta} \cdot \boldsymbol{u}_k \cdot \boldsymbol{n} - \mathbf{G}_k \cdot (\boldsymbol{u}_m - \boldsymbol{u}_k) \cdot \boldsymbol{n} \qquad (3.92)$$

$$\boldsymbol{n} \cdot \mu_k \nabla \boldsymbol{u}_k \cdot \boldsymbol{\lambda} = \mu_k \mathbf{A}_{k\eta} \cdot \boldsymbol{u}_k \cdot \boldsymbol{\lambda} - \mathbf{G}_k \cdot (\boldsymbol{u}_m - \boldsymbol{u}_k) \cdot \boldsymbol{\lambda} \qquad (3.93)$$

If assuming velocity is same for two-phase fluid in free flow region, then $\boldsymbol{u}_m = \boldsymbol{u}_k$ and we can get:

$$p_k^\omega - \boldsymbol{n} \cdot \left(\boldsymbol{I} p_k^\eta - \mu_k \nabla \boldsymbol{u}_k \right) \cdot \boldsymbol{n} = \mu_k \mathbf{A}_{k\eta} \cdot \boldsymbol{u}_k \cdot \boldsymbol{n} \tag{3.94}$$

$$\boldsymbol{n} \cdot \mu_k \nabla \boldsymbol{u}_k \cdot \boldsymbol{\lambda} = \mu_k \mathbf{A}_{k\eta} \cdot \boldsymbol{u}_k \cdot \boldsymbol{\lambda} \tag{3.95}$$

Note that Eq. (3.95) has the same form with BJS condition, which is Cauchy boundary condition.

(1) The validation on model analysis and interface condition

The accuracy and university of Eqs. (3.79)–(3.84) have been validated above. Thus whether this coupling model is accurate depends on the interface condition. According to some simplified result, the result of new interface condition has the same form as the typical boundary condition, which has showed the accuracy of new boundary condition. But the much stronger and direct validation require physical and numerical experiment.

At present, there are few physical experiments about the coupling flow problem. Single phase flow experiment made by Beavers and Joseph is a representative one. Corresponding two-phase flow have not been developed. The main reason is that it is very difficult to control and measure experiment parameters. With the development of hydromechanics, some scholars begin to apply LDA and PIV to measuring flow field close to the coupling interface but so far it is still restricted in single phase flow. On one hand, the above measurement device can only measure velocity and it cannot distinguish the data of two fluids. On the other hand, error is easy to generate due to the plugging of launcher by particles in porous media. Single phase flow can be regarded as a limitation of two-phase. Thus theoretically the interface condition of two-phase flow is suitable for single phase flow. At first the interface condition of coupling two-phase flow is simplified and transferred into the condition of single phase flow. Then a comparison with the experiment result of Beavers–Joseph is made.

At first, assuming there is k phase fluid exiting in flow region thus volume percentage $C_k = S_k = 1$. Put it into Eqs. (3.87), (3.88), (3.90) and (3.91), then we can get:

$$\boldsymbol{n} \cdot (\boldsymbol{u}_k - \boldsymbol{v}_k) = 0 \tag{3.96}$$

$$\boldsymbol{\lambda} \cdot (\boldsymbol{u}_k - \boldsymbol{v}_k) = u_s \tag{3.97}$$

$$p_k^\omega - \boldsymbol{n} \cdot \left(\boldsymbol{I} p_k^\eta - \mu_k \nabla \boldsymbol{u}_k \right) \cdot \boldsymbol{n} = \mu_k \left(\mathbf{A}_{k\eta} \cdot \boldsymbol{u}_k - \mathbf{A}_{k\omega} \cdot \boldsymbol{v}_k \right) \cdot \boldsymbol{n} \tag{3.98}$$

$$\boldsymbol{n} \cdot \mu_k \nabla \boldsymbol{u}_k \cdot \boldsymbol{\lambda} = \mu_k \left(\mathbf{A}_{k\eta} \cdot \boldsymbol{u}_k - \mathbf{A}_{k\omega} \cdot \boldsymbol{v}_k \right) \cdot \boldsymbol{\lambda} \tag{3.99}$$

Considering the experiment device of Beavers–Joseph in Fig. 3.11, when the flow state reach equilibrium, the pressure distribution is same between the upper

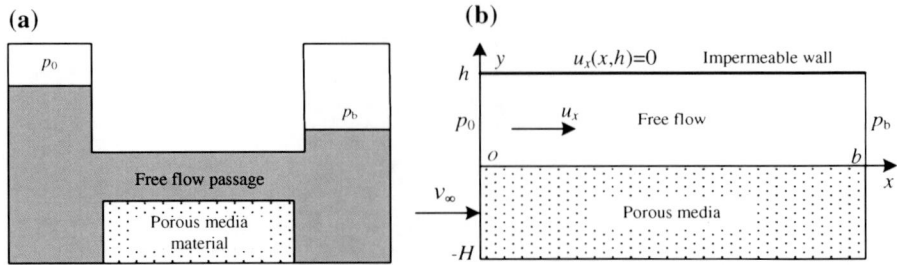

Fig. 3.11 Scheme of Beavers–Joseph experimental instrument (**a**) and its flow model (**b**)

free flow region and lower porous media region and flux Q in porous media is constant. As to free flow region, its upper boundary is impermeable wall where $u = 0$. u only changes in y direction when normal velocity and stress interface condition meet continuous condition. Thus we can get the following model:

$$\mu_k \frac{d^2 u}{dy^2} = \frac{dp}{dx}, \quad 0 \leq y \leq h \tag{3.100}$$

$$Q = -\frac{K_w}{\mu_k} \frac{dp}{dx}, \quad -H \leq y \leq 0 \tag{3.101}$$

$$u_B - Q = u_s, \quad \text{at } y = 0 \tag{3.102}$$

$$\mathbf{n} \cdot \nabla u_k \cdot \lambda = \lambda \cdot \left(\mathbf{A}_\eta \cdot \mathbf{u}_k - \mathbf{A}_\omega \cdot \mathbf{v}_k \right) \tag{3.103}$$

where $\mathbf{u}_k = u\lambda$, $\mathbf{v}_k = Q\lambda$. Put it into Eq. (3.103) and we can get the following boundary condition

$$\frac{du}{dy}\Big|_{y=0+} = \frac{\beta_1}{\sqrt{K_\omega}} u - \frac{\beta_2}{\sqrt{K_\omega}} Q \tag{3.104}$$

where $\beta_1 = \sqrt{K_\omega}(\lambda \cdot \mathbf{A}_\eta \cdot \lambda)$ and $\beta_2 = \sqrt{K_\omega}(\lambda \cdot \mathbf{A}_\omega \cdot \lambda)$, then

$$\frac{du}{dy}\Big|_{y=0+} = \frac{\beta_1}{\sqrt{K_\omega}} (u - \gamma Q) \tag{3.105}$$

Equation (3.105) is same with Eq. (3.6), where $\alpha = \beta_1$, $\gamma = \beta_2/\beta_1$. We can get:

$$\alpha = \sqrt{K_\omega} \left[(1 + \xi)\delta^{-1}(1 - \phi)\lambda \cdot \mathbf{A} \cdot \lambda \right] \tag{3.106}$$

$$\gamma = \phi^{-2} - \frac{\delta^2 \lambda \cdot \mathbf{D} \cdot \mathbf{K}_\omega^{-1} \cdot \lambda}{(1 + \xi)(1 - \phi)\lambda \cdot \mathbf{A} \cdot \lambda} \tag{3.107}$$

In Eq. (3.106), $(1 + \xi)$ means α is relevant to not only the structure characteristic of porous media but also the flow state while Eq. (3.107) shows γ is the function of fluid volume percentage. As to single phase flow, the percentages of fluid volume are 1 and ϕ respectively in free flow region and porous media region. If the porosity of porous media ϕ is very small, the velocity of porous flow Q is also very small. In this case, only when γ is very large, will porous flow velocity begin to work. But when ϕ is very large, Q is also very large when $\gamma = O(1)$. The porous media material in the experiment of Beavers–Joseph belongs to this case.

As to the above model, it is very easy to obtain analytical solution. Velocity in free flow region is

$$u = \frac{1}{2\mu_k}\frac{dp}{dx}y^2 - \frac{\alpha\sigma^2 - 2\alpha\sigma}{2\mu_k(1 + \alpha\sigma)}\frac{dp}{dx}hy - \frac{K_\omega(\sigma^2 + 2\alpha\sigma\gamma)}{2\mu_k(1 + \alpha\sigma)}\frac{dp}{dx} \tag{3.108}$$

where $\sigma = h/\sqrt{K_\omega}$.

As to Eq. (3.108), make an integration in y direction and we can get the corresponding total flux M:

$$M = \int_0^h u\,dy = -\frac{h^3}{12\mu_k}\frac{dp}{dx} - \frac{h^3}{12\mu_k}\frac{dp}{dx}\frac{3(\sigma + 2\alpha\gamma)}{\sigma(1 + \alpha\sigma)} \tag{3.109}$$

The first term at the right of Eq. (3.109) is the total flux M_p of the outlet end in Poiseuille flow problem. The sole difference between Poiseuille flow and coupling flow is the set of lower boundary condition. The lower and upper boundaries of Poiseuille flow are nonslip wall $(u = 0)$ while the lower boundary condition of coupling model is velocity slip condition $(u = u_B = 0)$. To simplify our study, Beavers and Joseph defines the parameter $\Phi = (M - M_p)/M_p$ to express the difference of these two models. We can get:

$$\Phi = \frac{3(\sigma + 2\alpha\gamma)}{\sigma(1 + \alpha\sigma)} \tag{3.110}$$

We can find the analytical result almost matches well with the experiment result of Beavers–Joseph by adjusting α and γ, better than the previous BJ condition. As Fig. 3.12 shows, black solid line is the fit line under BJ condition and $\alpha = 0.8$. In new fit line $\gamma = 1.5$ while α changes from 0.6 to 2.2. Obviously as to one certain media material, the parameter γ should be kept constant, which can reflect the geometry characteristic in interface region of porous media. As to Foam et al. material in Fig. 3.12, its porosity can be as high as 0.78. Thus there exists an apparent flow transition its coupling interface region. Based on Eq. (3.74) we can judge D is positive, thus we can estimate:

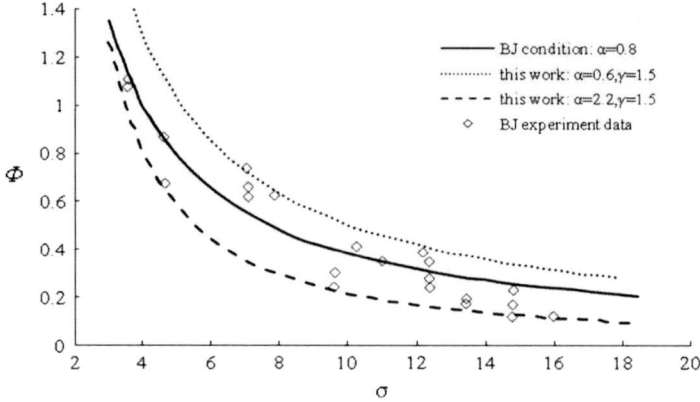

Fig. 3.12 Comparison between analytical solution for fractional excess flow rate and Beavers–Joseph experimental data(Foametal)

$$\gamma = 0.78^{-2} - \frac{\delta^2 \lambda \cdot D \cdot K_\omega^{-1} \cdot \lambda}{(1 + \xi)(1 - \varepsilon_{k\omega})\lambda \cdot A \cdot \lambda} \le 1.64 \qquad (3.111)$$

where $\gamma = 1.5$.

All the experiment data can be included between two dotted lines by adjusting α. As mentioned above, α is relevant to the structure characteristic of porous media and its flow state. Thus as to one certain experiment material, results are different when the experiment condition is different. This is why experimental data of Beavers–Joseph distributes zonally. From this perspective, new interface condition has more apparent physical significance, whose fitting effect is better. Other physical properties of Foam et al. as porous media material can be seen in Table 3.2.

To validate the accuracy of new interface condition, we also analyzed and contrast other experiment data in Beavers–Joseph experiment. The results are showed in Figs. 3.13 and 3.14. The physical properties of these five porous media materials are listed in Table 3.2.

Table 3.2 Material properties in Beavers–Joseph

Material	Permeability K_ω, m^2	Average diameter d, m	Porosity ϕ
Foam et al.	7.1×10^{-9}	$-*$	0.78
Foam et al. A	9.7×10^{-9}	4.06×10^{-4}	0.78
Foam et al. B	3.94×10^{-8}	8.64×10^{-4}	0.78
Foam et al. C	8.2×10^{-8}	1.14×10^{-3}	0.79
Aloxite 1	6.45×10^{-10}	3.30×10^{-4}	0.58
Aloxite 2	1.6×10^{-9}	6.86×10^{-4}	0.52

*Note Data about the average grain diameter of Foam et al. is absent in (Beavers and Joseph 1967)

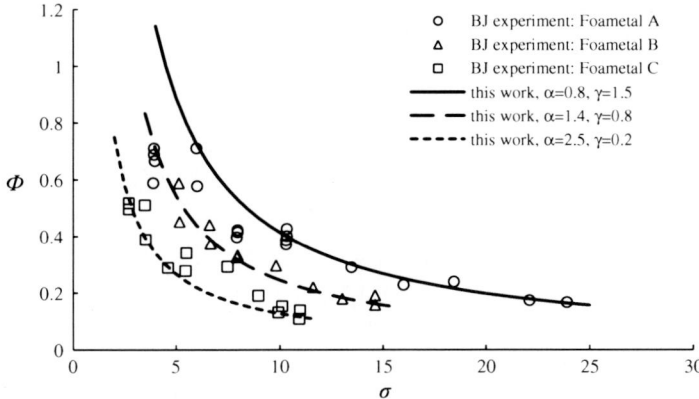

Fig. 3.13 Comparison between analytical under new interface condition solution and Beavers–Joseph experimental data (foam mental A–C)

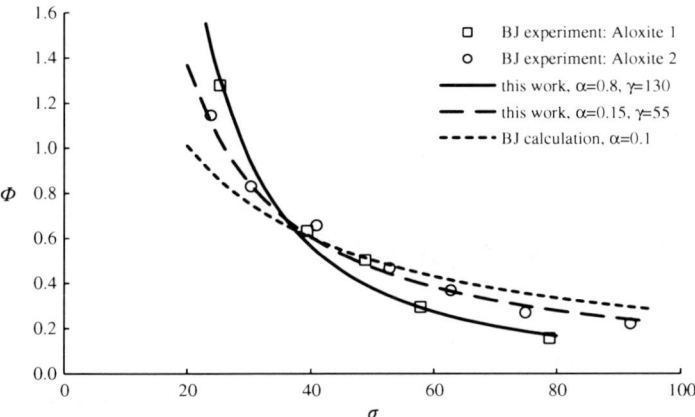

Fig. 3.14 Comparison between analytical under new interface condition solution and Beavers–Joseph experimental data (Aloxite 1–2)

As to the three kinds of porous media materials in Fig. 3.13, the porosity is $\phi \approx 0.79$, seen from Eq. (3.103), $\gamma \leq 1.6$. Meanwhile this figure shows the analytical solution obtained under new interface condition matches well with experimental data when γ is small. Figure 3.14 shows that the results based on new interface condition can still match well with the experimental data under a large γ, while at this moment BJ condition is not satisfied. The black solid line in Fig. 3.14 shows the best fit with BJ condition, which has a larger deviation compared with the experiment. Thus new coupling interface condition has an apparent advantage.

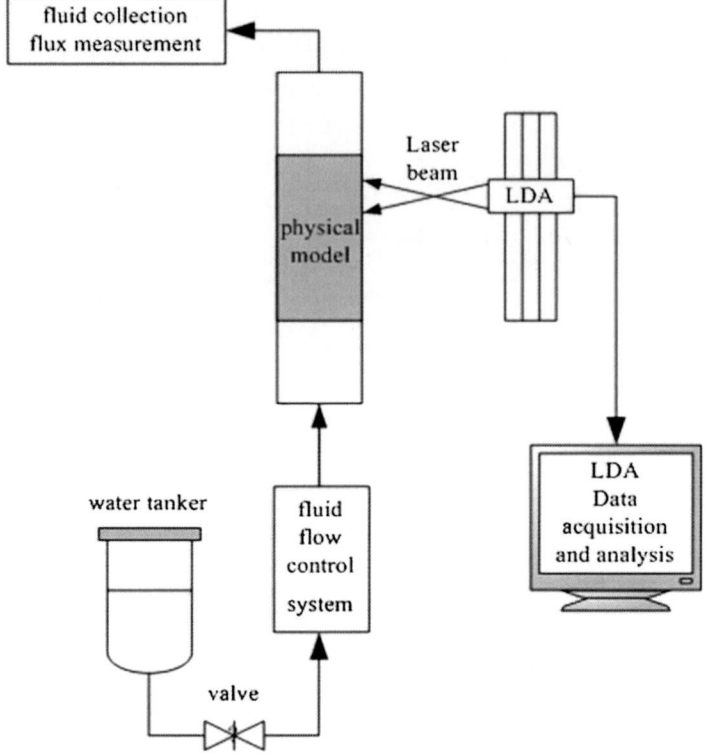

Fig. 3.15   Schematic of LDA experimental setups

Fig. 3.16   Scheme of the
experimental physical model

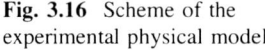

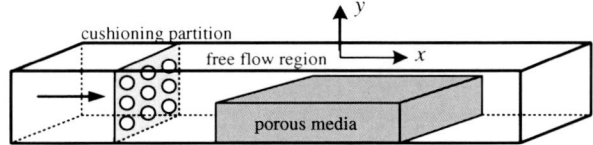

To validate the accuracy of the new interface condition further, we designed and made a single phase experimental instrument based on LDA, as shown in Figs. 3.15 and 3.16. We applied Flow-Explore 62N09 produced by Dantee Company to measure, collect and process experimental data.

Porous media in our model is filled with uniform rock particles by compaction whose porosity is 0.45 and the permeability is 5×10^{-9} m^2. In order to use LDA, organic glass is employed to close the experimental model and the height of free

Fig. 3.17 Comparison of velocity profile between theoretical and experimental results

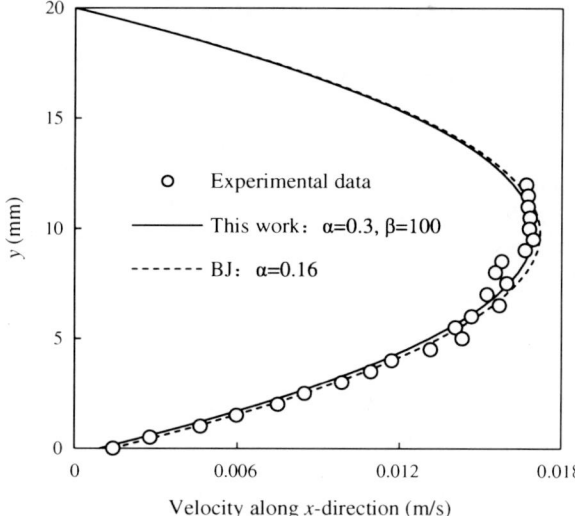

flow region is 20 mm. Pressure gradient is −0.33 Pa/m in experiment. Experimental fluid is fresh water and its viscosity is 0.001 Pa s.

Based on Eq. (3.105), we can get velocity distribution as follows:

$$u_s = \left(1 + \frac{\alpha}{\sqrt{K}}y\right)u_B + \frac{1}{2\mu}(y^2 + 2\beta\alpha\sqrt{K}y)\frac{\mathrm{d}p}{\mathrm{d}x} \tag{3.112}$$

where

$$u_B = -\frac{K}{2\mu}\left(\frac{\sigma^2 + 2\beta\alpha\sigma}{1 + \alpha\sigma}\right)\frac{\mathrm{d}p}{\mathrm{d}x} \tag{3.113}$$

Figure 3.17 shows both new interface condition and BJ condition can match up with LDA experiment data very well. We can get better result by adjusting β, especially for high flow velocity region in central section. Our experiment also shows velocity slip exists in the interface coupling of seepage flow and free flow, which is larger than Darcy velocity u_d. β is much larger because the heterogeneity of surface structure in porous media filled with rock particles. This result is similar to the aluminum sand model experiment of Beavers–Joseph. The new interfacial condition is verified further by this experiment.

3.3 Numerical Simulation of Free Flow in Vugs

3.3.1 Two Impressible Fluids Model

To adapt to complex geometry model of DFVN, we adopt upwind finite element method to make a numerical solution to two-phase flow in vugs system. Two-phase free flow mathematic model is macroscopic two-phase flow model. As to impressible fluid, its basic equation can be written as follows:

$$\frac{\partial C_k}{\partial t} + \nabla \cdot (C_k \boldsymbol{u}_k) = 0, \quad (k = \mathrm{w}, \mathrm{n}) \tag{3.114}$$

$$\frac{\partial \boldsymbol{u}_k}{\partial t} + \boldsymbol{u}_k \nabla \cdot \boldsymbol{u}_k = -\frac{1}{\rho_k} \nabla p + \boldsymbol{g} + \frac{1}{\rho_k} \nabla \tau_k - \lambda_k (\boldsymbol{u}_\mathrm{w} - \boldsymbol{u}_\mathrm{n}) \tag{3.115}$$

$$C_\mathrm{w} + C_\mathrm{n} = 1 \tag{3.116}$$

where $\lambda_k = b_k / (C_k \rho_k)$ is still defined as frictional coefficient between two-phase interface; $\tau_k = \mu_k \left[\nabla \boldsymbol{u}_k + (\nabla \boldsymbol{u}_k)^\mathrm{T} \right]$.

We assume the pressure of the two-phase fluid is equal, which is suitable for most of practical engineering.

3.3.2 Time Discrete Formulation

As to time term, we adopt ICE method (Harlow and Amsden 1971) (Implicit Continuous Fluid Eulerian Method) proposed by Harlow and Amsden to make a discretization and its corresponding time discrete format can be listed as follows:

$$\frac{C_k^{n+1} - C_k^n}{\Delta t} + \nabla \cdot \left(C_k^n \boldsymbol{u}_k^{n+1} \right) = 0, \quad (k = \mathrm{w}, \mathrm{n}) \tag{3.117}$$

$$\frac{\boldsymbol{u}_k^{n+1} - \boldsymbol{u}_k^n}{\Delta t} + \boldsymbol{u}_k^n \cdot \nabla \boldsymbol{u}_k^n = -\frac{1}{\rho_k} \nabla p^{n+1} + \boldsymbol{g} + \frac{1}{\rho_k} \nabla \cdot \tau_k^n - \lambda_k^n \left(\boldsymbol{u}_\mathrm{w}^n - \boldsymbol{u}_n^n \right) \tag{3.118}$$

$$C_\mathrm{w}^{n+1} + C_n^{n+1} = 1 \tag{3.119}$$

where superscript n is time step size number, $t = n\Delta t$.
Solve Eq. (3.118) and we can get

$$\boldsymbol{u}^{n+1} = -\frac{\Delta t}{\rho_k^*} \nabla p^{n+1} + \psi_k^n \tag{3.120}$$

where

$$\rho_k^* = d_k/\rho_n + c_k/\rho_w, \quad \psi_k^n = d_k \Psi_n^n + c_k \Psi_w^n \tag{3.121}$$

$$\Psi_n^n = u_k^n - \Delta\left(u_k^n \cdot \nabla u_k^n - \frac{1}{\rho_k}\nabla \cdot \tau_k^n - g\right) \tag{3.122}$$

$$\begin{pmatrix} d_n & c_n \\ d_w & c_w \end{pmatrix} = \begin{pmatrix} 1+\lambda_n^n\Delta t & -\lambda_n^n\Delta t \\ \lambda_w^n\Delta t & 1-\lambda_w^n\Delta t \end{pmatrix}^{-1} \tag{3.123}$$

Add up Eq. (3.117) of two-phase fluids and combine with Eq. (3.119), we can get

$$\nabla \cdot \left(C_w^n u_w^{n+1} + C_n^n u_n^{n+1}\right) = 0 \tag{3.124}$$

Equation (3.124) is called mass conservation equation.

3.3.3   Operator Splitting Method

We can get flow parameters at $t = (n+1)\Delta t$ by solving Eqs. (3.117), (3.120), and (3.124) simultaneously. Here we will adopt operator splitting method and separate it into two steps.

First, neglect pressure gradient term and get velocity $\tilde{u}_k$ approximately by applying Eq. (3.120),

$$\tilde{u}^{n+1} = \psi_k^n \tag{3.125}$$

Equation (3.120) minus Eq. (3.125), we can get

$$u_k^{n+1} = \tilde{u}_k^{n+1} - \frac{1}{\rho_k^*}\nabla\psi \tag{3.126}$$

where ψ is defined as follows:

$$p^{n+1} = \frac{\psi}{\Delta t} \tag{3.127}$$

At this moment, substitute Eq. (3.126) into Eq. (3.124) and we can get Poisson equation concerning about ψ

$$\nabla \cdot \left[\left(\frac{C_w^n}{\rho_w^*} + \frac{C_n^n}{\rho_n^*} \right) \nabla \psi \right] = \nabla \cdot \left(C_w^n \tilde{u}_w + C_n^n \tilde{u}_n \right) \tag{3.128}$$

The second step is to solve Poisson equation and get ψ. Then put ψ into (3.127) and Eq. (3.126), we can get p^{n+1} and u_k^{n+1}. And put u_k^{n+1} into Eq. (3.117) to get C_k^{m+1}.

3.3.4 Upwind Finite Element Numerical Calculation Format

Equation (3.127) and (3.128) are discretized numerically with upwind finite method. The computation domain is discretized by applying Delaunay triple network unit. Delaunay unit is shown in Fig. 3.18, where pressure p is only defined in the central spot. In other words, pressure in the whole unit is constant. However, velocity u_k, percentage of fluid volume C_k, and potential function ψ are all defined in angular point of triangle elements. We can make an approximation by the following interpolation function:

$$u_k \approx \sum_{i=1}^{3} N_i u_{ki}, \quad C_k \approx \sum_{i=1}^{3} N_i C_{ki}, \quad \psi \approx \sum_{i=1}^{3} N_i \psi_i \tag{3.129}$$

where N_i is interpolation function, which can be determined by

$$\begin{cases} N_1 = \frac{a_1 + b_1 x + c_1 y}{2A} \\ N_2 = \frac{a_2 + b_2 x + c_2 y}{2A} \\ N_3 = \frac{a_3 + b_3 x + c_3 y}{2A} \end{cases} \tag{3.130}$$

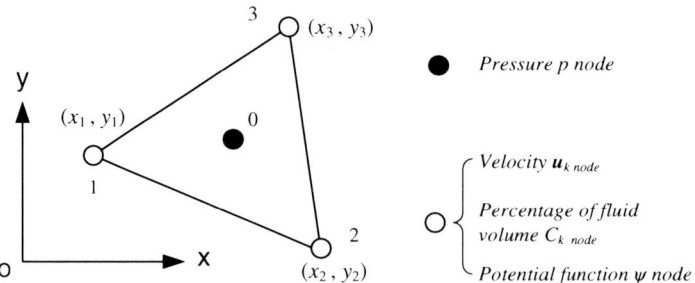

Fig. 3.18 Schematic of triangle element

The area of triangle unit A can be defined as follows:

$$A = \frac{1}{2} \begin{vmatrix} 1 & x_1 & y_1 \\ 1 & x_2 & y_2 \\ 1 & x_3 & y_3 \end{vmatrix} \tag{3.131}$$

where $a_1 = x_2 y_3 - x_3 y_2$, $b_1 = y_2 - y_3$, $c_1 = x_3 - x_2$. Their subscript is rotated by 1, 2, 3.

We can get the equivalent integration of Eq. (3.125) by weighted residual method.

$$
\begin{aligned}
\int_\Omega \boldsymbol{u}^* \cdot \tilde{\boldsymbol{u}}_k \mathrm{d}\Omega = & \int_\Omega \boldsymbol{u}^* \cdot (d_k \boldsymbol{u}_\mathrm{n} + c_k \boldsymbol{u}_\mathrm{w}) \mathrm{d}\Omega \\
& - \Delta t \int_\Omega \boldsymbol{u}^* \cdot (d_k \boldsymbol{u}_\mathrm{n} \cdot \nabla \boldsymbol{u}_\mathrm{n} + c_k \boldsymbol{u}_\mathrm{w} \cdot \nabla \boldsymbol{u}_\mathrm{w}) \mathrm{d}\Omega \\
& + \frac{\Delta t}{\rho_n} \left(\int_\Omega \tau_\mathrm{n}^n \nabla \cdot (\boldsymbol{u}^* d_k) \mathrm{d}\Omega - \int_\Gamma \boldsymbol{u}^* d_k \tau_\mathrm{n}^n \cdot \boldsymbol{n} \mathrm{d}\Gamma \right) \\
& + \frac{\Delta t}{\rho_\mathrm{w}} \left(\int_\Omega \tau_\mathrm{w}^n \nabla \cdot (\boldsymbol{u}^* c_k) \mathrm{d}\Omega - \int_\Gamma \boldsymbol{u}^* c_k \tau_\mathrm{w}^n \cdot \boldsymbol{n} \mathrm{d}\Gamma \right) \\
& - \Delta t \boldsymbol{g} \cdot \int_\Omega \boldsymbol{u}^* (d_k + c_k) \mathrm{d}\Omega
\end{aligned} \tag{3.132}
$$

where Γ is the outer boundary of solution region Ω and $\boldsymbol{n}$ is its unit outer normal vector.

To avoid the vibration of solutions, we adopt upwind weight Petrov–Galerkin finite element method Brooks and Hughes (1982) and Hughes et al. (1986) to process the convective term. In calculation, we will apply modified function

$$W_{jk} = \mathrm{e}^{-a_{k1}(x - x_j) - a_{k2}(y - y_j)} N_j \tag{3.133}$$

where

$$a_{k1} = k \frac{U_{kx}^n}{|\mathbf{U}_k^n|}, \quad a_{k2} = k \frac{U_{ky}^n}{|\mathbf{U}_k^n|} \tag{3.134}$$

where k is nonnegative constant and if $k = 0$, $W_{jk} = N_j$ and the above method will reduce to standard Galerkin method. U_{kx}^n and U_{ky}^n are the velocity components of $\mathbf{U}_k^n$ respectively. Velocity vector $\mathbf{U}_k^n$ can be defined as follows:

$$\mathbf{U}_k^n = d_k \boldsymbol{u}_\mathrm{n}^n + c_k \boldsymbol{u}_\mathrm{w}^n \tag{3.135}$$

Substitute Eq. (3.133) into Eq. (3.132) and we can get finite element numerical computational format in tangible unit. Arrange it as matrix form as follows:

$$\boldsymbol{M}_k^e \tilde{\boldsymbol{u}}_{ke} = \boldsymbol{M}_k^e \left(d_k \left(\boldsymbol{u}_{ne}^n + c_k \boldsymbol{u}_{we}^n \right) \right) - \left(\mathbf{F}_k^n \right)_e \Delta t \tag{3.136}$$

where

$$\boldsymbol{M}_k^e = \int_{\Omega_e} \boldsymbol{W}_k^{\mathrm{T}} \cdot \boldsymbol{N} \mathrm{d}\Omega_e \tag{3.137}$$

$$
\begin{aligned}
\left(\mathbf{F}_k^n \right)_e &= \boldsymbol{D}_k^e \left(d_k \boldsymbol{u}_{ne}^n \boldsymbol{u}_{ne}^n + c_k \boldsymbol{u}_{we}^n \boldsymbol{u}_{we}^n \right) + \mathbf{K}^e \left[(d_k \mu_n / \rho_n) \boldsymbol{u}_{ne}^n + (c_k \mu_w / \rho_w) \boldsymbol{u}_{we}^n \right] \\
&\quad - \mathbf{B}^e \left[d_k \left(\tau_{ne}^n \cdot \boldsymbol{n} \right) / \rho_n + c_k \left(\tau_{we}^n \cdot \boldsymbol{n} \right) / \rho_w \right] - S^e (d_k + c_k) \boldsymbol{g}
\end{aligned} \tag{3.138}
$$

$$\boldsymbol{D}_k^e = \int_{\Omega_e} \boldsymbol{W}_k^{\mathrm{T}} \cdot \left(\boldsymbol{N}^{\mathrm{T}} \cdot \nabla \boldsymbol{N} \right) \mathrm{d}\Omega_e \tag{3.139}$$

$$\mathbf{K}^e = \int_{\Omega_e} (\nabla \boldsymbol{N})^{\mathrm{T}} \cdot (\nabla \boldsymbol{N}) \mathrm{d}\Omega_e + \int_{\Omega_e} (\nabla^{\mathrm{T}} \boldsymbol{N}) \cdot (\nabla \boldsymbol{N}) \mathrm{d}\Omega_e \tag{3.140}$$

$$\mathbf{B}^e = \int_{\Omega_e} \boldsymbol{N}^{\mathrm{T}} \cdot \boldsymbol{N} \mathrm{d}\Omega_e, \quad S^e = \int_{\Omega_e} \boldsymbol{N}^{\mathrm{T}} \mathrm{d}\Omega_e \tag{3.141}$$

where e means an element.

As to Eq. (3.136), make a circulation for all the elements and we can get the following matrix equation to solve variation $\tilde{\boldsymbol{u}}_k$.

$$\boldsymbol{M}_k \tilde{\boldsymbol{u}}_k = \boldsymbol{M}_k \left(d_k \boldsymbol{u}_n^n + c_k \boldsymbol{u}_w^n \right) - \mathbf{F}_k^n \Delta t \tag{3.142}$$

where

$$\boldsymbol{M}_k = \sum_e \boldsymbol{M}_k^e, \quad \mathbf{F}_k^n = \sum_e \left(F_k^n \right)^e \tag{3.143}$$

Similarly as to Eqs. (3.117) and (3.126)–(3.128), standard Galerkin finite element method is adopted to make a numerical calculation. Corresponding finite element calculation format is as follows:

$$\boldsymbol{p}_e^{n+1} = S\psi / (|\Omega_e| \Delta t) \tag{3.144}$$

$$\boldsymbol{MC}_k^{n+1} = \boldsymbol{MC}_k^n - Q_k \Delta t \tag{3.145}$$

$$\boldsymbol{Mu}_k^{n+1} = \boldsymbol{M\mathfrak{u}}_k^n - \mathbf{E}_k \tag{3.146}$$

$$T\psi = -\boldsymbol{R} + \boldsymbol{P} \tag{3.147}$$

where

$$\mathbf{M} = \sum_e \mathbf{M}^e, \quad Q_k = \sum_e \mathbf{R}^e C_{ke}^n u_{ke}^{n+1}, \quad E_k = \sum_e \left(\mathbf{R}^e \psi_{ke}^n / \rho_k^* \right) \tag{3.148}$$

$$\mathbf{R} = \sum_e \mathbf{R}^e \left(C_{ne}^n \tilde{u}_{ne} + C_{we}^n \tilde{u}_{we} \right) \tag{3.149}$$

$$\mathbf{T} = \sum_e \mathbf{T}^e \left(\frac{C_{ne}^n}{\rho_n^*} + \frac{C_{we}^n}{\rho_w^*} \right), \quad \mathbf{P} = \sum_e \mathbf{P}^e \left(\frac{C_{ne}^n}{\rho_n^*} + \frac{C_{we}^n}{\rho_w^*} \right) \tag{3.150}$$

The unit characteristic matrix of the above equations are:

$$\mathbf{M}^e = \int_{\Omega^e} \mathbf{N}^T \cdot \mathbf{N} d\Omega_e, \quad \mathbf{R}^e = \int_{\Omega^e} \mathbf{N}^T \cdot \nabla \mathbf{N} d\Omega_e \tag{3.151}$$

$$\mathbf{T}^e = \int_{\Omega_e} (\nabla^T \mathbf{N}) \cdot (\nabla \mathbf{N}) \cdot \mathbf{N} d\Omega_e, \quad \mathbf{P}^e = \int_{\Omega_e} (\mathbf{N}^T \cdot \mathbf{N})(\nabla \psi \cdot \mathbf{n}) d\Omega_e \tag{3.152}$$

p^{n+1} can be obtained by solving Eq. (3.144), whilst C_k^{n+1}, u_k^{n+1} and ψ can be obtained by solving and combining Eqs. (3.145)–(3.147). We employ concentrated mass matrix technology for mass matrix M_k and M in the calculation. The whole flowchart can be seen in Fig. 3.19.

3.3.5 Examples and Analysis of Numerical Validation

(1) One-dimensional Burger equation
As Fig. 3.20 shows, 1-D water–oil displacement is considered here. Gravity is neglected and assume that fluid flow is laminar flow and the velocities of two-phase fluids are equal. Thus the friction between two-phase will be zero. Additionally, interfacial tension is neglected and the pressures of two-phase fluids are equal. Thus the basic mathematical model Eqs. (3.114)–(3.116) can be simplified as typical one-dimensional Burger equation.

$$\frac{\partial C_k}{\partial t} + u_k \frac{\partial C_k}{\partial x} = 0 \tag{3.153}$$

This equation is a classical convective equation and analytical solution exists. Equation (3.153) is solved with upwind finite element numerical method and the numerical method and code are validated by comparing with analytical solutions. Assume $u_w = u_n = 5$ m per day in the calculation. The total length of research region is $L = L_w + L_n = 100$ m. At initial time $L_w = 50$ m. Time step size is

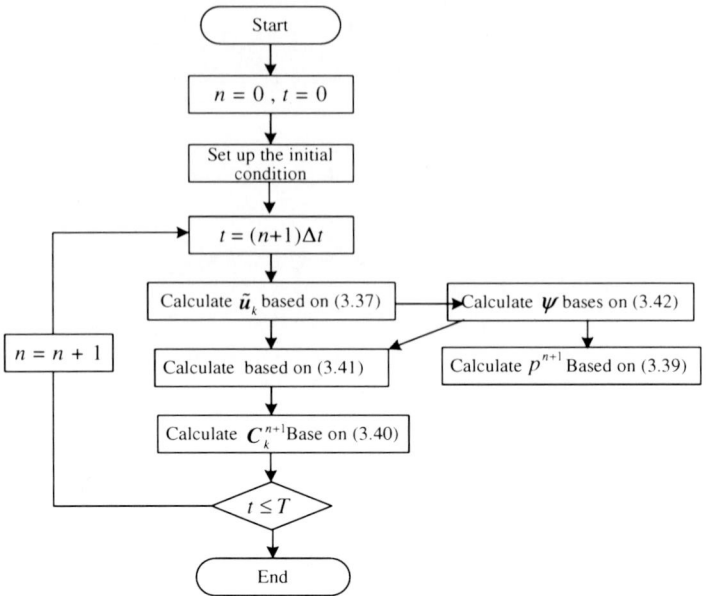

Fig. 3.19 Upwind finite element calculation flow chart

Fig. 3.20 Schematic of uniform velocity water flooding model

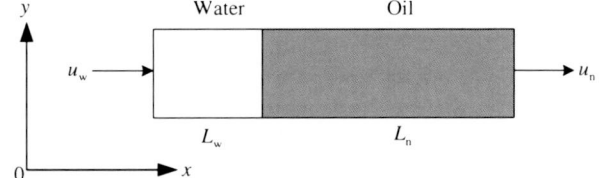

$\Delta t = 0.1$ day. Uniform network step size is $\Delta x = 0.5$ m. Parameter of shape function $k = 0.5$.

Figure 3.21 represents the distribution of the volume percentage of water phase and the comparison of analytical solutions at different time. Numerical results show that upwind finite element numerical format has ideal numerical stability and high precision. We can find that Eq. (3.153) is similar to Buckley–Leverett equation in forms.

(2) Two-dimensional cavity flow simulation

As for two-phase flow problem shown in Fig. 3.22, we will make a flow simulation by adopting different boundary conditions and initial conditions. The length of research region is $L = L_o + L_n = 100$ m. at initial time $L_o = 20$ m. time step size is $\Delta t = 0.1$ day. The initial pressure in whole research region is 1 atm, velocity of injection water on the left end is $u_o = 10$ m per day. Right end is set as constant pressure boundary (defined as 1 atm). The viscosity of water and oil are

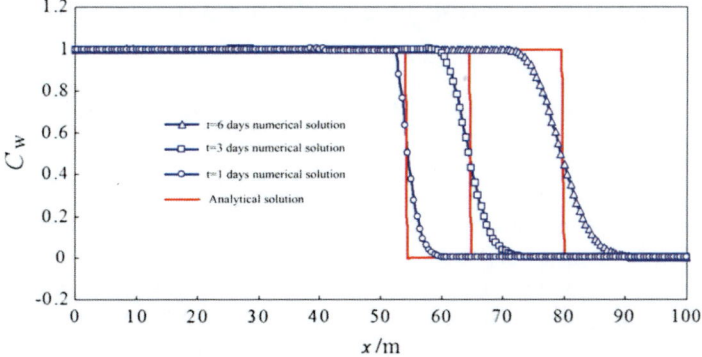

Fig. 3.21 Comparison between numerical solution and analytical solution

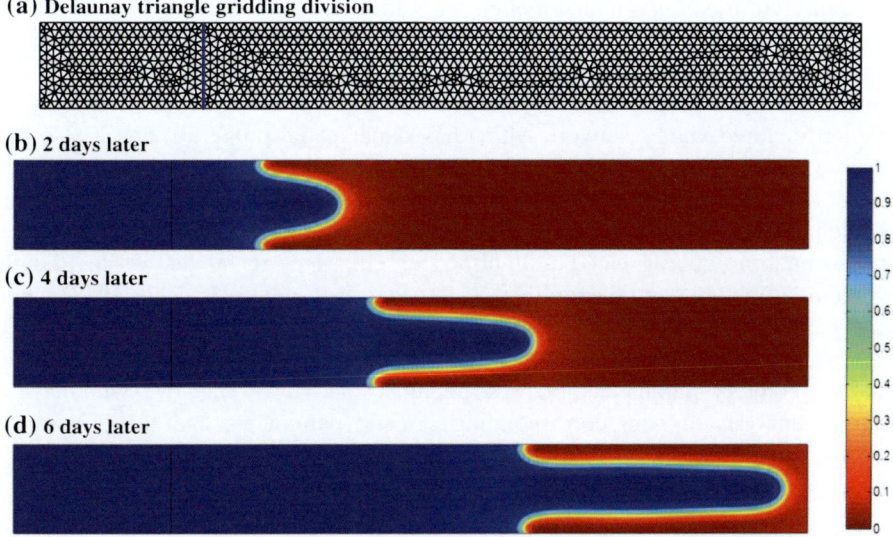

Fig. 3.22 Grid division and the volume distribution of water phase at different time

$\mu_\omega = 1$ mPa·s and $\mu_n = 5$ mPa·s respectively. Calculation network is shown in Fig. 3.22a. Research region can be divided into 2384 triangle units. Parameter of shape function is $k = 0.5$. We will neglect the velocity difference and friction between two phases.

Figure 3.22 shows the distribution of volume percentage of water phase. And results show that this method has ideal stability and robustness. Similar to porous media, apparent fingering phenomenon exists in two-phase free flow.

3.4 Numerical Simulation of Coupling Seepage Flow–Free Flow

3.4.1 The Establishment of Discrete Fractured-Vuggy Network Geometry Model

As classical discrete media model, the initial problem of discrete vugs network is to establish corresponding discrete media geometry model. At present, relevant research work mainly focuses on the establishment of discrete fracture network model. Popov and Qin et al. made an analysis (Popov et al. 2009; Qin et al. 2010) on pure cavity without considering fractures. In this section, we will add vug systems based on discrete fracture network model. Discrete fractured-vuggy network geometry model will be built by Monte Carlo method, which can lay a foundation for flow simulation research based on discrete fractured-vuggy network model.

In discrete media geometry model, we will assume that the seepage action in fractured-vuggy can be predicted by the geometry shape of fractured-vuggy, single fracture, and hydrodynamic characteristic in vugs. Spatial statistic properties (including the permeability of a small fracture) can be measured and used for generating fractured-vuggy network which has same spatial properties and it also can be used to solve flow laws in network. The other basic assumption is that a single fracture and a single vug both have regular geometry shapes. As to two-dimensional problems, a fracture is a linear segment which has different trace lengths, aperture, and inclination. A vug can be simplified as ellipse or rectangle. As to triangle problem, a fracture can be simplified as Baecher disk model (Baecher 1983) while vugs can be considered as spheroid or hexahedron.

To build discrete media geometry model, we must make a practical measurement on the geometry parameters of rock fractures and vugs. Then we will make a statistic analysis to gain corresponding statistic parameters and the probability density function it obeys. Based on these, equivalent discrete media geometry model in statistic degree is constructed.

1. The geometry parameter description on fractures and its statistic laws
Geometry parameters of fractures mainly include a shape, an aperture, occurrence, geometry size, and frequency (or density).

(1) Fracture shape
Fracture shape can be simplified as linear segment in a two-dimensional problem. In a three-dimensional problem, there exists two models at present. One is Baecher disk model and the other one is Veneziano polygon model, whose geometry shape is excessively complex in three-dimensional space. Thus most of scholars apply Baecher disk model which is relatively simple. Here we also adopt Baecher disk model.

(2) Fracture aperture
Fracture aperture is the distance between fracture surface. This parameter is the main contributor for porosity and permeability and it complies with logarithmic

normal distribution. Fracture aperture has an effect on flow laws of fluids in fractures. As to small fractures, seepage theory can be applied because it meets cube
formula. And as to big fracture systems, when hydraulic gradient is relatively big,
flow condition transfers from laminar flow to turbulent flow where cube formula
does not work and it need to be modified. Thus during the process of model
building, we make a division in order to approach to reality.

(3) Fracture occurrence
Generally, it is defined by two variables, strike and inclination. As Fig. 3.23 shows,
azimuth angle α and inclination angle β can be applied to describe fracture
occurrence. As it may be bunching in one or several statistics dominant. Thus we
need to make a distribution for fracture occurrence and then make a statistic
analysis on every group. The frequently used probability distribution are uniform
distribution, normal distribution, Arnold semi-sphere normal distribution, Bingham
distribution, and Fisher distribution. Literature (Zhang 2005) makes a comparison
on local geological data and it thinks Fisher distribution and Bingham have a good
matching.

    Fisher distribution is also called sphere normal distribution. If uniform direction
of occurrence is consistent with polar axis of reference spherical surface, then its
probability density function is,

$$f(\varphi_n, \theta_n) = \frac{1}{2\pi} \frac{\eta \sin \theta_n e^{\eta \cos \theta_n}}{e^{\eta} - e^{-\eta}} \tag{3.154}$$

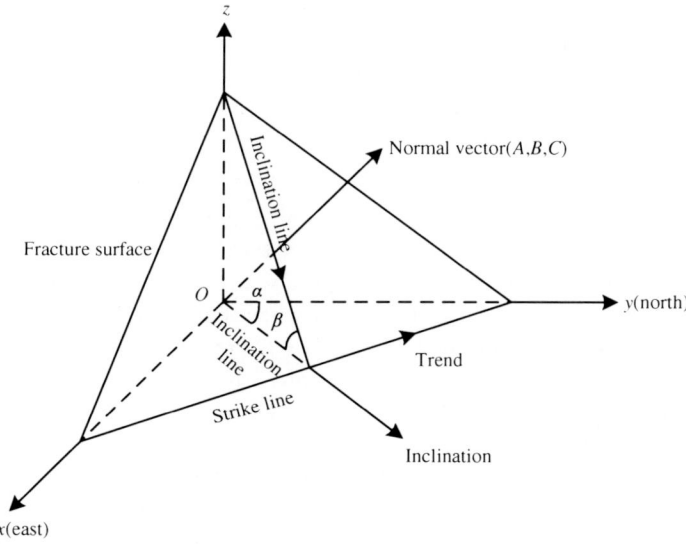

Fig. 3.23 Relationship between spatial location and orientation of fracture surface

Fig. 3.24 Schematic of
two-dimensional fracture line
in x–y axis

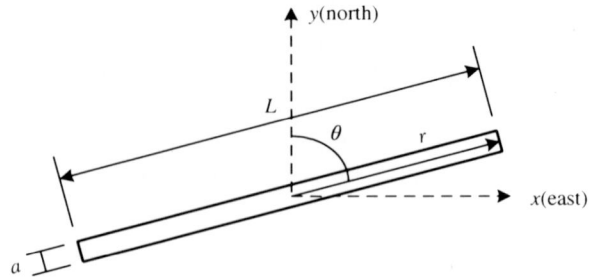

where (φ_n, θ_n) is a variable in spherical axis system and η is intensive extent
parameter of climax gather in fracture system. When $\eta = 0$, it is normal distribution. When η is very big, climaxes are concentrated in a small range in dominant
occurrence direction. As to two-dimensional problems, fracture occurrence can be
defined by azimuth θ as Fig. 3.24 shows.

(4) The geometric dimensioning of fracture surface
It is very difficult to gain the specific information under present technique condition.
We can only measure the trace length of a fracture. Generally, we assume that trace
length complies with logarithmic normal distribution or exponent distribution.
When the fracture surface is circular, the trace length can be fitted by an exponent
curve. When the fracture surface is generated, the distribution of fractural disk
diameters is needed and it is generally regarded the same as trace length. Although
the diameter distribution will not lead to the same distribution of trace length, the
effect of measurement errors is needed to be considered.

(5) Fracture frequency
Fracture frequency reflects the density of fracture and it can be defined as follows:
the number of central spots of the same group fracture surface in unit volume is
called volume frequency f_3 and the number of central spots of the same group
fractural trace length unit area is called area frequency f_2. The number of the same
group trace length which are intersected with unit line segment is called line frequency f_1.

(6) Porosity, permeability, and connectivity in fractures
A fracture pore is a variety of secondary pore which is formed by rock fracturing.
There is no big void space in Fractures. But when they are connected with primary
pore, porosity and permeability will increase dramatically. The void volume of
fractures can be calculated by other attributes such as the size of fracture and
effective fracture aperture. The permeability and connectivity of fractures reflect the
ability to flow in fractures. Generally, the permeability of fracture is very high,
which is mainly because the sole pores are connected by fractures, which have
become seepage channels. As to fractured medium, its permeability is contributed
by matrix and fractures. Fracture connectivity is the number of mutual intersected
fractures in unit area or volume, which will influence the flow condition in fractures

of the whole fluids. But in engineering practice, it is very difficult to make a quantitative description on the fracture connectivity.

In some reservoir stratum whose fractural porosity is very high, the fracture is both the primary seepage channel and the primary reservoir space. Especially in carbonatite stratum, these secondary pores are very obvious. In addition, some non-reservoir rock such as granite, when its fracture develop, it can work as the reservoir for oil and gas. This kind of reservoir is very common at the Archean Erathem in the south sea and the east sea and the pacific in Bohai Bay Basin.

2. Geometry description parameter on vugs and its statistic laws
Similarly, the geometry description parameters of a vug include shape, geometrical size, spatial azimuth, and frequency (or density).

(1) The shape of a vug
The shapes of natural vugs are very complicated. According to the results from local geologic survey, we can see vugs as spheroid in order to describe conveniently. Literature (Popov et al. 2009a, b; Qin et al. 2010) will simplify it into ellipse in two-dimensional problems.

(2) The geometry size of vugs
Under present technical condition, it is very difficult to gain the specific information about the geometry size of vugs. Through the above simplification, the size of vugs can be described by radius a, b, c in three principal axes of spheroid as Fig. 3.25a shows three principal radii which comply with logarithmic normal distribution, exponent normal distribution, or uniform distribution. As to two-dimensional problems, there are only two principal radii a and b as Fig. 3.25b shows.

(3) The spatial location of vugs
As to spheroid, since principal c is vertical to the plane where principal a and b exist, thus its spatial location can be determined uniquely. As Fig. 3.25a shows it can be described with azimuth angle θ and inclination angle φ. Its frequently used probability distributions are uniform distribution, normal distribution, Arnold hemisphere normal distribution, Bingham distribution, and Fisher distribution. Azimuth angle θ can be used to describe plane problem as Fig. 3.25b shows.

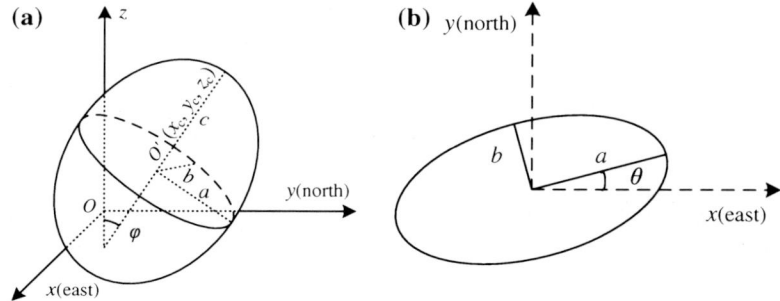

Fig. 3.25 The size of vugs and its spatial orientation

(4) The frequency of vugs
Similar to the frequency of fractures, the frequency of vugs reflects the density of vugs. The number of central spots of the same group vugs in unit volume is called volume frequency v_3. The number of central spots of the same group vugs ellipsoid in unit area is called area frequency v_2.

3. The method of Monte Carlo random model building
After getting the prior probability model about geometry parameters of fractures and vugs, we will make a random selection from these probability distributions in order to get the specific parameters of fractures or vugs system. Then we will make a combination between these parameters and it will become the computer fulfillment of geometry models on fractures and vugs. Based on random sampling in known probability distribution, Monte Carlo method will be adopted to edit a program. In this method, random number from 0 to 1 will be generated first. Based on this condition all kinds of random numbers which are matched to all distributions will be generated.

(1) The generation of random number from 0 to 1
At present congruence method is the most universal one, which is put forward by D. H.L in 1951 (Lehmer 1951). If integers N and M divide one positive integer and we can get the same remainder, we will call it congruence concerning about m. It can be written as N = M (mod m). The general formula which can be used to generate the random number can be written as follows:

$$x_n = (ax_{n-1} + c)(\bmod\ m), \quad R_n = \frac{x_n}{m} \tag{3.155}$$

where x_n is a random variable which is corresponding to random number. a is multiplier, commonly a constant number. c is an increment, commonly a constant number. m is a modulus. $\{R_n\}$ is a random array between 0 and 1. When n = 1, x_0 is a seed number, commonly system time of the computer.

The array we gain from Eq. (3.155) will repeat itself periodically. Thus the random we get by mathematic method is not the true one and we call it false random number. In order to gain a long enough circle, every parameter should attempt to utilize the word length of the computer. As to a 32-bit computer, the frequently used constant number will be listed as follows:

$$\begin{cases} m = 2^{31} - 1 \\ a = 2^{16} + 1 \\ c = (0.5 + \sqrt{3})/m \end{cases} \tag{3.156}$$

(2) The method of random sampling
As to random distribution function $F(x)$, direct sampling method is shown as follows:

$$X_F = \inf_{F(t) \geq R} t \tag{3.157}$$

where X_F is the individual in simple subsample $\{X_1, X_2, \ldots X_k\}$ coming from distribution function. R is the random number which is corresponding to random array $\{R_1, R_2, \ldots R_k\}$

As to continuous distribution function $F(x)$, if its inverse function exists, we can make a direct sampling as follows:

$$X_i = F^{-1}(R_i) \tag{3.158}$$

The fundamental step is:
① generate a uniform distribution random R_i;
② gain the random number which submitted to the given distribution by $X_i = F^{-1}(R_i)$.

Based on the above work, we make a program FracVugGen corresponding to two and three discrete fractured-vuggy network geometry model. You can see its flowchart in Fig. 3.26. We use MATLAB to edit the program and analyze the results by measurement data. We also can input the distribution laws of fractured-vuggy geometry parameter to fulfill model building. You can see its basic step as follows:

① at first, define a global coordinate system and determine a generated field in this system. This generated field should be larger than analysis field and research field in order to avoid the influence from boundary effect. In two dimension condition, we assume it rectangle and in three dimension condition we assume it hexahedron. Only if the central spot stays in the inner boundary, the boundary effect will work. Else those vugs whose parts of space stretche into the boundary will be neglected.

② then divide fractures and vugs into groups. Determine the geometry characteristic parameter of every fracture and vug. Calculate the number of every fracture and vug. We assume that the central spots location of fractures and vugs is submitted to uniform distribution, thus the number of fractures and vugs are equal to corresponding space region multiplied by frequency.

③ based on the above work, at first generate the location coordinate of the vug in generated domain and generate the specific geometry characteristic parameters by applying Monte Carlo method based on prior model, in order to determine the size and location of vugs.

④ as to the fractures in every group, generate the specific geometry parameter by applying Monte Carlo method based on the prior models: the location of central spots, occurrence (or azimuth), radius (or trace length), and aperture. After all the fracture parameters are generated, we will gain a discrete fracture network model, meanwhile it is allowed that determine the geometry parameters of the fracture according to observed value.

⑤ generate the research field boundary according to the actual sharp and size of the analysis field. Get the intersection lines or intersection points between fractures

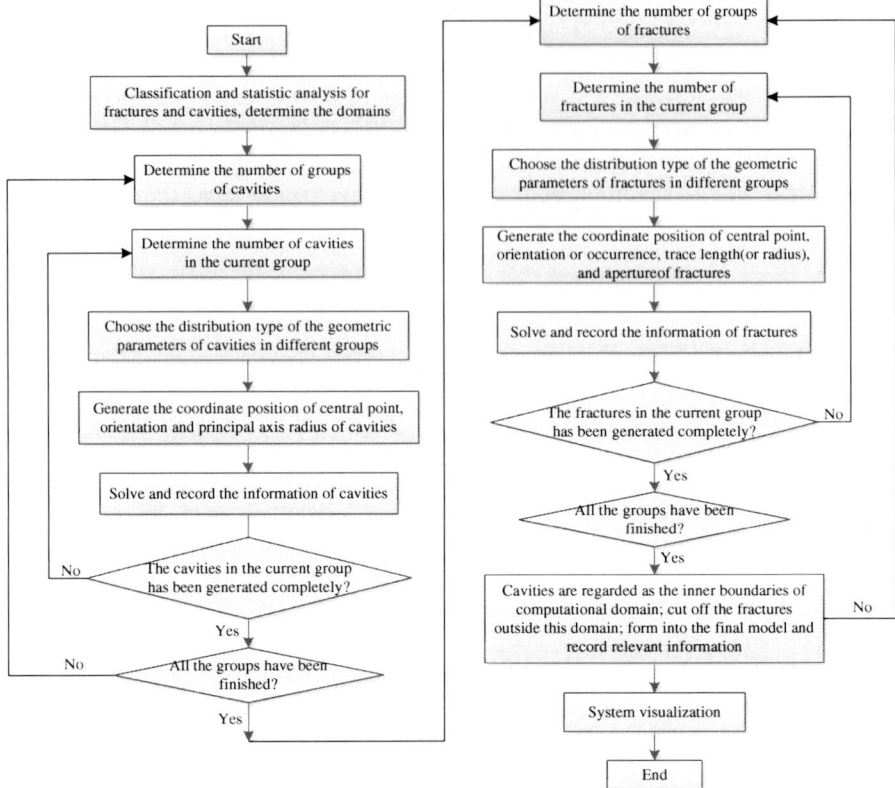

Fig. 3.26 FracVugGen program flow chart

and between fractures and boundaries(vugs included). As vugs are free flow space, it is regarded as internal boundary). Cut those parts outside the analysis domain and inside vugs, a discrete media model including vugs and fractures will be gained.

4. Cases studies

(1) Two-dimensional case 1

Assuming the research domain is square and the length of the side is a dimensionless unit length. This assumption is also applicable in the following cases. Figure 3.27a is an outcrop photograph data in one certain geological spot of Ta He oil field in Xin Jiang. To simplify the study, we transfer the length of sides into dimensionless unit length. Through statistics, the location of vugs is submitted to uniform random distribution between 0 and $\frac{\pi}{2}$. The principal axes a and b are submitted to normal distribution (0.2, 2). The average value is 0.2 and the standard deviation is 0. The frequency is $v_2 = 15$. Figure 3.27b shows a sample of the generated discrete fractured-vuggy network model. The modeling result shows the robustness.

(a)   (b)

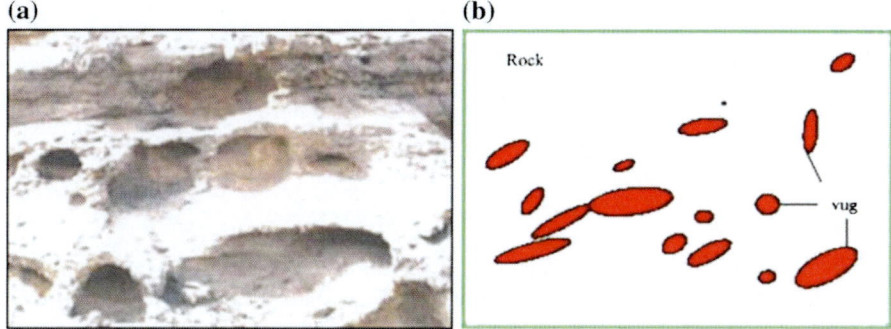

Fig. 3.27 Field outcrop and its random simulation in 2-D case 1

(a)   (b)

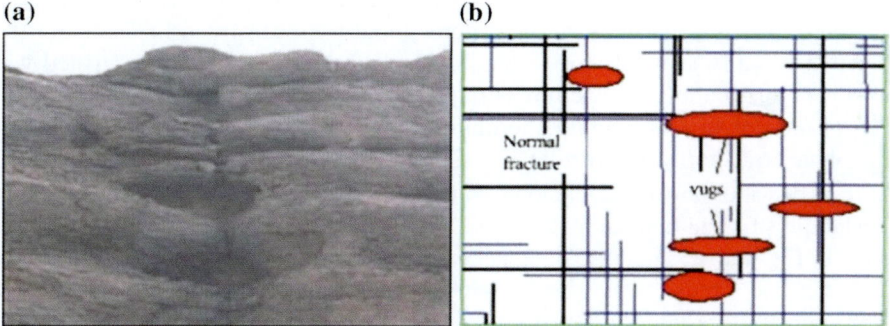

Fig. 3.28 Field outcrop and its random simulation in 2-D case 2

(2) Two-dimensional case 2

Figure 3.28a is another spot outcrop data. This rock contains vugs and vertical fracture system. Make a division statistics and analyze on it. The location of the vug is 0. The frequency is $v_2 = 5$. Other data is the same as that of in case 1. The fractures are optimized into four groups in two catalogs (large and small). The statistics has been input as Table 3.3 (the location is satisfied with normal distribution and the trace length is satisfied with exponent distribution. The aperture is

Table 3.3 Input data of fractures 2-D in case 2

Group	Frequency	Azimuth		Trace		Aperture	
		Mean	Standard deviation	Mean	Standard deviation	Mean	Standard deviation
1	10	0	0	0.8	0	0.0001	0
2	10	$\pi/2$	0	0.8	0	0.0001	0
3	10	0	0	0.8	0	0.001	0
4	10	$\pi/2$	0	0.8	0	0.001	0

Table 3.4 Input data of fractures 2-D in case 3

Group	Frequency	Azimuth		Trace		Aperture	
		Mean	Standard deviation	Mean	Standard deviation	Mean	Standard deviation
1	40	$\pi/3$	0	0.8	0	0.0001	0
2	10	$\pi/2$	0	0.8	0	0.001	0

satisfied with logarithmic normal distribution), the results simulated by FracVugGen program can be seen in Fig. 3.28b. The thick line stands for a fracture system with large aperture. The thin line stands for the small fracture system. The ellipse domains represent vugs. The modeling result shows the robustness.

(3) Two-dimensional case 3
In this case, there are two groups of different fractures with different occurrence and a group of vugs. And the data of fractures input has been listed in Table 3.4 (the location is satisfied with normal distribution and the trace length is satisfied with exponent distribution, the aperture is satisfied with logarithmic normal distribution), the location of vugs is submitted to uniform random distribution. The orientation is distributed randomly between 0 and $\frac{\pi}{2}$. The principal a and b are submitted to normal distribution (0.2, 2), among which the mean value is 0.2 and the standard deviation is 0. The frequency is $v_2 = 6$. The discrete media model generated can be seen in Fig. 3.29a. The thick line stands for a fracture system with large aperture. The thin line stands for the small fracture system. The ellipse domain means vugs. The simulation results show that random model is appropriate to the generation of two-dimensional complicated fracture network model.

(4) A three-dimensional case
In this case, the location of a vug is submitted to uniform random distribution. The orientation is constant. The principal a, b, c are distributed randomly between 0.1 and 0.3. The frequency is $v_2 = 5$. The parameters input of two groups of

(a)   **(b)**

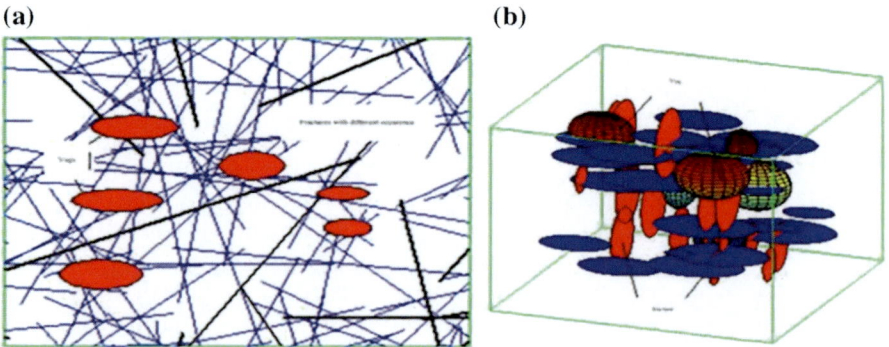

Fig. 3.29 The result from random simulation

Table 3.5 Input data of fractures 3-D in case 4

Group	Frequency	Dip angle		Radius		Aperture	
		Mean	Standard deviation	Mean	Standard deviation	Mean	Standard deviation
1	40	0	0	0.25	0.0075	0.001	0
2	10	$\pi/2$	0	0.25	0.0075	0.001	0

orthogonality fractures have been listed in Table 3.5 (the fracture inclination is satisfied with normal distribution, the radius is satisfied with uniform distribution, the aperture is satisfied with logarithmic normal distribution), the model results can be seen in Fig. 3.29b. The results show that the random model is appropriate for the generation of three-dimensional complicated fracture network model.

3.4.2 Two Phases Flow Mathematic Modeling of the Coupling of Seepage Flow–Free Flow

We will fulfill this simulation by combining new boundary condition. In the coupling flow simulation, we will adopt alternate solution method in order to decrease the complexity and calculation. In this method, we first view the physical quantities in free flow solution domain as the initial value in the coupling boundary, which is generally the calculated value of the last time step. Once we got the numerical solution of this seepage domain, the numerical value in the coupling interface can be substituted into the solution in free flow region as the boundary condition. If the solution method is steady and accurate, we can continue the alternate solution method.

We adopt boundary condition equation (3.87)–(3.91) in the coupling interface. Considering the low speed flow in reservoir, we can neglect the inertia item in free flow region of vugs and friction item between interfaces. Thus the normal line in the coupling interface is satisfied with the continuity condition of velocity and momentum. In the tangent line, it will degenerate into BJS condition, where the slip coefficient α is 1.0. In the following, two numerical cases will be given to validate the accuracy.

1. A case on the single vug media model
Consider a model which has one injection well and one production well shown in Fig. 3.30. There is a large rectangle vug in the center of reservoir model. As to matrix, this case will consider homogeneity and heterogeneity respectively. As to homogeneous matrix system, we assume that its porosity is $\phi = 0.2$ and its permeability is $K_m = 1$ μm^2. As to heterogeneous matrix rock, its permeability can be seen in Fig. 3.30b. At the initial time, the model is saturated by oil. The saturations of both the residual oil and the irreducible water are zero. The injection well will inject water with a constant speed $q_w = 0.01$ PV/d. The production well will

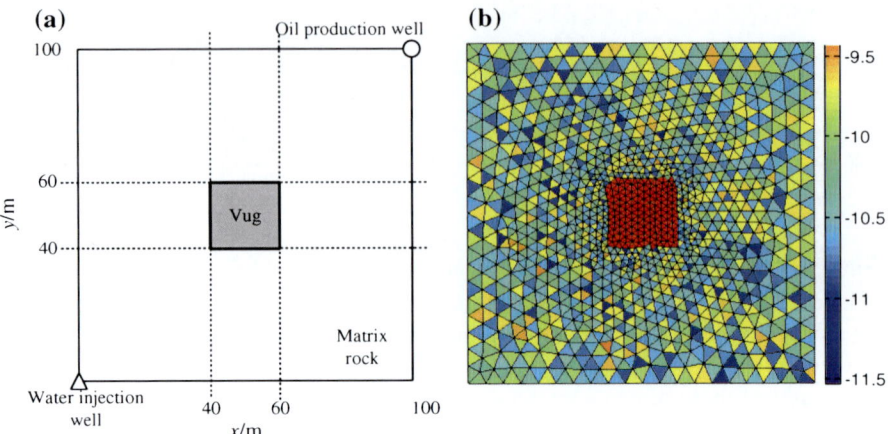

Fig. 3.30 Single vug model and its logarithmic distribution of matrix permeability

recover the oil at the same speed. For simplification, the effect of gravity and capillary force are neglected, the relative permeability for water phase is $k_{rw} = S_w^2$. The relative permeability for oil phase is $k_{rw} = (1 - S_w)^2$. The viscosity of water is $\mu_w = 1$ mPa·s, the viscosity of oil is $\mu_n = 5$ mPa·s, the densities of water and oil are $\rho_w = 1000$ kg/m^3 and $\rho_o = = 800$ kg/m^3 respectively. Delaunay triangle mesh is shown in Fig. 3.30b, where the mesh density of vug domain is 2.5 times for the mesh density of matrix rock domain and the time step length is 0.1d.

Figures 3.31 and 3.32 compares the distribution of water saturation and water phase pressure at different time. The data results show that the pressure in vugs is nearly constant because of its infinite conductivity, no matter in homogeneous matrix system, or heterogeneous matrix system. Thus this system can be seen as equipotential body. The forward speed in leading edge of oil and water is higher than matrix rock system. The simulation in the free flow region has neglected the effect of inertia item, thus the computation format in the whole region is standard Galerkin finite element numerical discrete format.

2. A case on the discrete fractured-vuggy network model
For the complex fractured-vuggy network reservoir model the thickness is 10 m, as shown in Fig. 3.33. The porosity of uniform and homogeneous matrix is $\phi = 0.2$, the permeability is $K_m = 0.1$ μm^2, the aperture is $a = 1$ mm and permeability is $K_f = \frac{a^2}{12} = 8.33 \times 10^4$ μm^2. At the initial time, the model is saturated by oil. The saturations of both the residual oil and the irreducible water are zero. The injection well will inject water with a constant speed $q_w = 0.01$ PV/d and the production well will recover the oil at the same speed.

For simplicity, the effect of gravity and capillary force are neglected, the relative permeability for water phase is $k_{rw} = S_w^2$. The relative permeability for oil phase is $k_{rw} = (1 - S_w)^2$. The viscosity of water is $\mu_w = 1$ mPa·s, the viscosity of oil is $\mu_n = 5$ mPa·s, the densities of water and oil are $\rho_w = 1000$ kg/m^3 and $\rho_w =$

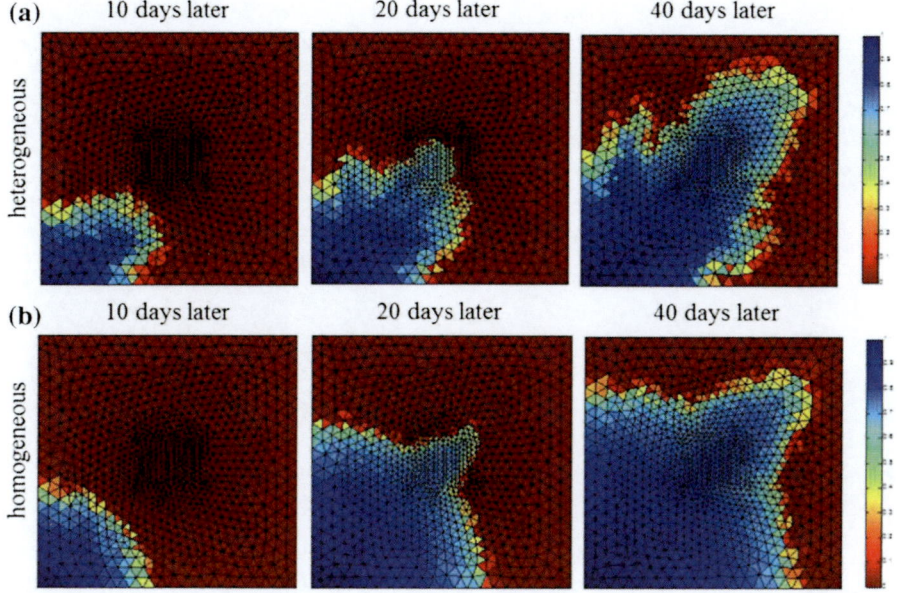

Fig. 3.31 Comparison of the distribution of water saturation

Fig. 3.32 Comparison of the distribution of water phase pressure

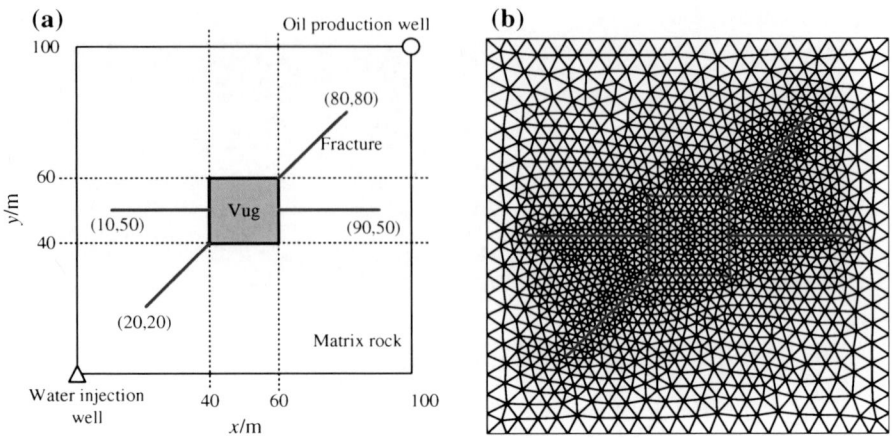

Fig. 3.33 Discrete fractured-vuggy model and its grid division

800 kg/m^3 respectively. Delaunay triangle mesh is shown in Fig. 3.30b, where the mesh density of fractured-vuggy domain is 2.5 times for the mesh density of matrix domain. The number of triangle element of matrix rock and vug domain is 2780. The number of fracture line element is 62.

Figure 3.34 compares the water saturation and the water phase pressure distribution at different time in discrete fracture-vug network model. According to the results, fractures and vugs can be seen as equipotential body. They have outstanding conductivity function. When the fractured-vuggy reservoir is developed by injecting water, we should avoid producing oil along the fractured-vuggy network, or it will lead to water channeling and logging, which will decrease the recovery efficiency. We should drive oil to the two sides of the fractures, which will stretch the water line and enlarge the displacement area and enhance the displacement efficiency. Thus the injection well should be arranged along the direction of fractures and the producing well should be arranged vertical to fractured-vuggy network.

3. A case study on one certain fractured-vuggy unit in Ta He oil field
In the following part, we have taken a certain single well model of one fractured-vuggy unit in Ta He oil field as an example and applied a discrete fractured-vuggy network model to simulate the characteristics of productivity and moisture content.

(1) A description on the single well model
This single well is drilled in a structure north high spot in Aixieke number 2, located in northeast 21°. The horizontal distance is 2.3 km. On January 1, 1999, this well began to be drilled and this project was completed on May 17, 1999. The depth of this well designed is 5587.7 m and its completed depth is 5612.7 m. The completed position is Lower Ordovician series. The initial productivity is 200 t/d and the original formation pressure is 55.04 Mpa. The calculated pressure

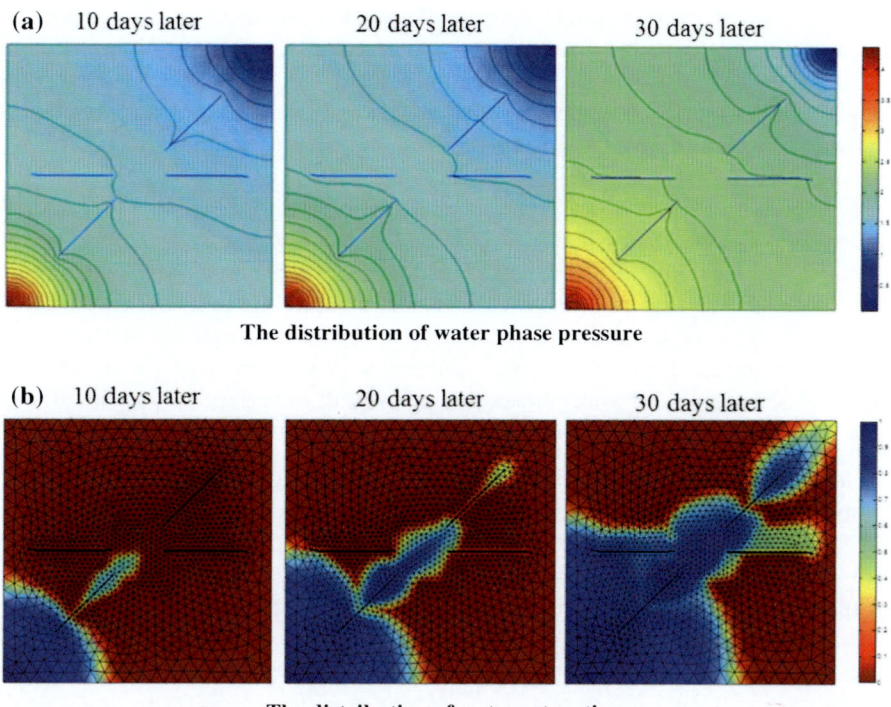

(a) 10 days later   20 days later   30 days later

The distribution of water phase pressure

(b) 10 days later   20 days later   30 days later

The distribution of water saturation

Fig. 3.34 The distribution of water phase pressure and saturation in fractured-vuggy medium

coefficient is 1.1. Thus the pressure in this formation is constant. The skin factor is −3.7, which means that the measurement of removing blocking is obvious after acid fracturing. The formation permeability is 3.3 μm², which means that the flow resistance is very small when oil flows out. Thus we can know that the fracture grow extraordinarily after formation acid fracturing. Number 16(1) formation (depth from 5414.0–5420.0 m) is the relative developed belt and number 16(2) formation (depth from 5420–5441.0 m) is fractured-vuggy developed belt. Up to Feb 13, 2008, the accumulated liquid productivity capacity is 359,613 t, the accumulated oil productivity capacity is 148,113 t and the accumulated water productivity capacity is 211,500 t.

During the process of fractured-vuggy formation, geologic structure such as fault will have an effect on the distribution of fractured-vuggy. Research finds that there is a very good corresponding relationship between small scale fractures and faults. The shorter the distance between fracture and fault is, the larger the density of the small scale fracture will be. Conversely, it will be smaller. In the fourth district Ta He oil field, there are three principal azimuthal faults developed: northeast, south–north, and northwest. The density of small scale fracture in northeast is the largest and the fracture density within 100 m far from northeast fault is the highest. Beyond 100 m, the density of small scale fracture reaches a stable level of 0.4 piece per meter. The density of south–north small scale fracture takes a second place. The density within 115 m far from south–north fracture is relatively higher. Beyond 115 m, the density of small scale fracture reaches a stable level of 0.2 piece per meter. Northwest small scale fractures developed least. There are small scale fractures developing within 165 m far from northwest fault, while not beyond 165 m.

Fractures in the single unit develop relatively well and there exist macroscopic large fractures and vugs which can be seen as fractured-vuggy geological formation. The permeability of formation is good and it presents an apparent uniform reservoir characteristic in well test curve. As a result of limited capacity of the fractured-vuggy media, the initial productivity capacity is high but the stable production period is short. The water content will increase and production will decrease seriously after water breakthrough. It will experience a long-term stage when the water content is high and oil production is low, as Fig. 3.35 shows.

(2) The numerical simulation on discrete fractured-vuggy network model
Number 16 formation is a typical fractural one. At this moment, we just make a DFN numerical simulation on formation 16(1). The depth of this formation is 5414.0–5420.0 m and its average effective thickness is 6.0 m. The average porosity of the formation matrix rock is around 4 %. The permeability is 3.5×10^{-3} μm^2. The average water saturation is 20 %. The porosity of the fracture is around 0.5 %. The macroscopic fracture probability is 8–30 %. Its aperture is around 1000 μm.

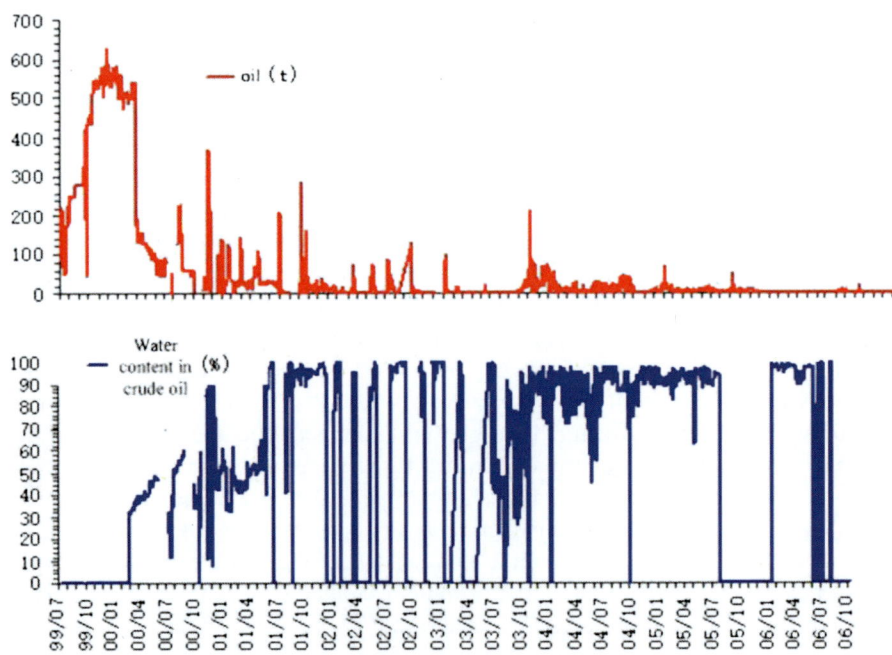

Fig. 3.35 Curves of oil production and water cut for single well per day

Table 3.6 Physical properties of the reservoir

Parameter	Value	Unit	Data source
Oil density ρ_o	0.8635	g/m^3	Well TK404PVT
Oil layer thickness h	6	m	Logging interpretation
Porosity ϕ	0.05	(–)	Logging interpretation
Volume coefficient B_o	1.1	(–)	Well TK404PVT
Compressibility C_o	0.001065	1/MPa	Well TK404PVT
Oil viscosity μ_o	24.09	mPa·s	Well TK404PVT

The inclination angle is larger than 80°, which can be seen as vertical fracture. Oil saturation is 82 %. The other basic parameters on the reservoir are shown in Table 3.6. The relative permeability curve is shown in Fig. 3.36. Due to its fractures and considering a simple calculation, the effect of capillary is neglected. The density of water is 1000 kg/m^3. The viscosity of water is 1.0 mPa·s.

According to the geological statistic data, we build a corresponding discrete fractured-vuggy network model as Fig. 3.37a shows. The mark of water injection well is TK407. The corresponding triangle gridding is shown in Fig. 3.37b.

We have applied the discrete fractured-vuggy network numerical simulation method to make a historic fitting on the production performance. In this unit there is a single producing well TK404 and it belongs to flowing production at the initial stage. As time goes by, water from injection well will bypass the fault and in-pour into this unit. The injected water will intrude into this place from a small area between northeast and south–north fault. Figure 3.38 shows the distribution of water saturation of a single unit at different moment.

The fitting result of oil production indicates a good precision as Fig. 3.39 shows. Thus we can use it as a model for the next historic fitting. By comparing with the

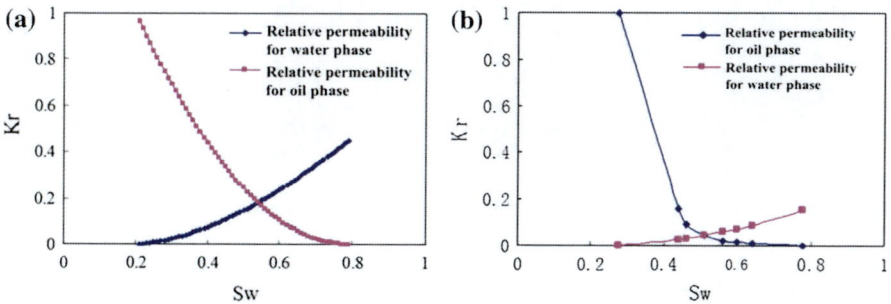

Fig. 3.36 Relative permeability curves of oil and water

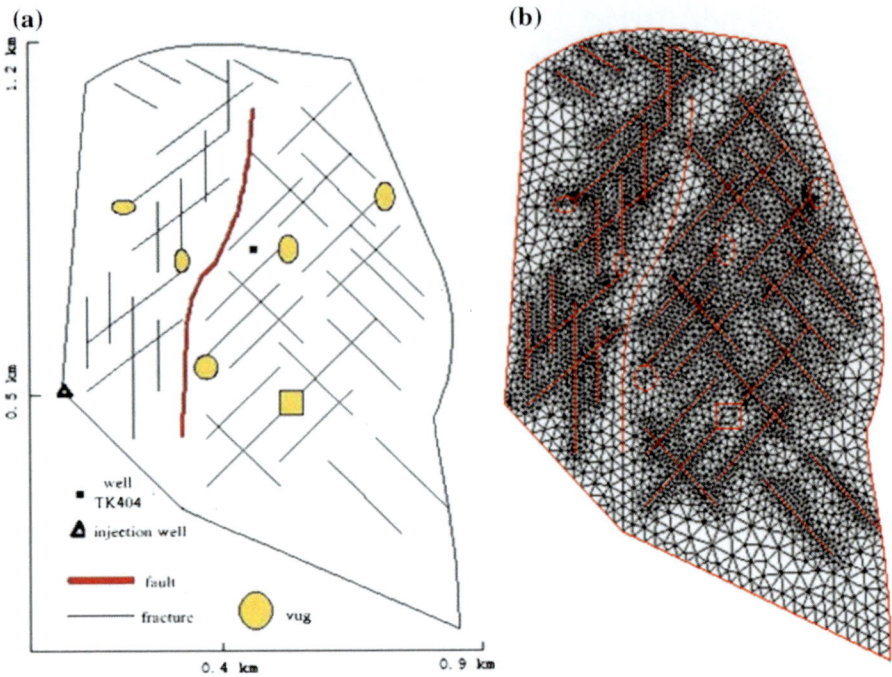

Fig. 3.37 Geological statistic model and grid division of TK404 well

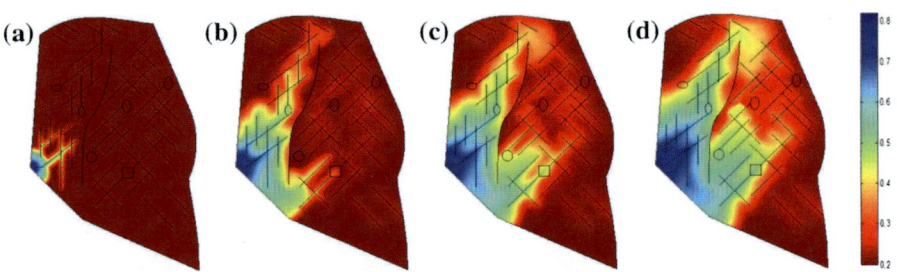

Fig. 3.38 The distribution of water saturation of single well at different time

true water content curve, we can find a good fitting precision in the early and late stages in the discrete fractured-vuggy network model. But this model will have water breakthrough much earlier than the actual situation. If adjusting the fracture occurrence and parameters, we can gain a better simulation effect (Fig. 3.40).

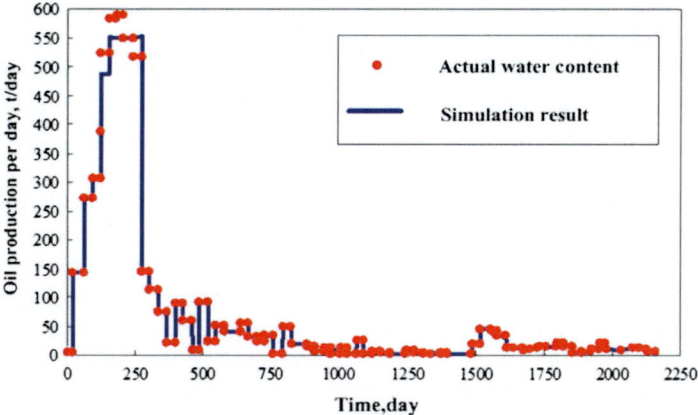

Fig. 3.39 Comparison of oil production per day for single well

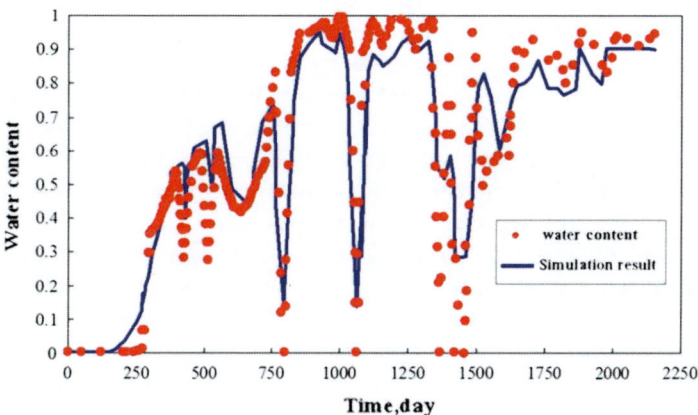

Fig. 3.40 Comparison of water cut for single well

3.5 Conclusions

(1) In this chapter, we come up with discrete fractured-vuggy network model based on the structural characteristics of fractured-vuggy media. This model add vug system into discrete fracture model to describe the flow characteristics of coupling seepage flow and free flow in fractured-vuggy medium. This new model is a typical mathematic model which overcomes the weakness of the current mathematic model. In this model, free flow exists in vug system, which can be described by Navier–Stokes equation. Seepage flow exists in fracture and matrix rock system. Based on geological statistic principle, we formed a

set of random model building method on discrete fractured-vuggy network and presented the corresponding cases.

(2) Based on the volume average method, we deduce two phases flow numerical model which couples seepage flow and free flow by two timescale upgrades. Then we build the corresponding coupling interface condition. By comparing with Beavers–Joseph experimental data, we validate the accuracy of this coupling interface condition. The coupling mathematic model we build is a typical two phases flow Darcy mathematic model in seepage region while it is a typical macroscopic two phases model in free flow region. As to the coupling flow mathematic model, we can build discrete fractured-vuggy network two phases flow mathematic model by applying discrete fracture model in seepage region. We make an elaborate research on the flow simulation theory and method of macroscopic discrete fractured-vuggy network model. As to its complexity, we form the corresponding unstructured gridding division method. As to two-dimensional problem, we will adopt Delaunay triangle gridding. As to three-dimensional problem, we will adopt Delaunay tetrahedron gridding. Based on the discrete fracture model, we build two phases flow mathematic model in the seepage region. We build two phases free flow mathematic model in vugs by applying two fluids model. Then we make this numerical value discretized by adopting upwind Petrov–Galerkin definite element method and the algorithm is verified by 1-D Burger equation

(3) Combining the coupling interface condition and applying alternate solution method, we fulfill the numerical simulation on two phases flow which couples seepage flow and free flow. Then corresponding numerical cases are given. The calculation results show that this system has an infinite conductivity capacity and it can be seen as equipotential body which is easy to form a dominant transport. Based on the geological statistic data of a single well in one certain fractured-vuggy unit in Ta He oil field, we build the corresponding discrete fracture-vug network model. Then we applied the simulation method we build in this chapter to make a simulation on the production performance. The results show that this model has a higher fitting precision. By adjusting the fracture occurrence and other parameters, we can gain better results.

References

Abdassah D, Ershaghi I (1986) Triple-porosity systems for representing naturally fractured reservoirs. SPE Form Eval 1:113–127

Arbogast T, Brunson DS, Bryant SL, Jennings JW (2004) A preliminary computational investigation of a macro-model for vuggy porous media. Dev Water Sci 55:267–278

Baca RG, Arnett RC, Langford DW (1984) Modelling fluid flow in fractured-porous rock masses by finite-element techniques. Int J Numer Meth Fl 4(4):337–348

Baecher GB (1983) Statistical analysis of rock mass fracturing. J Int Assoc Math Geol 15:329–348

Bear J (1972). Dynamics of fluids in porous media. American Elsevier, New York

Beavers GS, Joseph DD (1967) Boundary conditions at a naturally permeable wall. J Fluid Mech 30:197–207

Beavers GS, Sparrow EM, Magnuson RA (1970) Experiments on coupled parallel flows in a channel and a bounding porous medium. J Fluids Eng 92:72

Beavers GS, Sparrow EM, Masha BA (1974) Boundary condition at a porous surface which bounds a fluid flow. AIChE J 20:596–597

Brooks AN, Hughes TJR (1982) Streamline upwind/Petrov-Galerkin formulations for convection dominated flows with particular emphasis on the incompressible Navier-Stokes equations. Comput Methods Appl Mech Eng 32:199–259

Chandesris M, Jamet D (2006) Boundary conditions at a planar fluid–porous interface for a Poiseuille flow. Int J Heat Mass Transf 49:2137–2150

Chandesris M, Jamet D (2009) Derivation of jump conditions for the turbulence k6103 model at a fluid/porous interface. Int J Heat Fluid Flow 30:306–318

Chang X, Yao J, Dai W, Wang Z (2004) The study of well test interpretation method for a triple medium reservoir. J Hydrodyn 19:339–346

Dagan G (1979) The generalization of Darcy's Law for nonuniform flows. Water Resour Res 15:1–7

Erzeybek S, Akin S (2008) Pore network modeling of multiphase flow in fissured and vuggy carbonates. In: SPE improved oil recovery symposium

Ger W, Mikeli, 1999. On the interface boundary condition of beavers, Joseph and Saffman. SIAM J Appl Math 60:1111–1127

Goyeau B, Lhuillier D, Gobin D, Velarde MG (2003) Momentum transport at a fluid–porous interface. Int J Heat Mass Transf 46:4071–4081

Harlow FH, Amsden AA (1971) A numerical fluid dynamics calculation method for all flow speeds. J Comput Phys 8:197–213

Huang ZQ, Yao J, Wang YY, Lv XR (2011a) Numerical Simulation of the waterflooding development in fractured reservoir based on discrete fracture model. Chin. J. Comput. Phys. 28:41–49

Huang ZQ, Yao J, Wang YY, Tao K (2011b) Numerical study on two-phase flow through fractured porous media. Sci. China Technol. Sci. 54:2412–2420

Hughes TJR, Franca LP, Balestra M (1986) A new finite element formulation for computational fluid dynamics: V. Circumventing the babu08 ka-brezzi condition: a stable Petrov-Galerkin formulation of the stokes problem accommodating equal-order interpolations. Comput Methods Appl Mech Eng 59:85–99

Jager W, Mikeli A (2009) Modeling effective interface laws for transport phenomena between an unconfined fluid and a porous medium using homogenization. Transp Porous Media 78:489–508

Jones IP (1973) Low Reynolds number flow past a porous spherical shell. In Mathematical proceedings of the Cambridge philosophical society, vol 73, No 01. Cambridge University Press, pp 231–238

Jamet D, Chandesris M (2009) On the intrinsic nature of jump coefficients at the interface between a porous medium and a free fluid region. Int J Heat Mass Transf 52:289–300

Le Bars M, Worster MG (2006) Interfacial conditions between a pure fluid and a porous medium: implications for binary alloy solidification. J Fluid Mech 550:149–173

Lehmer DH (1951) Mathematical methods in large-scale computing units. In: Proceedings of 2nd Symposium on Large-Scale Digital Calculating Machinery, Harvard University Press, Cambridge, pp 141–146

Liu Q, Prosperetti A (2010) Pressure-driven flow in a channel with porous walls. In: 63rd annual meeting of the APS division of fluid dynamics, pp 77–100

Liu J, Bodvarsson GS, Wu YS (2003) Analysis of flow behavior in fractured lithophysal reservoirs. J Contam Hydrol 62–63:189–211

Louis C, Wittke W (1971) Experimental study of water flows in jointed rock massif, Tachien project, Formosa. Geotechnique 21(1):29

Lucia FJ (2007) Carbonate reservoir characterization. Springer, Berlin Heidelberg

Mat MD, Ilegbusi OJ (2002) Application of a hybrid model of mushy zone to macrosegregation in alloy solidification. Int J Heat Mass Transf 45:279–289

Mosthaf K, Baber K, Flemisch B, Helmig R, Leijnse A, Rybak I, Wohlmuth B (2011) A coupling concept for two-phase compositional porous-medium and single-phase compositional free flow. Water Resour Res 47:447

Neale G, Nader W (1974) Practical significance of Brinkman's extension of Darcy's law: coupled parallel flows within a channel and a bounding porous medium. Can J Chem Eng 52(4): 475–478

Noorishad J, Mehran M (1982) An upstream finite element method for solution of transient transport equation in fractured porous media. Water Resour Res 18:588–596

Ochoa-Tapia JA, Whitaker S (1995a) Momentum transfer at the boundary between a porous medium and a homogeneous fluid–I. Theoretical development. Int J Heat Mass Transf 38:2635–2646

Ochoa-Tapia JA, Whitaker S (1995b) Momentum transfer at the boundary between a porous medium and a homogeneous fluid—II. Comparison with experiment. Int J Heat Mass Transf 38:2647–2655

Popov P, Efendiev Y, Qin G (2009a) Multiscale modeling and simulations of flows in naturally fractured karst reservoirs. Commun Comput Phys 6:162–184

Popov P, Qin G, Bi L-F, Efendiev Y, Kang Z-J, Li J-L (2009b) Multiphysics and multiscale methods for modeling fluid flow through naturally fractured carbonate karst reservoirs. SPE Reserv Eval Eng 12:218–231

Qin G, Bi L, Popov P, Efendiev Y, Espedal M (2010) An efficient upscaling procedure based on Stokes-Brinkman model and discrete fracture network method for naturally fractured carbonate karst reservoirs. In: International Oil and Gas Conference and Exhibition in China. Society of Petroleum Engineers

Rhodes CA, Rouleau WT (1966) Hydrodynamic lubrication of partial porous metal bearings. J Fluids Eng 88:53–60

Richardson S (1971) A model for the boundary condition of a porous material. Part 2. J Fluid Mech 49

Saffman PG, Saffman PG (1971) On the boundary condition at the surface of a porous medium. Stud Appl, Math

Slattery JC (1967) Flow of viscoelastic fluids through porous media. Aiche J 13:1066–1071

Taylor GI (1971) A model for the boundary condition of a porous material. Part 1. J Fluid Mech 49:319–326

Velazquez RC, Vasquez-Cruz MA, Castrejon-Aivar R, Arana-Ortiz V (2005) Pressure transient and decline curve behaviors in naturally fractured Vuggy carbonate reservoirs. SPE Reserv Eval Eng 8:95–112

Whitaker S (1986) Flow in porous media II: the governing equations for immiscible, two-phase flow. Transport Porous M 1(2):105–125

Whitaker S (1999) The method of volume averaging, theory and applications of transport in porous media. Kluwer Academic, The Netherlands

Wilson CR, Witherspoon PA (1974) Steady state flow in rigid networks of fractures. Water Resour Res 10:328–335

Wu YS, Ge JL (1983) The fluid flow problem in triple medium fracture-vug reservoirs. Chin J Theor Appl, Mech

Wu Y, Liu HH, Bodvarsson GS (2004) A triple-continuum approach for modeling flow and transport processes in fractured rock. J Contam Hydrol 73:145–179

Wu YS, Qin G, Ewing RE, Efendiev YY, Kang Z, Ren Y (2006) A multiple-continuum approach for modeling multiphase flow in naturally fractured Vuggy petroleum reservoirs. In: International oil and gas security conference, China

Wu Y-S, Ehlig-Economides C, Qin G, Kang Z, Zhang W, Ajayi B, Tao Q (2007) A triple-continuum pressure-transient model for a naturally fractured vuggy reservoir. In: 2007 Society of petroleum engineers (SPE) annual technical conference and exhibition, Anaheim, California, U.S.A., 11–14 November 2007

Yao J, Zisheng W (2007) Theory and method for well test interpretation in fractured-vuggy carbonate reservoirs. China University of Petroleum Press, Shandong Dongying

Yang J, Yao J, Wang ZS (2005) Study of pressure transient characteristic for triple medium composite reservoirs. J Hydrodyn 20:418–425

Yao J, Dai W, Wang Z (2004) Study on well testing interpretation method in triple medium reservoir of variable well-bore storage. J Univ Pet 28:46–51

Yao J, Huang Z, Li Y, Wang C, Xinrui LV, Yao J, Huang Z, Li Y, Wang C, Xinrui LV (2010a) Discrete fracture-vug network model for modeling fluid flow in fractured vuggy porous media. In: International oil and gas security conference, exhibition, China

Yao J, Huang ZQ, Wang ZS, Li YJ, Wang C (2010b) Mathematical model of fluid flow in fractured Vuggy reservoirs based on discrete fracture-vug network. Acta Pet Sin 31:815–819

Yao J, Wang ZS, Zhang Y, Huang ZQ (2010c) Numerical simulation method of discrete fracture network for naturally fractured reservoirs. J Univ Pet 31:284–288

Zhang Y (2005) Rock hydraulics and engineering. China WaterPower Press, Beijing

Zhang Q, Prosperetti A (2009) Pressure-driven flow in a two-dimensional channel with porous walls. J Fluid Mech 631:1–21

Zhang D, Yao J, Wang Z, Zhan A (2008) Study on well testing model and pressure characteristics of triple medium reservoirs. Xinjiang Pet Geol 29:222–226

Zheng SQ, Li Y, Zhang WM, Zhang HF, Zhao YY (2009) Composite medium model and fluid flow mathematical model for fractured Vuggy reservoir. Pet Geol Oilfield Dev Daqing 28:63–66

Chapter 4
Equivalent Medium Model

Abstract Numerical simulation of fluid flow in fractured karst reservoirs is still a challenging issue. The multiple-porosity model is the major approach until now. However, the multiple-porosity assumption in this model is unacceptable for many cases. In the present work, an efficient numerical model has been developed for fluid flow in fractured karst reservoirs based on the idea of equivalent continuum representation. First, based on the discrete-fracture model and homogenization theory, the effective absolute permeability tensors for each grid blocks are calculated. And then an analytical procedure to obtain a pseudo-relative permeability curves for a grid block containing fractures and cavities has been successfully implemented. Next, a full-tensor simulator has been designed based on a hybrid numerical method (combining mixed finite element method and finite volume method). A simple fracture system has been used to demonstrate the validity of our method. Lastly, we have used the fracture and cavity statistics data from a TAHE outcrop in west China, effective permeability values and other parameters from our code, and an equivalent continuum simulator to calculate the water flooding profiles for a more realistic system.

Keywords Fractured karst reservoirs · Effective permeability tensor · Discrete-fracture model · Full-tensor simulator

4.1 The Research Status and Trends

The equivalent continuum model (ECM) was first proposed by Snow (1970), it is a mathematical model which describes flow in a fractured media based on the equivalent continuum theory. Since then, several scholars have made a large number of researches on (largely researched) this method (Feng et al. 2007; Liu et al. 2000; JCS Long 2012; Oda 1986; Tian 1984).

In the equivalent continuum model, the whole research region (most of the research work) is regarded as a hypothetical continuum. Every point in the system is at local equilibrium as the fluid exchanges between fracture and matrix. This

© Petroleum Industry Press and Springer-Verlag Berlin Heidelberg 2017   143
J. Yao and Z.-Q. Huang, *Fractured Vuggy Carbonate Reservoir Simulation*,
Springer Geophysics, DOI 10.1007/978-3-662-55032-8_4

method focuses on the macro-flow characteristic expressed by fractured media. First, the permeability of fractures is distributed to the whole fractured media uniformly, then the fractured media is regarded as an anisotropic porous media, and the equivalent continuum model can be established based on the flow theory of porous media. However, the equivalent is just for flow.

The outstanding advantage of ECM is that the flow can be analyzed based on the anisotropic continuum media theory. So, it has solid foundation on both theory and method. (Therefore, both theory and practically tried methods contribute to the solid foundations of ECM). Moreover, it does not need to know the exact position and hydraulic characteristic for each fracture. Therefore, (This proves that) ECM is a valuable tool for the engineering problem which has difficulty in getting exact data for each fracture. Although the ECM is easy to implement, it still has two difficulties: one is the validity of the ECM; the other is the calculation of the equivalent parameters for fractured media.

The porous media, which is composed of solid particles and voids between particles, is discontinuous at microscale. However, analyzing flow in porous media with the continuum theory has never been in doubt, it is because that (simply because), the respective element volume (REV) of porous media is sufficiently small. Mechanical property of any material can be obtained by special experiment. The property obtained by experiment has no significant relationship with the size of sample as the size of sample is bigger than
a specific value. On the contrary, the property obtained will be fluctuant as the size of sample is smaller than a specific value. For the permeability coefficient of a porous media, the relationship between the value and volume V can be plotted as a curve, as shown in Fig. 4.1. Because the REV of porous media is small and its structure is uniform, the reliable permeability can be obtained by using several small samples. Therefore, it is reasonable to treat the porous media as continuum media.

For the controversy about validity of treating the fractured media as continuum media, several scholars have given the criterion, respectively: Louis (1972) regarded that the equivalent media model can be applied when the number of fracture is more than 1000 within the range of engineering; Zhou et al. (2004) regarded that the equivalent media model can be applied when the radio between the average fracture interval and building size is lesser than 0.05; Wilson et al. (1983) regarded

Fig. 4.1 The relationship curve for permeability coefficient K and representative element volume REV (V_0)

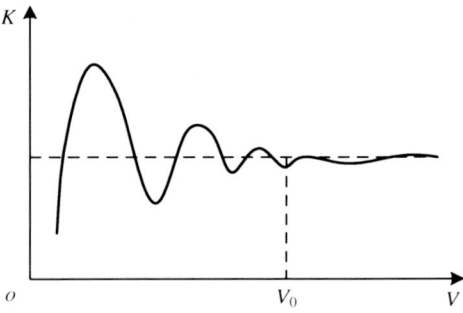

that the equivalent media model can be applied when the radio between the maximum joint spacing and building size is lesser than 0.02. However, all of these criteria are deduced by specific project or theory, which means it is difficult to be applied in practical.

Long et al. (1985) pointed out two conditions that need to be met as the fractured reservoir can be viewed as continuum media: (1) the REV is existent and the equivalent permeability changes with the size of sample negligibly; (2) the equivalent permeability tensor has symmetry. The method that determines whether the permeability tensor with symmetry is measuring directional permeability or not. Assuming that the permeability at potential gradient direction is K_J and the permeability at flow direction is K_f, the media has symmetrical permeability tensor if $(K_J)^{-1/2}$ and $(K_f)^{-1/2}$ can constitute the ellipse in polar coordinates. Obviously, the ellipse turns into circle for homogeneous isotropic media, and the flow direction is consistent with the gradient direction.

For fractured reservoir, the equivalent permeability depends on the density and the distribution of fractures as well as the fracture network. In terms of permeability, the value of REV for fractures rock mass is several orders of magnitude larger than the value of REV for porous rock mass. Sometimes it is even nonexistent. So it is not always feasible to view the fractured reservoir as equivalent media. Moreover, Youtian Zhang summarizes up the essential condition for fractured reservoir when analyzing flow problem with the tool of equivalent media model.

For the equivalent permeability of fracture-vug media, Neale and Nader (1973) are the pioneers of the related research. In their study, Navier–Stokes equation was employed in the spherical cavity, and the Darcy equation was used to describe the flow in porous medium. They studied the impact of spherical vugs on the permeability in homogeneous isotropic porous media based on flow equivalent principle. However, the systematic theory has not been developed as the research object is too simple. Recently, Arbogast et al. (2004), Arbogast and Gomez (2009), Arbogast and Lehr (2006) studied the permeability of fracture-vug media. They described the macro-flow in fracture-vug media by applying Darcy–Stokes equations. Using the Beavers–Joseph–Saffman boundary condition, the fracture-vug media is coupled to the free-flow region. And based on the homogenization theory, they gain a macroscopic equivalent Darcy flow equation and the theoretical formula of equivalent permeability tensor. Moreover, Arbogast et al. point out that the equivalent permeability distributions obtained by theory computation are totally different for the same medium model when they used different domain decomposition methods, although oversample technology is used to process the computation. As shown in Fig. 4.2, the values of equivalent permeability and the corresponding distributions calculated at 2×2 grid blocks and 3×3 grid blocks are different. The fundamental reason is that the artificial domain decomposition destroyed the topological structure of original fracture-vug media, while the problem is difficult to solve by the methods we have now.

Popov et al. (2009) think that the real vugs are always companied with different degrees of filling. Then the Stokes–Brinkman equations are more effective when describing the coupling flow in fracture-vug media and can avoid the explicit

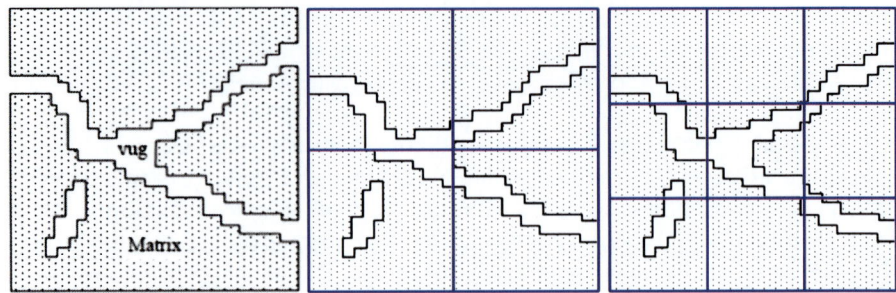

Fig. 4.2 The example diagrams in the research of Arbogast

formulation of the boundary conditions at the interface between fluid and porous. In their study, the fractures are treated as free-flow regions as same as vugs. This approach provided an accurate model. But it was not practical due to a large number of grids required because of two different length scales.

Toward this, Huang et al. (2010, 2011) and Qin et al. (2010) brought the concept of discrete fracture in the research of fractured-vuggy media, and studied the impact of the size, shape, position, and vugs combination on equivalent permeability. But all existing researches are about single-phase flow, the permeability calculated is equivalent permeability. The equivalent relative permeability of different phases is needed to simulate fluid flow in real reservoir. However, few works have been studied. Therefore, Huang et al. (2013) proposed a method to calculate equivalent relative permeability curve based on discrete fracture-vug network model and preferential flow assumption, while the method can only be adapted to the reservoir which has a high degree of fracture-vug network connectivity.

4.2 The Equivalent Process of Fractured Medium

4.2.1 Brief Introduction to Permeability Tensor

The permeability has directionality. This directionality reflects the anisotropy of a reservoir and the permeability should be represented by tensor.

Low-order tensor is widely applied in various disciplines. Scalar (e.g., mass, density) is zero-order tensor; vector (e.g., velocity, displacement) is one-order tensor; two-order tensor is defined as a physical quantity that has one value and two directions and has nine components. The mathematical meaning can be understood as a matrix whose size of value and directions are changed simultaneously. Another important property of tensor is that tensor (it) is irrelevant to coordinate system, which means one tensor expresses one (the) same physical meaning in different coordinates, although this tensor may have different forms in different coordinates.

This property that provides theoretical foundation for permeability tensor can be expressed by permeability elliptic.

Permeability tensor in Darcy's law is the element that connects pressure gradient U with flow rate Q:

$$Q = KU \tag{4.1}$$

For the 3-D reservoir, permeability tensor is two-order tensor and it has the form

$$K = \begin{bmatrix} k_{xx} & k_{xy} & k_{xz} \\ k_{yx} & k_{yy} & k_{yz} \\ k_{zx} & k_{zy} & k_{zz} \end{bmatrix} \tag{4.2}$$

where the first subscript represents the direction of flow; the second subscript represents the direction of pressure gradient. For example, k_{xx} represents the flow rate that the pressure gradient in x-direction creates in the x-direction. Likewise, k_{yz} represents the flow rate that the pressure gradient in z direction creates in the y-direction. After expansion, Darcy's law will have the following form:

$$
\begin{aligned}
Q_x &= -\left[k_{xx}\left(\tfrac{\partial P}{\partial x}\right) + k_{xy}\left(\tfrac{\partial P}{\partial y}\right) + k_{xz}\left(\tfrac{\partial P}{\partial z}\right) \right] \\
Q_y &= -\left[k_{yx}\left(\tfrac{\partial P}{\partial x}\right) + k_{yy}\left(\tfrac{\partial P}{\partial y}\right) + k_{yz}\left(\tfrac{\partial P}{\partial z}\right) \right] \\
Q_z &= -\left[k_{zx}\left(\tfrac{\partial P}{\partial x}\right) + k_{zy}\left(\tfrac{\partial P}{\partial y}\right) + k_{zz}\left(\tfrac{\partial P}{\partial z}\right) \right]
\end{aligned}
\tag{4.3}
$$

The physical meaning of permeability tensor can be understood in this way: if a certain pressure gradient is put on a rock in a certain direction. Fluid outflows not only from this direction, but also from other direction. The flow in this direction is called the main flow; the flow in other directions of this pressure gradient is very small and it is called the secondary flow. Because the main direction permeability is usually much larger than the secondary direction permeability, the main flow is much larger than the secondary flow. So in most practical problems, permeability can (or must) be assumed to be diagonal tensor. This assumption is conditional. According to the property that tensor is irrelevant to the coordinate, there is always a direction that can make the non-diagonal tensor to be zero. This direction is called the main permeability direction. The permeability tensor in this direction has the following form:

$$K = \begin{bmatrix} k_{xx} & 0 & 0 \\ 0 & k_{yy} & 0 \\ 0 & 0 & k_{zz} \end{bmatrix} \tag{4.4}$$

If the coordinate is in parallel with the main permeability direction, the permeability has the form of Formula 4.4. In other coordinates, the permeability has the form of Formula 4.2. So in most cases, neglecting the non-diagonal element of

Fig. 4.3 The schematic
diagram of stratified reservoir

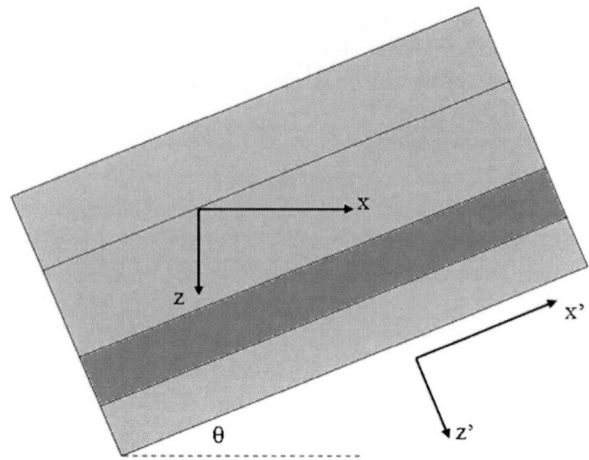

permeability tensor can cause serious error. Fancy the thought that when the
coordinate and the main permeability are at an angle of 45°, the error of the flow
calculated with Darcy's law can be 45 %.

There are some examples of permeability tensor in several special situations.

(1) Diagonal permeability tensor is adopted in black oil model, like Formula 4.4.
    At this time, the angle between the coordinate direction and the main per-
    meability direction is zero or very small.
(2) As we can see from the stratified reservoir in Fig. 4.3, if it is known to us that
    the permeability in x' direction is k_a, then the permeability in z' direction is k_h.
    In this 2-D section, if the x-z coordinate is parallel to the direction of x'-z'
    coordinate, the permeability tensor is

$$K = \begin{bmatrix} k_a & 0 \\ 0 & k_h \end{bmatrix} \tag{4.5}$$

While, if the x-z coordinate is not parallel to the direction of x'-z' coordinate, the
permeability tensor is

$$K = \begin{bmatrix} k_{xx} & k_{xy} \\ k_{yx} & k_{yy} \end{bmatrix} = \begin{bmatrix} k_a \cos^2 \theta + k_h \sin^2 \theta & (k_a - k_h) \cos \theta \sin \theta \\ (k_a - k_h) \cos \theta \sin \theta & k_a \sin^2 \theta + k_h \cos^2 \theta \end{bmatrix} \tag{4.6}$$

For a strong anisotropic reservoir, the effective permeability calculated by
equivalent continuum model should be full tensor, like Formula 4.2. In some
special situations, when the fracture is perpendicular to the boundary, like Fig. 4.4,
equivalent permeability tensor has the following form:

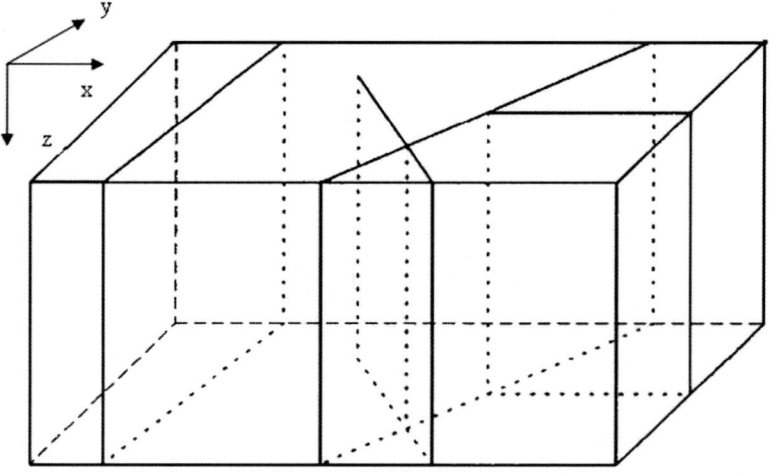

Fig. 4.4 Typical vertical fracture grid block

$$K = \begin{bmatrix} K_{xx} & K_{xy} & 0 \\ K_{yx} & K_{yy} & 0 \\ 0 & 0 & K_{zz} \end{bmatrix} \tag{4.7}$$

4.2.2   The Equivalent Permeability Tensor of Fractured Medium

For the calculation of equivalent permeability tensor of fractured medium, researchers have done many researches and put forward many calculation methods. (To find the equivalent permeability tensor of a fractured medium, researchers have explored far and wide and have put forward many calculation methods). These methods can be roughly divided into two categories: the analytical method which is based on the geometric information of the fracture and the numerical method which is based on flow simulation. The analytical method based on geometric information requires statistical analysis of the fracture geometric information: generalizing group of all the fractures and then applying analytical formula to calculate the equivalent permeability tensor. This kind of method is very effective, however it ignores connectivity between the fractures so the result is inaccurate. With the continuous improvement of numerical method and development of computer technology, the numerical method based on flow simulation is attracting more and more attention. The main idea of this method is: to implement single flow simulation on a small scale, then calculate the equivalent permeability tensor according to equivalent flow assumption. For this kind of method, boundary condition is especially important for the calculation of equivalent permeability. The most

Fig. 4.5 A 2-D grid with
vertical fracture

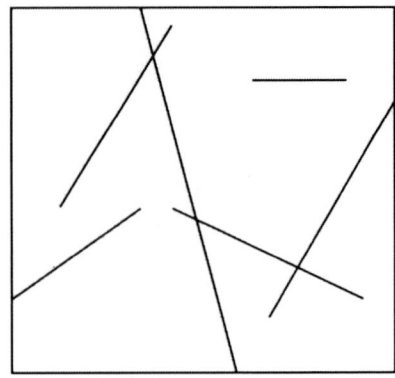

commonly used boundary conditions are periodic boundary condition and closed
fixed pressure condition boundary condition.

(1) The calculation method of periodic boundary condition
The equivalent permeability tensor of grid reflects the effect that bedrock and
fracture have on seepage. Thus, the calculation of equivalent permeability tensor
should consider the effect of the fluids flow in the bedrock or the fracture and the
channeling between the two media. Let us take the calculation of the equivalent
permeability tensor of the grid in the Fig. 4.5 for example, to illustrate the basic
principle of the calculation of periodic boundary condition. A grid of fractured
reservoir is shown in Fig. 4.5. There are six fractures in the grid and all the fractures
satisfy the following conditions: (1) The fractures are distributed randomly, the
fractures can intersect with each other and may end in the grid or stretch across
several grids. (2) When one fracture stretches across several grids, the solution is to
segment the fracture according to the grids before analysis. (3) If one fracture
intersects with the boundary, we can shorten the fracture inward at the intersection
of the fracture and the boundary. (4) The fractures have a mutual height and height
equal to the grid height. This grid block's horizontal equivalent permeability tensor
has the following form:

$$K = \begin{bmatrix} k_{xx} & k_{xy} \\ k_{yx} & k_{yy} \end{bmatrix} \tag{4.8}$$

As shown in Fig. 4.5, there are two kinds of medium in the 2-D grid: bedrock
and fracture. Fracture permeability is far larger than bedrock permeability and when
fluids flow through the two kinds of medium, the fluids satisfy Darcy's law and
conservation of mass. Replace the practical grids with equivalent grids. Equivalent
grids are homogeneous and anisotropic. Equivalent permeability is K (tensor).
When the fluids flow through equivalent grids, the fluids also satisfy Darcy's law
and conservation of mass. A fluid of unit viscosity is assumed to flow through the

equivalent grid, a pressure gradient J (vector) is exerted on the grid. We can get average flow rate Q(vector) according to Darcy's law

$$Q = -K \times J \qquad (4.9)$$

If the pressure gradient exerted on the grid $J = (1,0)^T$, we can get the average flow rate at this pressure gradient according to Formula 4.9:

$$Q = -(k_{xx}, k_{yx})^T \qquad (4.10)$$

That is to say, the first column of permeability tensor corresponds with the average flow rate at unit pressure gradient. At this pressure gradient, we can calculate the flow Q at the grid boundary when fluids flow through real grids, and we can calculate k_{xx} and k_{yx} according to Formula 4.10. Likewise, we can calculate the other two elements of this permeability tensor if we exert an unit pressure gradient $J = (1,0)^T$ in direction y.

Let us look at the rectangular grid block in the Fig. 4.6, it has four outer boundaries Γ_i ($i = 1, 2, 3, 4$), and n_i ($i = 1, 2, 3, 4$) is the outer normal vector of the boundaries. For example, exert an unit pressure gradient $J = (1,0)^T$ in direction x, now periodic condition boundary is

$$\begin{cases} p|_{\Gamma_2} = p|_{\Gamma_4} - 1, & \text{on } \Gamma_2 \text{ and } \Gamma_4 \\ p|_{\Gamma_1} = p|_{\Gamma_3}, & \text{on } \Gamma_3 \text{ and } \Gamma_1 \\ q|_{\Gamma_1} \cdot n_1 = -q|_{\Gamma_3} \cdot n_3, & \text{on } \Gamma_3 \text{ and } \Gamma_1 \\ q|_{\Gamma_2} \cdot n_2 = -q|_{\Gamma_4} \cdot n_4, & \text{on } \Gamma_2 \text{ and } \Gamma_4 \end{cases} \qquad (4.11)$$

Through solving the single-phase steady discrete fracture model of periodic boundary, and through the above calculation process, we can get the equivalent permeability tensor. We can refer to Chap. 1 for the specific establishment process of discrete fractured model and the solving process of finite element method. Some scholars adopt boundary element method to solve this problem.

Fig. 4.6 Grid block and its boundaries

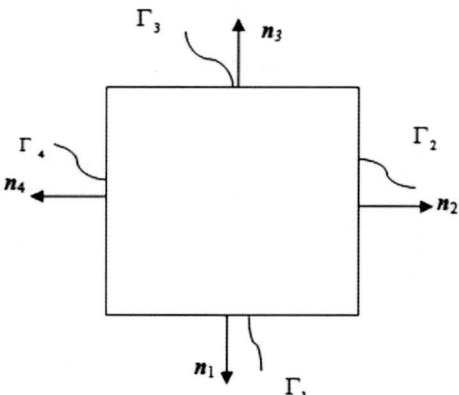

Periodic boundary condition guarantees us an accurate equivalent permeability tensor in full tensor form. But for the fracture that intersects with the grid boundary, we need to shorten the fracture inward at the intersection of the fracture and the boundary in order to make the fractures in the grid satisfy periodic distribution. This behavior reduces the flow conductivity of the fracture in certain degree.

(2) The calculation method of closed fixed pressure boundary condition
However, for practical fractured reservoir, the fracture distribution situation in each grid is not the same, and fractures often intersect with the boundaries. So the fractures in the grid cannot satisfy periodic distribution, the grid sub-tense intersects with the fracture and the situation is symmetrical. At this time, if we adopt periodic boundary condition, the results will have big deviation. To this, a new method based on discrete fracture model to calculate equivalent permeability tensor is put forward. This method adopts closed fixed pressure boundary condition and this method is applicable in various situations in which fractures are distributed complicatedly. First, build discrete fracture model in the coarse grid unit and solve the model by the finite unit method. Then calculate the average value of the pressure and the flow rate in the coarse grid unit. Finally, calculate the equivalent permeability in full tensor form according to Darcy's law (Fig. 4.7).

In coarse grid unit, deem the fractures as the inner boundary to divide unstructured grid and solve the discrete fracture model in the grid unit (refer to Chap. 2 for the specific solving process), then we can get the pressure of each nodes. Based on this, use interpolation to calculate the velocity and the pressure gradient of each triangular grid in the coarse grid unit, then solve the average value of the velocity and the volume average value of pressure gradient in the coarse grid unit:

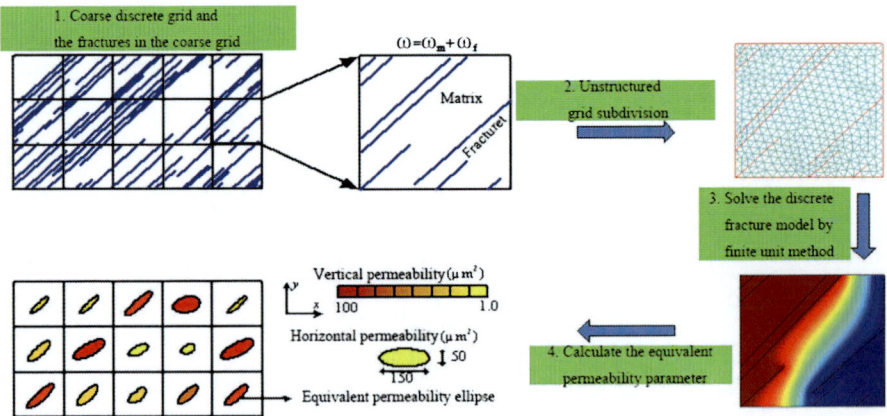

Fig. 4.7 The flow chart of the calculation of the equivalent permeability field in fractured reservoir

$$\langle u \rangle^j = \frac{1}{V_b} \int_V u^j dV = \frac{1}{V_b} \sum_{l=1}^{N} u_l V_l \tag{4.12}$$

where $j = x, y$ (represents the coordinate axis direction of the fixed pressure boundary); V_b represents the volume of the coarse grid unit; u_l and $(\nabla p)_l$ represent, respectively, the velocity and the pressure gradient in the number l triangular grid of the coarse grid unit; V_i represents the volume of number l triangular grid; N represents the number of the units in the coarse grid.

$\langle u \rangle^j$ and $\langle \nabla p \rangle^j$ consist of two parts: direction x and direction y. Combined with Darcy's law $\langle u \rangle = -\frac{k_{eff}}{\mu} \cdot \langle \nabla p \rangle$, we can get the following set of equations.

$$\begin{pmatrix} \langle \nabla p \rangle_x^x & \langle \nabla p \rangle_y^x & 0 & 0 \\ 0 & 0 & \langle \nabla p \rangle_x^x & \langle \nabla p \rangle_y^x \\ \langle \nabla p \rangle_x^y & \langle \nabla p \rangle_y^y & 0 & 0 \\ 0 & 0 & \langle \nabla p \rangle_x^y & \langle \nabla p \rangle_y^y \end{pmatrix} \begin{pmatrix} k_{xx} \\ k_{xy} \\ k_{yx} \\ k_{yy} \end{pmatrix} = -\mu \begin{pmatrix} \langle u \rangle_x^x \\ \langle u \rangle_y^x \\ \langle u \rangle_x^y \\ \langle u \rangle_y^y \end{pmatrix} \tag{4.13}$$

We can get equivalent permeability tensor by solving this set of equations. When we consider the closed fixed pressure boundary condition, the permeability we get may not satisfy symmetry. We can let $k_{xy} = k_{yx} = \sqrt{k_{xy} \cdot k_{yx}}$.

(3) Oversample technology

In view of that, the conventional methods just consider the grid during the small-scale flow simulation process and the influence of the peripheral grids on the grid is neglected. And these methods can be collectively referred to as local flow analytical method. The advantage of this kind of method is that they have a small amount of calculation. Its disadvantage is that, when handling reservoirs that contain large fractures, this kind of method cannot characterize the effective conductivities of the big-scale fractures which cut through several grid blocks and the influence of the distribution of the fractures in the peripheral grids. So the result is not accurate. Therefore, some scholars put forward global flow analytical method. Namely obtain discrete fractured geological model seepage field of the whole fractured reservoir, then analyze the seepage field in each grid unit. The accurate result of the whole seepage field can be obtained. But the amount of calculation of this technique is too huge to be applied to practical engineering.

Considering comprehensively about the advantages and the disadvantages of the two methods above, we put forward the concept of oversample cell: on the basis of target grid, expand the flow analysis area to the scale of several grids to give full consideration to the influence that the peripheral grids have on this flow. Although the amount of calculation increases, this technology can fully reflect the conductivity of long fractures.

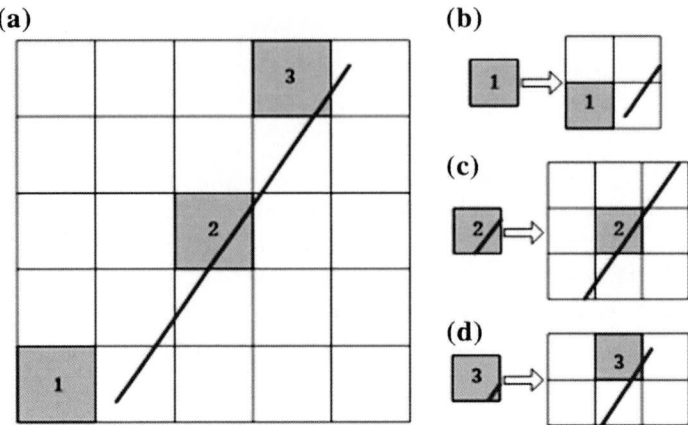

Fig. 4.8 The schematic of oversample technology

Oversample technology consists of peripheral grids of the object grid (shown in Fig. 4.8). For example, in 2-D systems, when object grid is in the corner, the oversample grids contain four grids; when the object grid is on the edge, the oversample grids contain six grids; when the object grid is inside, the oversample grids contain nice grids.

The specific calculation process is consistent with the previous calculation process of closed fixed pressure boundary on the whole. The main difference is that oversample technology requires obtaining the oversample cell of every coarse grid. It need to establish discrete fractured model in the oversample cell and do the single-phase numerical simulation for calculating the pressure and velocity field in the oversample cell. Finally, the volume average of the velocity and pressure gradient in the target primitive coarse grid are calculated and we can get the equivalent permeability of the target grid. Figure 4.9 is the flowchart of calculation.

(4) The analysis of the example calculation

Take the fractured reservoir in Fig. 4.9a as an example. First, employ 10 × 5 and 20 × 10 two different grid systems to discrete this model (Fig. 4.10a, b). The basic parameters of the model are listed in Table 4.1.

Figures 4.11 and 4.12 give the distribution result of equivalent permeability with respectively applying the proposed method and conventional method in two of different grid systems.

The results show that equivalent permeability field obtained without using oversample technology obviously loses the connectivity of long fracture, while the equivalent permeability field obtained by using oversample technology characterizes the long fracture's connectivity well.

Based on 20 × 10 grid system and the equivalent permeability calculated by the above two methods, respectively, establish equivalent continuum media model to

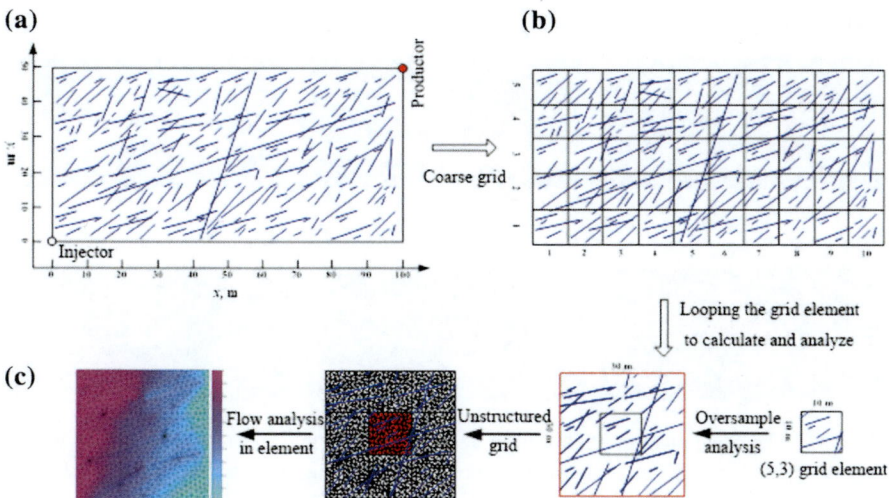

Fig. 4.9 Flowchart of the calculation of permeability field in fractured reservoir based on oversample technology. **a** The distribution of fracture in reservoir. **b** 5 × 10 grid system. **c** The schematic of oversample analysis for grid element

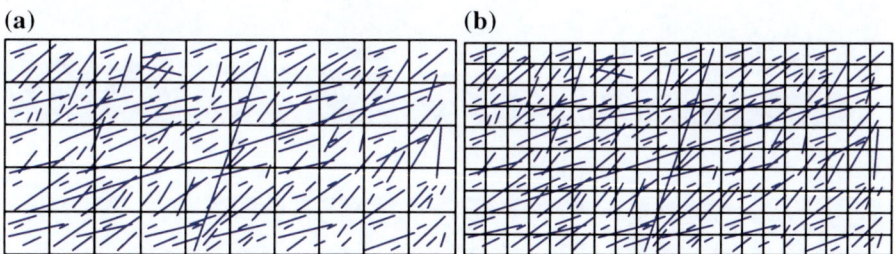

Fig. 4.10 Grid system of fractured reservoir. **a** 10 × 5 grid system. **b** 20 × 10 grid system

Table 4.1 The basic parameters of the fractured reservoir

Parameter name	Parameter value
Matrix permeability (μm^2)	1×10^{-3}
Fracture permeability (μm^2)	1×10^{4}
Fracture aperture (m)	1×10^{-3}
Fluid viscosity (mPa s)	1

numerically simulate the single-phase flow in fractured reservoir. Compare the result with the simulation result of discrete fracture model (shown in Fig. 4.13). Obviously, the result calculated by the proposed method is more realistic and

(a1)   **(b1)**   **(c1)**

(a2)   **(b2)**   **(c2)**

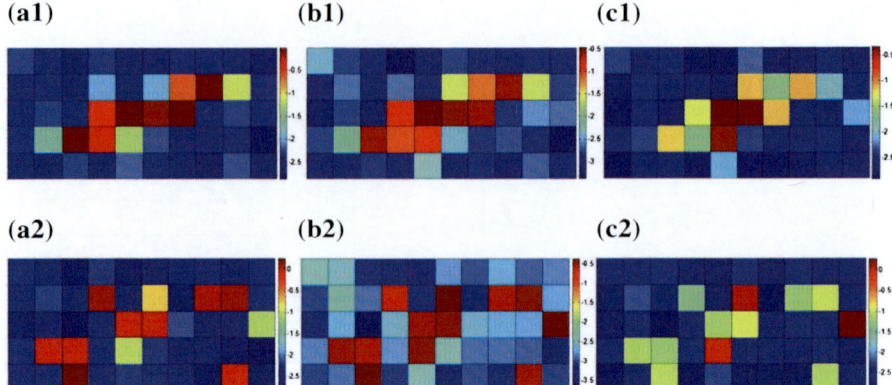

Fig. 4.11 The cloud pictures of equivalent absolute permeability in 10×5 grid system with two methods, μm^2. (a1) Using oversample, $\lg(k_{xx})$ (b1) Using oversample, $\lg(k_{xy})$ (c1) Using oversample, $\lg(k_{yy})$ (a2) Not using oversample, $\lg(k_{xx})$ (b2) Not using oversample, $\lg(k_{xy})$(c2) Not using oversample, $\lg(k_{yy})$

(a1)   **(b1)**   **(c1)**

(a2)   **(b2)**   **(c2)**

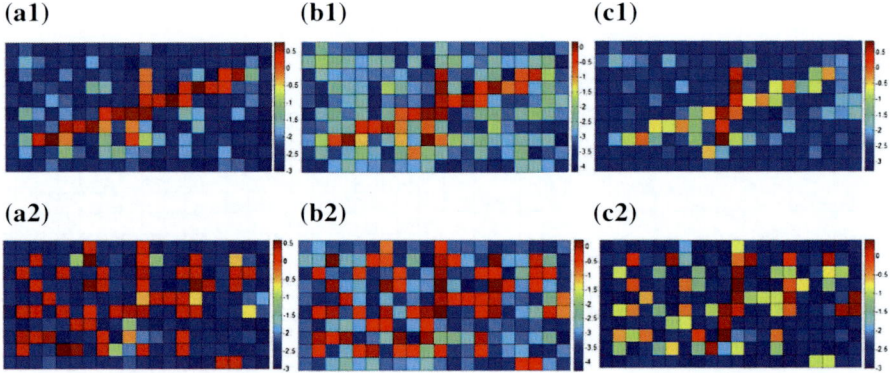

Fig. 4.12 The cloud pictures of equivalent absolute permeability in 20×10 grid system with two methods, μm^2. **a1** Using oversample, $\lg(k_{xx})$. **b1** Using oversample, $\lg(k_{xy})$. **c1** Using oversample, $\lg(k_{yy})$. **a2** Not using oversample, $\lg(k_{xx})$. **b2** Not using oversample, $\lg(k_{xy})$. **c2** Not using oversample, $\lg(k_{yy})$

consistent with the result of discrete fracture model. Figures 4.14 and 4.15, respectively, give the comparison of pressure curves in the conditions of $y = 20$ m and $x = 40$ m. As the figure shows, result calculated by the method presented in this paper fits better with discrete fracture model which also verify the correctness and effectiveness of this method.

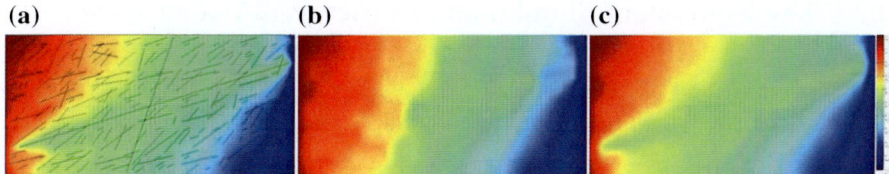

Fig. 4.13 The pressure distribution of 2-D example, MPa. **a** Discrete fracture model. **b** Not using oversample. **c** Using oversample

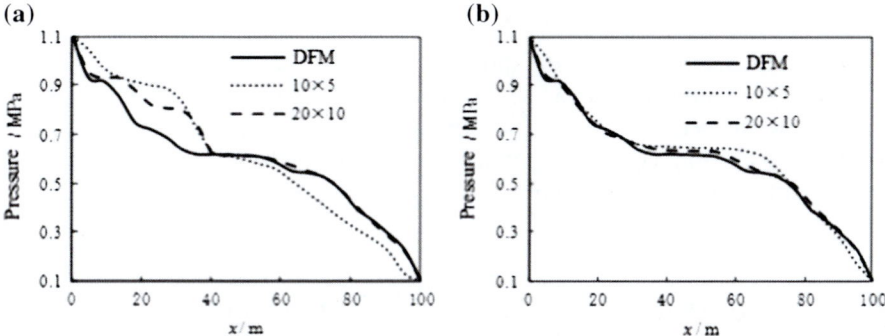

Fig. 4.14 The comparison of pressure distribution along $y = 20$ m. **a** Not using oversample. **b** Using oversample

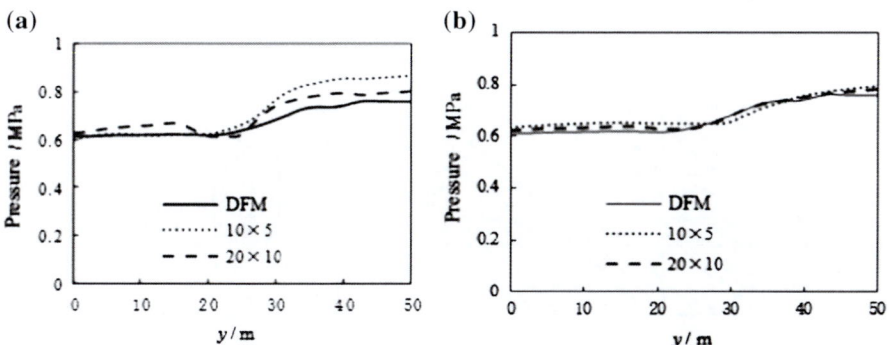

Fig. 4.15 The comparison of pressure distribution along $y = 40$ m. **a** Not using oversample. **b** Using oversample

4.3   The Equivalent Simulation of Fractured-Vuggy Media

The flow simulation of discrete fracture-vug network model has been introduced in Chap. 3. The simulation results indicate that this model can subtly simulate the real flow in fractured-vuggy media at macro-REV scale and provide important massage for identifying the flow rules in fractured-vuggy media. However, the model calculates expensively, which means it is difficult to implement flow simulation for the whole 3-D oil field under the existing computational condition. Therefore, we will describe another alternative numerical model for immiscible two-phase flow in fractured karst reservoirs with homogenization method based on discrete fracture-vug network model.

For the fractured-vuggy media shown in Fig. 4.16, the coarse grids are generated first, then the equivalent parameters of each coarse grid are obtained to describe macro-flow characteristic of fractured-vuggy media. The key problem in equivalent simulation is getting equivalent parameters of each coarse grid (such as equivalent permeability, relative permeability curve, capillary pressure curve, etc.).

4.3.1   Equivalent Absolute Permeability Calculation

There are two methods that can be applied to calculate the equivalent permeability of coarse grid: one is based on simple arithmetic, geometric average, or statistical average (Aarnes et al. 2009); The other one is flow-based method which has

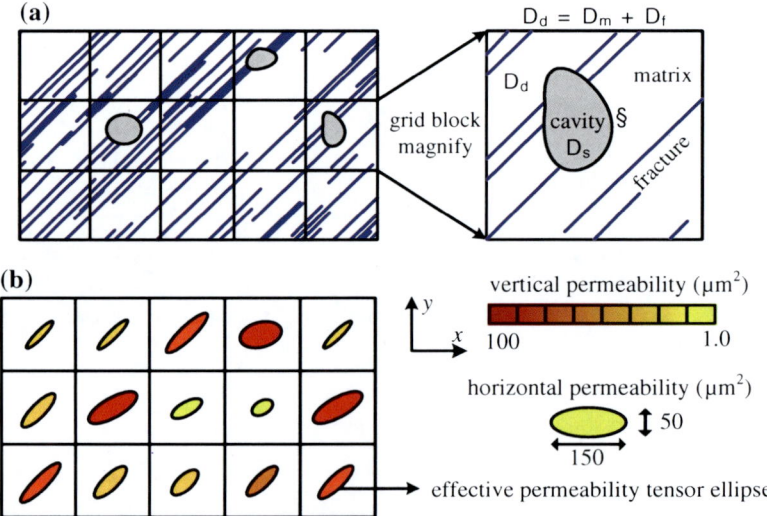

Fig. 4.16 Simple fracture-vug media and the permeability tensor map

comprehensive applicability (Efendiev and Durlofsky 2002), and its key point is the establishment and solution of flow mathematical model. Obviously, the former one cannot be applied to fractured-vuggy media.

1. Single-phase flow mathematical model of discrete fracture-vug network

As shown in Fig. 4.16, the grid-block problem that is used to obtain effective permeability tensor of a grid block can be described with discrete fracture-vug network model. The corresponding single-phase flow mathematical model can be obtained by simplifying the general coupled two-phase flow mathematical model proposed in Chap. 3.

(1) Free-flow region
To calculate the effective permeability tensor of coarse grid, we first study the steady flow. Then corresponding flow mathematical model can be described by Stokes equation

$$\nabla \cdot \boldsymbol{u} = 0 \tag{4.14}$$

$$-\mu \nabla^2 \boldsymbol{u} + \nabla p_s = \rho \boldsymbol{f} \tag{4.15}$$

where f is unit body force, m/s^2; the subscript s represents free-flow region.
For incompressible Newtonian fluid, the stress tensor $\boldsymbol{\sigma}$ is

$$\boldsymbol{\sigma} = -p_s \boldsymbol{I} + 2\mu \boldsymbol{S}(\boldsymbol{u}) \tag{4.16}$$

where $\boldsymbol{I}$ is unit tensor, $\boldsymbol{S}(\boldsymbol{u})$ is strain rate

$$\boldsymbol{S}(\boldsymbol{u}) = \frac{1}{2}(\nabla \boldsymbol{u} + \boldsymbol{u}\nabla) \tag{4.17}$$

The corresponding boundary condition: Dirichlet (velocity) and Neumann (traction) conditions are as follows:

$$\boldsymbol{u} = \boldsymbol{u}_{\mathrm{D}}, \quad \text{on } \Gamma_{\mathrm{D}} \tag{4.18}$$

$$\boldsymbol{n} \cdot \boldsymbol{\sigma} = \boldsymbol{t}_{\mathrm{N}}, \quad \text{on } \Gamma_{\mathrm{N}} \tag{4.19}$$

where $\boldsymbol{u}_D$ is the specific velocity on Dirichlet boundary; $\boldsymbol{t}_N$ is the specific force on Neumann boundary.

Besides above boundaries, no-slip boundary conditions are specified on the impermeable wall of the open fluid domain. Traction-free boundary conditions are imposed on the outlet surface. The conditions at the interface between the free-flow region and porous medium need to be handled carefully, details of such interfacial conditions will be presented below.

(2) Porous flow region

For the porous flow region, one has the classical Darcy law both for rock matrix and fractures, specific equation as follows:

$$\int_{\Omega} FEQ \, d\Omega = \int_{\Omega_m} FEQ \, d\Omega_m + e \times \int_{\Omega_f} FEQ \, d\Omega_f \qquad (4.20)$$

where flow control equation FEQ is single-phase flow mathematical model which can be written as

$$\nabla \cdot v = 0 \qquad (4.21)$$

$$\mu(K)^{-1} v + \nabla p_d = \rho f \qquad (4.22)$$

where subscript d represents porous flow region.

The boundary conditions of above mathematical model are

$$p_d = p_D, \quad \text{on } \Gamma_D \qquad (4.23)$$

$$n \cdot \frac{K}{\mu}(\nabla p_d - \rho f) = q_N, \quad \text{on } \Gamma_N \qquad (4.24)$$

where p_d is specific pressure on Dirichlet boundary; q_N is specific quantity of flow on Neumann boundary.

(3) Interfacial boundary conditions

The problem then remains in defining relevant boundary conditions at the interface between the two regions. It is clear that the mass and momentum must be balanced across the interface between the free-flow region and porous medium. Continuity of normal stress tensor and normal velocity (i.e., mass conservation) are robust and generally accepted boundary conditions expressed as

$$u \cdot n = v \cdot n, \quad \text{on } \Sigma \qquad (4.25)$$

$$n \cdot (-\sigma \cdot n) = n \cdot (p_d I \cdot n), \quad \text{on } \Sigma \qquad (4.26)$$

The tangential velocity and stress condition are generally written as

$$\lambda \cdot (u - v) = u_s \qquad (4.27)$$

$$u \cdot \lambda = \frac{\sqrt{\lambda \cdot K \cdot \lambda}}{\mu \alpha}(-\sigma \cdot n) \cdot \lambda, \quad \text{on } \Sigma \qquad (4.28)$$

The left term of Eq. 4.28 neglects the effect of flow velocity for the effect is too small to be considered when permeability of media is not high. For Newtonian fluid, Eqs. 4.26 and 4.28 can be simplified as follows:

$$2\mu n \cdot S(u) \cdot n = p_s - p_d, \quad \text{on } \Sigma \tag{4.29}$$

$$u \cdot \lambda = -2\frac{\sqrt{\lambda \cdot K \cdot \lambda}}{\alpha} n \cdot S(u) \cdot \lambda, \quad \text{on } \Sigma \tag{4.30}$$

2. The fundamentals of homogenization theory

Homogenization theory was proposed by Benssousan et al. (2011), when they studied macroscopic equivalent material parameter of composite material in 1970s. After that, the homogenization theory is widely used in the field of material science and solid mechanics. With strict mathematical theoretical background, the method that is called multi-scale homogenization method in some paper is widely applied in the field of heat and mass transfer in porous media and hydromechanics, etc. (Allaire 1992; Auriault 1991; Hornung 1997). From a mathematical point of view, the theory of homogenization is a limit theory which uses the asymptotic expansion and the assumption of periodicity to substitute the differential equations with rapidly oscillating coefficients, with differential equations whose coefficients are constant or slowly varying in such a way that the solutions are close to the initial equations. As the periodic dimension approaches zero, the homogenized effective or equivalent properties are obtained and their asymptotic behavior can be calculated.

A heterogeneous medium is said to have a regular periodicity if the functions denoting some physical quantity of the medium (either geometrical or some other characteristics) have the following property

$$F(x + NY) = F(x) \tag{4.31}$$

where

$$N = \begin{bmatrix} n_1 & 0 & 0 \\ 0 & n_2 & 0 \\ 0 & 0 & n_3 \end{bmatrix} \tag{4.32}$$

where $x = \{x_1, x_2, x_3\}$ is the coordinate vector of spatial point; n_i is any integer, $i = 1, 2, 3$; $Y = [Y_1, Y_2, Y_3]^{\mathrm{T}}$ is constant vector which determines the period of research region; F is the function of position vector and it can be scalar, vector or tensor.

In homogenization theory, assume that the period Y is very small compared with the whole research region and strong heterogeneity characteristic function will rapidly change in a small neighborhood of a point x. Accordingly, all physical quantities rely on two different scales: one on the global level or coarse scale x, which describes the slow variations; and the other on local level or fine scale y, which indicates the rapid oscillations. And then, we can introduce a small parameter ε, which is the ratio of two different scale unit vectors, so $\varepsilon y = x$ or $y = x/\varepsilon$. For strong heterogeneity media, any physical quantity ψ can be written as

$$\psi = \psi(\mathbf{x}, \mathbf{y}) = \psi\left(\mathbf{x}, \frac{\mathbf{x}}{\varepsilon}\right) = \psi(\mathbf{x}, \varepsilon) \tag{4.33}$$

To illustrate this technique, let us assume that $\Phi(\mathbf{x})$ is a rapidly oscillating quantity function of strong heterogeneity media and its variation is described in Fig. 4.17a. In order to study these oscillations, we should use the two-scale expansion, and the space can be enlarged as indicted in Fig. 4.17b. Our purpose is to find the slowly changing equivalent curve of $\Phi(\mathbf{x})$ on whole research region, as $\overline{\Phi}(\mathbf{x})$ shown in Fig. 4.17.

We assign a coordinate system $\mathbf{x} = (x_1, x_2, x_3)$ in $\mathbf{R}^3$ space to define the overall domain Ω. And then we assume that the region is periodically arranged by base cells whose identical dimensions are εY_1, εY_2, and εY_3, where Y_1, Y_2, and Y_3 are the sides of base cell in a local coordinate system $\mathbf{y} = (y_1, y_2, y_3) = \mathbf{x}/\varepsilon$. $\mathbf{Y} = [Y_1, Y_2, Y_3]^\mathrm{T}$ is the boundary of base cell. Assuming that the physical quantity at point $\mathbf{x}$ of coarse-scale research region is periodic of period $\mathbf{Y}$, the physical quantity Φ can be written as follows:

$$\Phi^\varepsilon(\mathbf{x}) = \Phi\left(\mathbf{x}, \frac{\mathbf{x}}{\varepsilon}\right) = \Phi(\mathbf{x}, \mathbf{y}) = \sum_i^\infty \varepsilon^i \Phi_i(\mathbf{x}, \mathbf{y}) \tag{4.34}$$

where $\varepsilon \to 0$, $\Phi_i(\mathbf{x}, \mathbf{y})$ is a smooth function to coordinate $\mathbf{x}$ and $\mathbf{Y}$-periodic in $\mathbf{y}$, which means that the periodic boundary condition should be imposed to base cell.

(a)   **(b)**

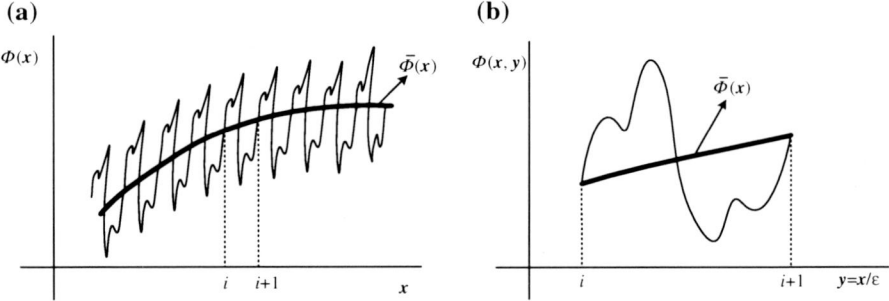

Fig. 4.17 The rapidly oscillating quantity function in strong heterogeneity media. **a** Rapidly oscillating function. **b** Changes on partial enlarged view

3. **The solving of equivalent permeability tensor**

Considering a fractured-vuggy media Ω and assuming that the research region is periodic of period Y, the element volume of base cell Y is $|Y|$. Setting Ω_s^ε is the free-flow region in base cell, Ω_d^ε is corresponding porous flow region, Σ^ε is the coupling interface, and n_s is unit normal vector, λ_s is the unit tangent vector. Then the single-phase flow mathematical model of discrete fracture-vug network in base cell Y can be written as

(1)   Free-flow region (Stokes equation).

$$-\mu\varepsilon^2\nabla^2 u^\varepsilon + \nabla p_s^\varepsilon = \rho f, \quad \text{in } \Omega_s^\varepsilon \tag{4.35}$$

$$\nabla \cdot u^\varepsilon = 0, \quad \text{in } \Omega_s^\varepsilon \tag{4.36}$$

(2)   Porous flow region (Darcy equation).

$$-\mu K^{-1} v^\varepsilon + \nabla p_s^\varepsilon = \rho f, \quad \text{in } \Omega_d^\varepsilon \tag{4.37}$$

$$\nabla \cdot v^\varepsilon = 0, \quad \text{in } \Omega_d^\varepsilon \tag{4.38}$$

(3)   Interfacial boundary condition.

$$u^\varepsilon \cdot n_s = v^\varepsilon \cdot n_s, \quad \text{on } \Sigma^\varepsilon \tag{4.39}$$

$$2\mu\varepsilon^2 n_s \cdot S(u^\varepsilon) \cdot n_s = p_s^\varepsilon - p_d^\varepsilon, \quad \text{on } \Sigma^\varepsilon \tag{4.40}$$

$$u^\varepsilon \cdot \lambda_s = -2\frac{\varepsilon\sqrt{\lambda_s \cdot K \cdot \lambda_s}}{\alpha} n_s \cdot S(u^\varepsilon) \cdot \lambda_s, \quad \text{on } \Sigma^\varepsilon \tag{4.41}$$

(4)   Outer boundary condition.

$$u^\varepsilon = 0, \quad \text{on } \partial\Omega \cap \partial\Omega_s^\varepsilon \tag{4.42}$$

$$v^\varepsilon \cdot n = 0, \quad \text{on } \partial\Omega \cap \partial\Omega_d^\varepsilon \tag{4.43}$$

The homogenization problem is to determine the behavior of the system as $\varepsilon \to 0$. In order to ensure the existence of limitation of pressure and velocity as $\varepsilon \to 0$, in the equations we have scaled both the viscosity μ and the permeability tensor K^ε by ε^2. In fact this is the usual scaling for deriving Darcy's law from Stokes flow in porous media. If we assume that the representative element volume (REV) is existing, then the pressure and velocity variable in above mathematical model can be written as the form of asymptotic expansion

$$u^\varepsilon(x) = \sum_{i=0}^{\infty} \varepsilon^i u_i(x,y) = u_0(x,y) + \varepsilon^1 u_1(x,y) + \varepsilon^2 u_2(x,y) + \dots \qquad (4.44)$$

$$v^\varepsilon(x) = \sum_{i=0}^{\infty} \varepsilon^i v_i(x,y) = v_0(x,y) + \varepsilon^1 v_1(x,y) + \varepsilon^2 v_2(x,y) + \dots \qquad (4.45)$$

$$p_s^\varepsilon(x) = \sum_{i=0}^{\infty} \varepsilon^i p_s^i(x,y) = p_s^0(x,y) + \varepsilon^1 p_s^1(x,y) + \varepsilon^2 p_s^2(x,y) + \dots \qquad (4.46)$$

$$p_d^\varepsilon(x) = \sum_{i=0}^{\infty} \varepsilon^i p_d^i(x,y) = p_d^0(x,y) + \varepsilon^1 p_d^1(x,y) + \varepsilon^2 p_d^2(x,y) + \dots \qquad (4.47)$$

Substituting the above four equations in Eqs. (4.35)–(4.43), and considering that $\nabla = \nabla_x + \varepsilon^{-1}\nabla_y$. We can see that the first term of the right of Eqs. (4.46) and (4.47) only changes at coarse scale while has no relationship with fine scale by comparing the coefficient of ε^{-1} in equations.

$$p^0(x) = p_s^0(x) = p_d^0(x), \quad \text{on } \Omega \qquad (4.48)$$

After that, we can obtain the corresponding cell problem by comparing the coefficient of ε^0 as follows:

$$-\mu\nabla_y^2 u_0 + \nabla_y p_s^1 = -\left(\nabla_x p_s^0 - \rho f\right), \quad \text{in } Y_s \qquad (4.49)$$

$$\nabla_y \cdot u_0 = 0, \quad \text{in } Y_s \qquad (4.50)$$

$$-\mu K^{-1} v_0 + \nabla_y p_d^1 = -\left(\nabla_x p_d^0 - \rho f\right), \quad \text{in } Y_d \qquad (4.51)$$

$$\nabla_y \cdot v_0 = 0, \quad \text{in } Y_d \qquad (4.52)$$

$$u_0 \cdot n_s = v_0 \cdot n_s, \quad \text{on } \Sigma \qquad (4.53)$$

$$2\mu n_s \cdot S(u_0) \cdot n_s = p_s^1 - p_d^1, \quad \text{on } \Sigma \qquad (4.54)$$

$$u_0 \cdot \lambda_s = -2\frac{\sqrt{\lambda_s \cdot K \cdot \lambda_s}}{\alpha} n_s \cdot S(u_0) \cdot \lambda_s, \quad \text{on } \Sigma \qquad (4.55)$$

Generally, the right term of Eqs. (4.49) and (4.51) can be written as

$$\nabla_x p_l^0 - \rho f = \sum_j e_j \left[\partial_{xj} p_l^0(x) - \rho f_{xj}\right], \quad l = s, d \qquad (4.56)$$

where e_j is the unit vector at j-direction in Cartesian coordinate system.

Separation of variables for $\boldsymbol{u}_0$, $\boldsymbol{v}_0$ and p_l^1, we obtain

$$\boldsymbol{u}_0(\boldsymbol{x},\boldsymbol{y}) = -\frac{1}{\mu}\sum_j \boldsymbol{e}_j\left[\partial_{xj}p_s^0(\boldsymbol{x}) - \rho\boldsymbol{f}_{xj}\right]\boldsymbol{w}_s^j \tag{4.57}$$

$$\boldsymbol{v}_0(\boldsymbol{x},\boldsymbol{y}) = -\frac{1}{\mu}\sum_j \boldsymbol{e}_j\left[\partial_{xj}p_d^0(\boldsymbol{x}) - \rho\boldsymbol{f}_{xj}\right]\boldsymbol{w}_d^j \tag{4.58}$$

$$p_l^1(\boldsymbol{x},\boldsymbol{y}) = -\frac{1}{\mu}\sum_j \boldsymbol{e}_j\left[\partial_{xj}p_l^0(\boldsymbol{x}) - \rho\boldsymbol{f}_{xj}\right]\pi_s^j \tag{4.59}$$

where $\boldsymbol{w}_l^j$ and π_l^j (l = s,d) are both physical field function with period $\boldsymbol{Y}$.

Substituting Eqs. (4.56)–(4.59) in Eqs. (4.49)–(4.55), we can obtain auxiliary equation for base cell

$$-\nabla_y^2\boldsymbol{w}_s^j + \nabla_y\pi_s^j = \boldsymbol{e}_j, \quad \text{in } \boldsymbol{Y}_s \tag{4.60}$$

$$\nabla_y \cdot \boldsymbol{w}_s^j = 0, \quad \text{in } \boldsymbol{Y}_s \tag{4.61}$$

$$\boldsymbol{K}^{-1}\boldsymbol{w}_d^j + \nabla_y\pi_d^j = \boldsymbol{e}_j, \quad \text{in } \boldsymbol{Y}_d \tag{4.62}$$

$$\nabla_y \cdot \boldsymbol{w}_d^j = 0, \quad \text{in } \boldsymbol{Y}_d \tag{4.63}$$

$$\boldsymbol{w}_s^j \cdot \boldsymbol{n}_s = \boldsymbol{w}_d^j \cdot \boldsymbol{n}_d, \quad \text{on } \Sigma \tag{4.64}$$

$$\boldsymbol{n}_s \cdot \boldsymbol{S}\left(\boldsymbol{w}_s^j\right) \cdot \boldsymbol{n}_s = \pi_s^j - \pi_d^j, \quad \text{on } \Sigma \tag{4.65}$$

$$\boldsymbol{w}_s^j \cdot \boldsymbol{\tau}_s = -2\frac{\sqrt{\boldsymbol{\tau}_s \cdot \boldsymbol{K} \cdot \boldsymbol{\tau}_s}}{\alpha}\boldsymbol{n}_s \cdot \boldsymbol{S}\left(\boldsymbol{w}_s^j\right) \cdot \boldsymbol{\tau}_s, \quad \text{on } \Sigma \tag{4.66}$$

After solving the above auxiliary problem, we can obtain the equivalent permeability tensor $\boldsymbol{K}$ at coarse scale by below equation

$$\boldsymbol{K} = \frac{1}{|\boldsymbol{Y}|}\left(\int_{Y_s}\boldsymbol{w}_s^j dy + \int_{Y_d}\boldsymbol{w}_d^j dy\right) \tag{4.67}$$

Obviously, the component is

$$K_{ij} = \frac{1}{|\boldsymbol{Y}|}\left(\int_{Y_s}\left(\boldsymbol{w}_s^j\right)_i dy + \int_{Y_d}\left(\boldsymbol{w}_d^j\right)_i dy\right) \tag{4.68}$$

From the definition of discrete fracture model, we can obtain

$$\int_{Y_d} \left(w_d^j\right)_i dy = \int_{Y_m} \left(w_m^j\right)_i dy + e \times \int_{Y_f} \left(w_f^j\right)_i dy, \quad m = \text{matrix, } f = \text{fracture} \quad (4.69)$$

The equivalent quantity of flow at macro-coarse scale can be written as classical Darcy's law when $\varepsilon \to 0$, specific equation as follows:

$$\mu(\mathbf{K})^{-1}\bar{\mathbf{u}} + \nabla p^0 = \rho f \tag{4.70}$$

$$\nabla \cdot \bar{\mathbf{u}} = 0 \tag{4.71}$$

where $\bar{\mathbf{u}}$ is macro-rate of flow.

Note that w^j in Eqs. (4.60)–(4.66) are the fine-scale velocities in the base cell, and unit vector in j-direction in Eqs. (4.60) and (4.62) can be written as

$$\nabla_y \left(\pi_l^j - y_j\right) = \nabla_y \pi_l^j - e_j \tag{4.72}$$

Above equation indicates that the rate of flow w^j results from the boundary base cell that is periodic when it is under unit pressure gradient effect in j-direction. Accordingly, above auxiliary problem for base cell has the same formula with Strokes–Darcy problem and is easy to solve. Moreover, we can get the equivalent absolute permeability tensor at coarse scale with Eq. (4.67).

4. Example verification

For real engineer problem, we will need Eqs. (4.70) and (4.71) to simulate the flow at coarse scale. To verify the validity for above method, we analyze a simple fracture-vug media (as shown in Fig. 4.18a) and make some numerical computation on it. First, the whole region is discretized into the 5 × 5 coarse grids as shown in Fig. 4.18b and the equivalent permeability tensor at every coarse grid block is calculated based on homogenization theory and numerical simulation. Then, Eqs. (4.70) and (4.71) are used to simulate the flow at coarse scale. Lastly, we compare the computation result with DFVN model at fine scale.

We plot the corresponding coarse-scale pressure in Fig. 4.19. We have compared this coarse-scale pressure with the averaged coarse-scale pressure obtained from the fine-scale solution. The numerical result indicates that the numerical computation result at coarse scale can reflect the characteristic of flow while the fine solution at fine scale can describe the fractures and vug more detailed. More comparison result is shown in Fig. 4.20, and a good match with fine-scale and coarse-scale solutions has been achieved. It is clear that the solution of macro-model based on equivalent permeability would be more accurate as ε tends to zero.

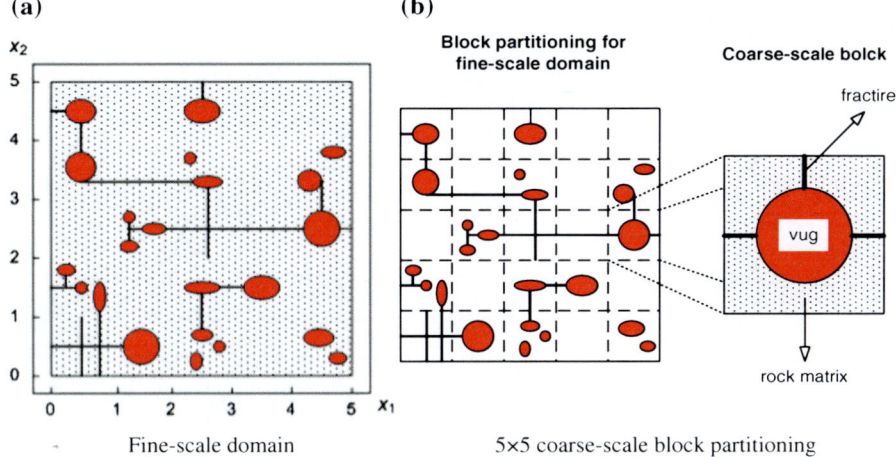

Fig. 4.18 Fine-scale domain **a** consisting of rock matrix and discrete fracture-vug networks. The coarse-scale block partitioning **b** used for numerical calculations of equivalent permeability

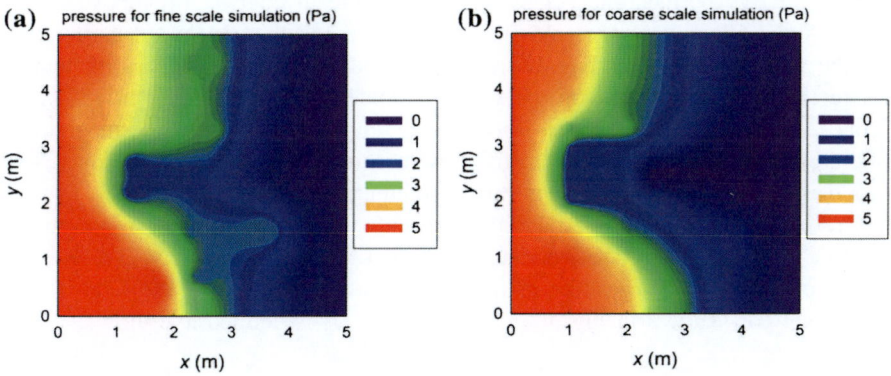

Fig. 4.19 Comparison of the fine-scale **a** and coarse-scale **b** pressure solutions

4.3.2   Computation of Pseudo-relative Permeability

For two-phase flow simulation of fracture-vug media at coarse scale, the key problem is how to describe the influence of fractures and vug for water and oil front. Toward this, the concept of pseudo-relative permeability at coarse scale is proposed based on preferential flow assumption for fracture-vug, and the corresponding solution has been formed. The pseudo-relative permeability is not a new concept, Hearn (1971) has proposed it when he studied the two-phase flow problem of layered formation. After that, Talleria et al. (1999) studied the application and limitation of it. Meanwhile, van Golf-Racht (1982) pointed out that it is possible to

Fig. 4.20 Comparison of the fine-scale (*solid curves*) and coarse (*dashed curves*) pressure profiles at various values of *y* and *x*

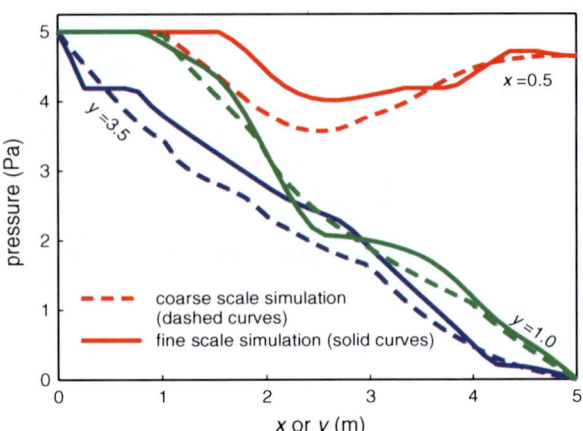

establish pseudo-relative permeability curve with experiment method for fractures media. Pruess et al. (1990) applied this concept to study equivalent medium parameter for double medium model and deduced some simple expressions.

van Lingen et al. (2001) developed a set of theory method to solve pseudo-relative permeability that includes fracture at coarse grid block. On this basis, Abdel-Ghani (2009) reinterpret the solution and the concept of parameter in it. The research result indicated that the method can adapt to formation that has fractures developed. Based on the work of Abdel-Ghani, Huang et al. (2011a, b) proposed a set of theory and method for solving pseudo-relative permeability of fracture-vug media. First, preferential flow assumption of fracture-vug network is proposed, which means wetting fluid preferentially flows in fractures and vug before absorbed into matrix with the method of imbibition. This assumption is realistic for fracture-vug network always having high conductivity, which is 3–7 orders of magnitude large than matrix. This is also verified in Tahe field. For the coarse network shown in Fig. 4.16, the porosity ϕ_b is defined as follows:

$$\phi_b = \phi_m + \phi_f + \phi_c = \phi_m + \frac{\sum e_i l_i}{V} + \frac{\sum (V_c)_j}{V} \tag{4.73}$$

where ϕ_m, ϕ_f and ϕ_c are the porosity of matrix, fracture and vug, respectively; e_i and l_i are the aperture and length of i-th fracture, respectively; $(V_c)_j$ is the volume of i-th vug; V is the volume of coarse grid. Note that both the inner porosity in fractures and cavities are taken as 1.

1. Residual oil saturation and irreducible water saturation

The effective residual saturations and end-point relative permeabilities of grid blocks are changed by the presence of discrete fracture-cavity networks. The effective residual oil saturation of a fractured karst gird block $S_{or,b}$ is calculated using the following arithmetic average:

$$S_{or,b} = \frac{\phi_m S_{or,m} + (\phi_f + \phi_c) S_{or,fc}}{\phi_m + \phi_f + \phi_c} \tag{4.74}$$

where $S_{or,m}$ is the residual oil saturation in the matrix, and $S_{or,fc}$ is the residual oil saturation in the discrete fracture-cavity network. Similarly, the effective connate water saturation of a gird block $S_{wc,b}$ is calculated as

$$S_{wc,b} = \frac{\phi_m S_{wc,m} + (\phi_f + \phi_c) S_{wc,fc}}{\phi_m + \phi_f + \phi_c} \tag{4.75}$$

where $S_{wc,m}$ is the connate water saturation in the matrix, and $S_{wc,fc}$ is the connate water saturation in the discrete fracture-cavity network.

After having above definition, the corresponding $k_{oe,b}$ and $k_{we,b}$ can be deduced. Relative permeability $k_{oe,b}$.

$$k_{oe,b} = \frac{k_{oe,m} k_m \phi_m + k_{oe,fc} k_{fc} (\phi_f + \phi_c)}{k_m \phi_m + k_{fc} (\phi_f + \phi_c)} \tag{4.76}$$

where $k_{oe,m}$ is the relative permeability of oil that corresponds to the residual oil saturation of matrix; $k_{oe,fc}$ is the relative permeability of oil that corresponds to the residual oil saturation of fracture-vug system; $k_m = \mathrm{trace}(K_m)/n$, where n is the space dimension, K_m is the permeability tensor of matrix; $k_{fc} = \mathrm{trace}(K_{fc})/n$, where K_{fc} is the permeability tensor of fracture-vug.

$$K = K_m + K_{fc} \tag{4.77}$$

As can be seen from above equation, three permeabilities are all symmetric positive definite two-order tensor. We assume that the relative permeability has a property of direction-invariance and it is universal in the multiphase study for porous media. Relative permeability $k_{we,b}$.

$$k_{we,b} = \frac{k_{we,m} k_m \phi_m + k_{we,fc} k_{fc} (\phi_f + \phi_c)}{k_m \phi_m + k_{fc} (\phi_f + \phi_c)} \tag{4.78}$$

where $k_{we,m}$ is the relative permeability of oil that corresponds to the irreducible water saturation of matrix; $k_{we,fc}$ is the relative permeability of oil that corresponds to the irreducible water saturation of fracture-vug system.

2. Pseudo-relative permeability curve

Based on preferential flow assumption, specific calculation process for pseudo-relative permeability curve is shown in Fig. 4.21. First, normalization processing is carried out for the real relative permeability curve of matrix and fracture-vug system, where the normalized relative permeability curve of fracture-vug system is viewed as classical X shape (as shown in Fig. 4.12a). To

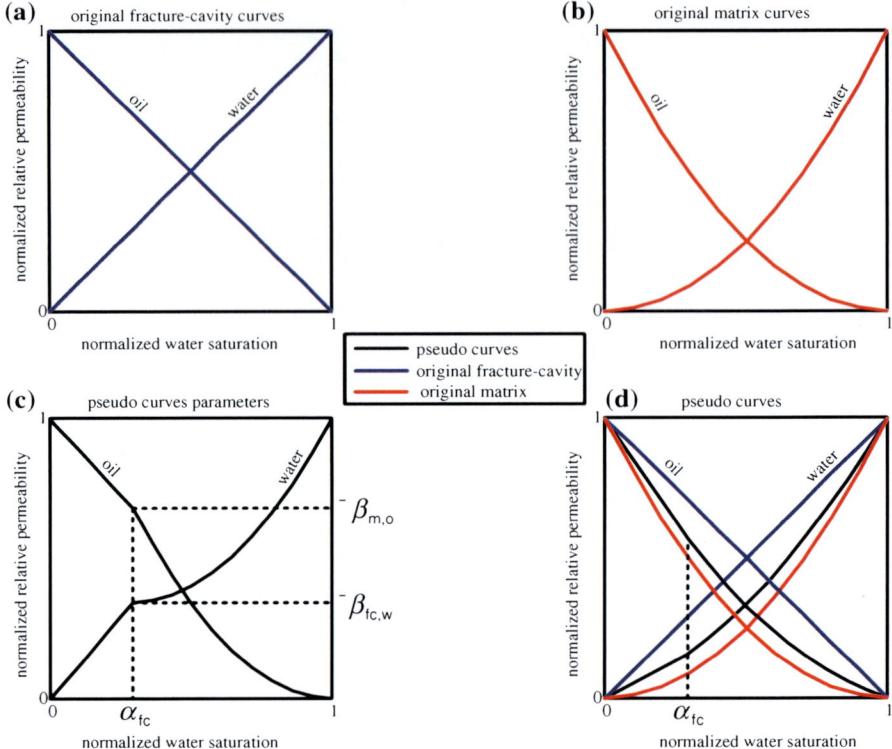

Fig. 4.21 Schematic of solving pseudo-relative permeability curve

solve parameters in Fig. 4.21c, we need couple relative permeability parameters of matrix and fracture-vug system as follows:

$$\alpha_{\text{fc}} = \frac{\left(1 - S_{\text{wc,fc}} - S_{\text{or,fc}}\right)\left(\phi_{\text{f}} + \phi_{\text{c}}\right)}{\left(1 - S_{\text{wc,fc}} - S_{\text{or,fc}}\right)\left(\phi_{\text{f}} + \phi_{\text{c}}\right) + \left(1 - S_{\text{wc,m}} - S_{\text{or,m}}\right)\phi_{\text{m}}} \tag{4.79}$$

α_{fc} represents the contribution of the fracture-cavity volume to the total mobile porosity in a grid block. $\beta_{\text{fc,w}}$ in Fig. 4.21, the contribution of fracture-cavity system to the maximum grid block relative permeability to water, is defined as

$$\beta_{\text{fc,w}} = \frac{k_{\text{fc}}k_{\text{we,fc}}\left(\phi_{\text{f}} + \phi_{\text{c}}\right)}{k_{\text{fc}}k_{\text{we,fc}}\left(\phi_{\text{f}} + \phi_{\text{c}}\right) + k_{\text{m}}k_{\text{we,m}}\phi_{\text{m}}} \tag{4.80}$$

$\beta_{\text{m,o}}$, the contribution of matrix to the maximum grid block relative permeability to oil, is defined as

$$\beta_{\mathrm{m,o}} = \frac{k_{\mathrm{m}} k_{\mathrm{oe,m}} \phi_{\mathrm{m}}}{k_{\mathrm{fc}} k_{\mathrm{oe,fc}} (\phi_{\mathrm{f}} + \phi_{\mathrm{c}}) + k_{\mathrm{m}} k_{\mathrm{oe,m}} \phi_{\mathrm{m}}} \tag{4.81}$$

And then the new normalized points from the original matrix curves are calculated as the following transformations

$$\begin{cases} S_{\mathrm{wn,b}}^* = S_{\mathrm{wn,m}} (1 - \alpha_{\mathrm{fc}}) + \alpha_{\mathrm{fc}} \\ k_{\mathrm{rw,b}}^* = k_{\mathrm{rw,m}} + \left(S_{\mathrm{wn,b}}^* - k_{\mathrm{rw,m}} \right) \beta_{\mathrm{fc,w}} \\ k_{\mathrm{ro,b}}^* = k_{\mathrm{ro,m}} + \left(1 - S_{\mathrm{wn,b}}^* - k_{\mathrm{ro,m}} \right) (1 - \beta_{\mathrm{m,o}}) \end{cases} \tag{4.82}$$

where $S_{\mathrm{wn,m}}$ represents the saturation of water phase in matrix.

Remark, the pseudo-relative permeability curve obtained by above method is between the relative permeability curve of matrix and fracture-vug, as shown in Fig. 4.21d.

Then the quantitative evaluation of the effective continuum capillary pressure is straightforward. Based on preferential flow assumption, there are two flow stages in a grid block containing fracture-cavity networks, i.e., the preferential flow stage in the fracture-cavity network and the second stage flow in matrix. So given a certain average water saturation of grid block $S_{\mathrm{w,b}}$, the corresponding water saturation $S_{\mathrm{w,m}}$ and $S_{\mathrm{w,fc}}$, in fracture-cavity system and matrix can be found from the following equation

$$S_{\mathrm{w,b}} = \frac{\phi_{\mathrm{m}} S_{\mathrm{w,m}} + (\phi_{\mathrm{f}} + \phi_{\mathrm{c}}) S_{\mathrm{w,fc}}}{\phi_{\mathrm{m}} + \phi_{\mathrm{f}} + \phi_{\mathrm{c}}} \tag{4.83}$$

The capillary pressure could be found from the capillary functions of the fracture-cavity system and matrix, respectively.

With this, we have successfully obtained the equivalent absolute permeability tensor and the pseudo-relative permeability curve of coarse grid.

4.4   Numerical Simulation Method for Equivalent Media

Generally, the equivalent permeability is a full rank tensor rather than diagonal tensor. Therefore, the full tensor numerical simulator should be applied to do corresponding flow numerical simulation. However, corresponding commercial software has not been developed, and we will develop an efficient full tensor numerical simulation technology by combining mixed finite element and finite volume method in this section. We apply the classical IMPES method to larger scale two-phase flow simulation: the pressure equation is implicitly discretized by mixed finite element method, and the saturation equation is explicitly solved by finite volume method.

4.4.1 Two-Phase Flow Mathematical Model in Large Scale

(1) Global pressure equation
First, by combining the continuity equation of water and oil phase and expanding it, we can obtain

$$\nabla \cdot (v_w + v_o) + \frac{\partial \phi}{\partial t} + \phi \frac{S_w}{\rho_w} \frac{\partial \rho_w}{\partial t} + \phi \frac{S_o}{\rho_o} \frac{\partial \rho_o}{\partial t} + \frac{v_w}{\rho_w} \cdot \nabla \rho_w + \frac{v_o}{\rho_o} \cdot \nabla \rho_o = q \quad (4.84)$$

For simplicity, the fluid and rock mass are assumed to be incompressible. Then above equation can be simplified as

$$v = -[K\lambda_w(\nabla p_w - \rho_w G) + K\lambda_o(\nabla p_o - \rho_o G)] , \quad \nabla \cdot v = q \quad (4.85)$$

where $v = v_w + v_o$, overall velocity of fluid; $q = q_w + q_o$, source term; K, permeability tensor of coarse grids; $G = g\nabla z$, acceleration of gravity term.
There are two pressure variables p_w and p_o in above equation. By introducing capillary pressure $p_c = p_o - p_w$ (assume that water is wetting and capillary pressure is a function of saturation S_w), we can eliminate a variable

$$v = -[K(\lambda_w + \lambda_o)\nabla p_o - K\lambda_w \nabla p_{cow}] + K(\lambda_w \rho_w + \lambda_o \rho_o)G \quad (4.86)$$

Obviously, this method will result in a strong coupling between pressure equation and saturation equation, which is difficult to be solved. So we apply another method to eliminate ∇p_c. We first assume that the capillary pressure p_{cow} is the monotone function of water saturation S_w. Then the overall pressure $p = p_o - p_{com}$ is introduced, where p_{com} is called modified pressure

$$p_{com}(S_w) = \int_1^{S_w} f_w(\tau) \frac{\partial p_c}{\partial S_w}(\tau)d\tau \quad (4.87)$$

where $f_w = \lambda_w/(\lambda_w + \lambda_o)$ is fractional flow function of water phase.
From above equation, we can know $\nabla p_{com} = f_w \nabla p_c$, then

$$v = -K\lambda\nabla p + K(\lambda_w \rho_w + \lambda_o \rho_o)G \quad (4.88)$$

where $\lambda = (\lambda_w + \lambda_o)$, total mobility.
Substituting above equation in continuity equation of Eq. 4.85, we can obtain

$$-\nabla \cdot [K\lambda\nabla p - K(\lambda_w \rho_w + \lambda_o \rho_o)G] = q \quad (4.89)$$

Obviously, the pressure equation is elliptic equation.

(2) Saturation equation for water phase

For saturation equation, we usually use water phase equation. Based on the Darcy's law, we can obtain

$$\lambda_o v_w - \lambda_w v_o = K\lambda_w \lambda_o \nabla p_c + K\lambda_w \lambda_o (\rho_w - \rho_o)G \qquad (4.90)$$

Substituting $v_o = v - v_w$ in above equation and the dividing it by λ, we can get the velocity of water phase

$$v_w = f_w[v + K\lambda_o \nabla p_c + K\lambda_o (\rho_w - \rho_o)G] \qquad (4.91)$$

Substituting above equation in continuity equation of water phase, we can obtain

$$\phi \frac{\partial S_w}{\partial t} + \nabla \cdot \{f_w(S_w)[v + K\lambda_o \nabla p_c + K\lambda_o (\rho_w - \rho_o)G]\} = q_w \qquad (4.92)$$

This is the classical parabolic equation.

4.4.2 Mixed FEM for Pressure Equation

The mixed finite element method (Aarnes et al. 2007; Durlofsky 1993; Yotov 1996) is applied to solve globe pressure equation Eq. (4.89). This method has advantages of finite element method and finite volume method: one is meeting the conservation of mass in each element; another one is that it can deal with the computation of full tensor permeability conveniently. The difference with standard Galerkin finite element is that the mixed finite element takes the pressure and velocity as direct physical variables, discretizes the Darcy equation and continuity equation respectively, and establish corresponding mixed computation format.

Specific procedure: searching the approximate solution (p, v) in space $L^2(\Omega) \times H_0^{1,\mathrm{div}}(\Omega)$ to satisfy the equivalent integral equations

$$\int_\Omega u \cdot \left(K\lambda\left(S_w^k\right)\right)^{-1} \cdot v^{k+1} \, d\Omega - \int_\Omega p^{k+1}\nabla \cdot u \, d\Omega$$
$$= \int_\Omega u \cdot \left(f_w\left(S_w^k\right)\rho_w + f_o\left(S_w^k\right)\rho_o\right)G \, d\Omega \qquad (4.93)$$

$$\int_\Omega l\nabla \cdot v^{k+1} \, d\Omega = \int_\Omega lq^{k+1} \, d\Omega \qquad (4.94)$$

For all $u \in H_0^{1,\mathrm{div}}(\Omega)$ and $l \in L^2(\Omega)$. The superscript k represents k-th time step.

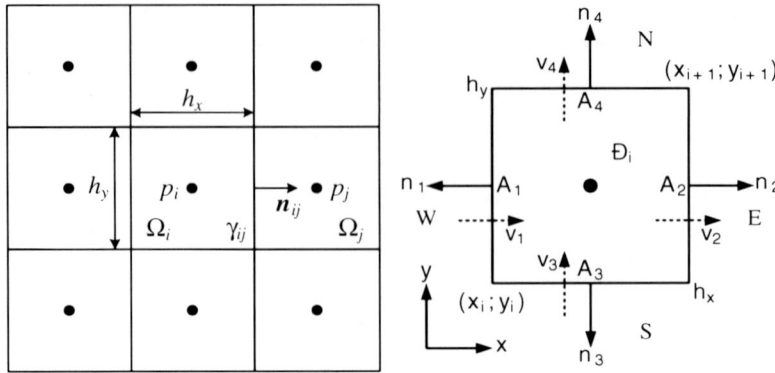

Fig. 4.22 Schematic of mixed finite element

Here, we use the low-order linear Raviart–Thomas space (RT_0 space), specific contents as follows:

$$P = \left\{ P \in L^2(\Omega) : p|_{\Omega_i} \text{ is constant } \forall \Omega_i \in \Omega \right\} \tag{4.95}$$

$$V = \left\{ \begin{array}{l} v \in H_0^{1,\mathrm{div}}(\Omega) : v|_{\Omega_i} \text{ have linear components } \forall \Omega_i, \\ \left. (v \cdot n_{ij}) \right|_{\gamma_{ij}} \text{ is constant } \forall \gamma_{ij} \in \Omega, \text{ and } v \cdot n_{ij} \text{ is continuous across } \gamma_{ij} \end{array} \right\} \tag{4.96}$$

where n_{ij} is the unit normal to γ_{ij} pointing from Ω_i and Ω_j, as shown in Fig. 4.22. The corresponding Raviart–Thomas mixed FEM thus seeks $(p, v) \in P \times V$ such that holds for all $u \in V$ and $l \in P$.

To express Eqs. (4.93) and (4.94) as a linear system, observe that functions in V are, for admissible grids, spanned by base functions $\{w_{ij}\}$ that are defined by

$$\{w_{ij}\} \in \mathcal{P}(\Omega_i)^d \cup \mathcal{P}(\Omega_j)^d \tag{4.97}$$

$$\int_{\gamma_{kl}} w_{ij} \cdot n_{kl} \, d\Gamma = \begin{cases} 1 & \text{if } \gamma_{kl} = \gamma_{ij} \\ 0 & \text{else} \end{cases} \tag{4.98}$$

where $\mathcal{P}(\Omega_i)$ and $\mathcal{P}(\Omega_j)$ are linear functions at Ω_i and Ω_j; superscript d is space dimension.

For pressure space P, we can definite the basis function space as below

$$U = \text{span}\{\psi_m\}$$

where

$$\chi_m = \begin{cases} 1 & \text{if } x \in \Omega_m \\ 0 & \text{else} \end{cases} \tag{4.99}$$

Accordingly, the approximate expression of pressure and velocity can be written as

$$\hat{p} = \sum_{\Omega_m} p_m \psi_m, \quad \hat{v} = \sum_{\gamma_{ij}} v_{ij} w_{ij} = Wv \tag{4.100}$$

Substituting above equation in Eqs. (4.93) and (4.94), and after integration by part, we can obtain

$$\sum_{\Omega_e} \left(\int_{\Omega_e} W_e^T \cdot [\kappa \lambda (S_W^k)]_e^{-1} \cdot W_e \, d\Omega \, v_e^{k+1} \right) - \sum_{\Omega_e} \left(\int_{\Omega_e} \nabla \cdot W_e^T \, d\Omega \, p_e^{k+1} \right)$$

$$= \sum_{\Omega_e} \left(\int_{\Omega_e} W_e^T \cdot \rho_e G_e \, d\Omega \right) \tag{4.101}$$

$$\sum_{\Omega_e} \left(\int_{\Omega_e} \nabla \cdot W_e \, d\Omega \, v_e^{k+1} \right) = \sum_{\Omega_e} \left(\int_{\Omega_e} q^{k+1} \, d\Omega \right) \tag{4.102}$$

where

$$\rho_e = f_w (S_W^k) \rho_w + f_o (S_W^k) \rho_o$$

Generally, we write above equations as follows:

$$\begin{bmatrix} B & -C^T \\ C & 0 \end{bmatrix} \begin{bmatrix} v^{k+1} \\ p^{k+1} \end{bmatrix} = \begin{bmatrix} g^{k+1} \\ q^{k+1} \end{bmatrix} \tag{4.103}$$

where

$$B = \sum_{\Omega_e} \left\{ \int_{\Omega_e} \boldsymbol{W}_e^{\mathrm{T}} \cdot \left[\kappa \lambda \left(S_{\mathrm{w}}^k \right) \right]_e^{-1} \cdot \boldsymbol{W}_e \ \mathrm{d}\Omega v_e^{k+1} \right\}$$

$$C = \sum_{\Omega_e} \left\{ \int_{\Omega_e} \nabla \cdot \boldsymbol{W}_e \ \mathrm{d}\Omega v_e^{k+1} \right\}$$

$$g^{k+1} = \sum_{\Omega_e} \left\{ \int_{\Omega_e} \boldsymbol{W}_e^{\mathrm{T}} \cdot \rho_e \boldsymbol{G}_e \ \mathrm{d}\Omega \right\}$$

$$q^{k+1} = \sum_{\Omega_e} \left\{ \int_{\Omega_e} q^{k+1} \ \mathrm{d}\Omega \right\}$$

To get the information of elements in above equations, we need to analyze the characteristic of element. The main idea of mixed finite element method is similar to block-centered finite volume method and differs from standard Galerkin finite element, which means pressure is defined at the center point of element to present the average pressure of whole element as shown in Eq. 4.99 and velocity is defined at the boundary of element as shown in Fig. 4.22b. For rectangular element, the velocity can be deduced by the following equation:

$$v = \sum_{A_m} v_m \boldsymbol{w}_m = (-\boldsymbol{w}_1, \boldsymbol{w}_2, -\boldsymbol{w}_3, \boldsymbol{w}_4) \begin{pmatrix} v_1 \\ v_2 \\ v_3 \\ v_4 \end{pmatrix} = \boldsymbol{W}v \qquad (4.104)$$

In the above equation, we define that all positive values of vector function are consistent with the positive direction of coordinate, as shown in Fig. 4.22b. Obviously, velocity is linear function for RT_0 space. Then we have

$$\boldsymbol{w}_m = \begin{pmatrix} a_{m1} + a_{m2}x \\ a_{m3} + a_{m4}y \end{pmatrix} \qquad (4.105)$$

where the coefficient can be solved by the characteristic of unit basis function below

$$\int_{A_i} \boldsymbol{w}_m \cdot \boldsymbol{n}_i \ \mathrm{d}A = \delta_{mi} \qquad (4.106)$$

As shown in Fig. 4.22b, for basis function, we can obtain

$$\int_{A_1} \boldsymbol{w}_1 \cdot \boldsymbol{n}_1 \, \mathrm{d}A = 1 \Rightarrow \int_0^{h_y} -(a_{m1} + a_{m2}x_i) \, \mathrm{d}y = 1$$

$$\int_{A_2} \boldsymbol{w}_1 \cdot \boldsymbol{n}_2 \, \mathrm{d}A = 0 \Rightarrow \int_0^{h_y} (a_{m1} + a_{m2}x_{i+1}) \, \mathrm{d}y = 0$$

$$\int_{A_3} \boldsymbol{w}_1 \cdot \boldsymbol{n}_3 \, \mathrm{d}A = 1 \Rightarrow \int_0^{h_x} -(a_{m3} + a_{m4}y_i) \, \mathrm{d}x = 0$$

$$\int_{A_4} \boldsymbol{w}_1 \cdot \boldsymbol{n}_4 \, \mathrm{d}A = 1 \Rightarrow \int_0^{h_x} (a_{m3} + a_{m4}y_{i+1}) \, \mathrm{d}x = 0$$

By solving above equations, we can get coefficient a_{mi} and the specific expression of basis function $\boldsymbol{w}_1$. In the similar way, we can get the velocity basis function at other boundary as follows:

$$\boldsymbol{w}_1 = \frac{1}{h_x h_y} \begin{pmatrix} x - h_x \\ 0 \end{pmatrix}, \quad \boldsymbol{w}_2 = \frac{1}{h_x h_y} \begin{pmatrix} x \\ 0 \end{pmatrix}$$

$$\boldsymbol{w}_3 = \frac{1}{h_x h_y} \begin{pmatrix} 0 \\ y - h_y \end{pmatrix}, \quad \boldsymbol{w}_4 = \frac{1}{h_x h_y} \begin{pmatrix} 0 \\ y \end{pmatrix} \tag{4.107}$$

Substituting above equation in Eq. (4.104), we will obtain the specific expression of unit approximate function for velocity. After substituting it in Eqs. (4.101) and (4.102), we will obtain the element characteristic matrix of Eq. (4.103) as follows:

$$\boldsymbol{B}_e = \int_{\Omega_e} \boldsymbol{W}_e^{\mathrm{T}} \cdot \left[\boldsymbol{K}\lambda(S_w^k) \right]_e^{-1} \cdot \boldsymbol{W}_e \, \mathrm{d}\Omega$$

$$= \int_{\Omega_e} \begin{bmatrix} -\boldsymbol{w}_1^e \\ \boldsymbol{w}_2^e \\ -\boldsymbol{w}_3^e \\ \boldsymbol{w}_4^e \end{bmatrix}_{4 \times 2} \begin{bmatrix} \lambda K_{11}^e & \lambda K_{12}^e \\ \lambda K_{21}^e & \lambda K_{22}^e \end{bmatrix}_{2 \times 2}^{-1} [-\boldsymbol{w}_1^e \quad \boldsymbol{w}_2^e \quad -\boldsymbol{w}_3^e \quad \boldsymbol{w}_4^e]_{2 \times 4} \, \mathrm{d}\Omega$$

$$= \frac{1}{(h_x h_y)^2} \int_0^{h_y} \int_0^{h_x} \begin{bmatrix} h_x - x & 0 \\ x & 0 \\ 0 & h_y - y \\ 0 & y \end{bmatrix}_{4 \times 2} \begin{bmatrix} K_{11} & K_{12} \\ K_{21} & K_{22} \end{bmatrix}_{2 \times 2} \begin{bmatrix} h_x - x & x & 0 & 0 \\ 0 & 0 & h_y - y & y \end{bmatrix}_{2 \times 4} \, \mathrm{d}x \mathrm{d}y$$

Then we can get

$$
\boldsymbol{B}_{\mathrm{e}} = \begin{bmatrix}
\dfrac{K_{11}h_x}{3h_y} & \dfrac{K_{11}h_x}{6h_y} & \dfrac{K_{12}}{4} & \dfrac{K_{12}}{4} \\[8pt]
\dfrac{K_{11}h_x}{6h_y} & \dfrac{K_{11}h_x}{3h_y} & \dfrac{K_{12}}{4} & \dfrac{K_{12}}{4} \\[8pt]
\dfrac{K_{21}}{4} & \dfrac{K_{21}}{4} & \dfrac{K_{22}h_y}{3h_x} & \dfrac{K_{22}h_y}{6h_x} \\[8pt]
\dfrac{K_{21}}{4} & \dfrac{K_{21}}{4} & \dfrac{K_{22}h_y}{6h_x} & \dfrac{K_{22}h_y}{3h_x}
\end{bmatrix}_{4\times 4}
\tag{4.108}
$$

Similarly, the characteristic matrix and array of other elements are as follows, respectively.

$$
\boldsymbol{C}_{\mathrm{e}} = \int_{\Omega_e} \nabla \cdot \boldsymbol{W}_{\mathrm{e}}\, \mathrm{d}\Omega = \int_{\Omega_e} \begin{bmatrix} -\nabla \cdot w_1^{\mathrm{e}} & \nabla \cdot w_2^{\mathrm{e}} & -\nabla \cdot w_3^{\mathrm{e}} & \nabla \cdot w_4^{\mathrm{e}} \end{bmatrix}_{1\times 4}\, \mathrm{d}\Omega
$$

$$
= \frac{1}{h_x h_y} \int_{\Omega_e} \begin{bmatrix} -1 & 1 & -1 & 1 \end{bmatrix}_{1\times 4}\, \mathrm{d}\Omega = \begin{bmatrix} -1 & 1 & -1 & 1 \end{bmatrix}_{1\times 4}
$$

$$
\tag{4.109}
$$

$$
\boldsymbol{g}_{\mathrm{e}} = \int_{\Omega_e} \boldsymbol{W}_{\mathrm{e}}^{\mathrm{T}} \cdot \rho_{\mathrm{e}} \boldsymbol{G}_{\mathrm{e}}\, \mathrm{d}\Omega = \int_{\Omega_e} \begin{bmatrix} -w_1^{\mathrm{e}} \\ w_2^{\mathrm{e}} \\ -w_3^{\mathrm{e}} \\ w_4^{\mathrm{e}} \end{bmatrix}_{4\times 2} \begin{bmatrix} \rho g_x \\ \rho g_y \end{bmatrix}_{2\times 1} \mathrm{d}x\mathrm{d}y
$$

$$
= \frac{\rho_{\mathrm{e}}}{h_x h_y} \int_0^{h_y} \int_0^{h_x} \begin{bmatrix} h_x - x & 0 \\ x & 0 \\ 0 & h_y - y \\ 0 & y \end{bmatrix}_{4\times 2} \begin{bmatrix} g_x \\ g_y \end{bmatrix}_{2\times 1} \mathrm{d}x\mathrm{d}y
\tag{4.110}
$$

$$
= \frac{\rho_{\mathrm{e}}}{2} \begin{bmatrix} g_x h_x \\ g_x h_x \\ g_y h_y \\ g_y h_y \end{bmatrix}_{4\times 1}
$$

$$
q_{\mathrm{e}} = \int_{\Omega_e} q^{k+1}\, \mathrm{d}\Omega = A_{\mathrm{e}} q^{k+1}
\tag{4.111}
$$

Looping all the elements, constituting overall matrix Eq. (4.103) with element characteristic matrix and applying Gaussian elimination method to solve the equation, we can obtain the value of elements at whole coarse grid system and the rate of flow at the boundary of elements.

4.4.3   FVM for Saturation Equation

In this section, we describe the finite volume method used for the approximation of the saturation equation. Only a short description of the method employed will be given. The interested reader is referred to Afif and Amaziane (2002a, b, 2003) for more details. The saturation discretization in the i-th grid block based on finite volume method is given as

$$\int_{\Omega_i} \phi \frac{\partial S}{\partial t} \, d\Omega + \int_{\partial \Omega_i} \left[f_w (v + K\lambda_o \cdot \nabla p_c + K\lambda_o \cdot (\rho_w - \rho_o)G) \right] \cdot n_i \, d\Gamma = \int_{\Omega_i} q_w \, d\Omega$$

$$(4.112)$$

For convenience, we dropped the subscript w for water saturation S_w. Applying divergence theory, we write above equation as

$$\int_{\Omega_i} q_w \, d\Omega = \int_{\Omega_i} \frac{\phi}{\Delta t} \left(S_i^{k+1} - S_i^k \right) d\Omega$$

$$+ \sum_{\gamma_{ij}} \left[f_w(S)_{ij} \left(v \cdot n_{ij} + K\lambda_o \cdot \nabla p_c \cdot n_{ij} + K\lambda_o \cdot (\rho_w - \rho_o)G \cdot n_{ij} \right) \right]$$

$$(4.113)$$

Applying time discretization θ-rule for temporal discretization, we obtain

$$\frac{\phi_i}{\Delta t} \left(S_i^{k+1} - S_i^k \right) + \frac{1}{|\Omega_i|} \sum_{\gamma_{ij}} \left[\theta F_{ij} \left(S^{k+1} \right) + (1 - \theta) F_{ij} \left(S^k \right) \right] = q_w \left(S_i^k \right) \qquad (4.114)$$

where

$$F_{ij}(S) = \int_{\gamma_{ij}} \left[f_w(S)_{ij} \left(v \cdot n_{ij} + K\lambda_o \cdot \nabla p_c \cdot n_{ij} + K\lambda_o \cdot (\rho_w - \rho_o)G \cdot n_{ij} \right) \right] d\Gamma$$

$$(4.115)$$

Here, $F_{ij}(S)$ is the numerical approximation of the flux over edge γ_{ij}.

For a first-order scheme, it is common to use upstream weighting for the fractional flow

$$f_w(S)_{ij} = \begin{cases} f_w(S_i) & \text{if } v \cdot n_{ij} \geq 0 \\ f_w(S_j) & \text{if } v \cdot n_{ij} < 0 \end{cases} \qquad (4.116)$$

An explicit scheme, i.e., $\theta = 0$, is employed. Such scheme is quite accurate but need impose stability restrictions on the time step, i.e., the CFL condition,

$$\Delta t \leq \frac{\phi_i |\Omega_i|}{v_i^{in} \max\{f_w'(S)\}_{0 \leq S \leq 1}} \qquad (4.117)$$

where

$$v_i^{in} = \max(q_i, 0) - \sum_{\gamma_{ij}} \min(v_{ij}, 0) \qquad (4.118)$$

$$\frac{\partial f_w}{\partial S} = \frac{\partial f_w}{\partial S^*} \frac{\partial S^*}{\partial S} = \frac{1}{1 - S_{wc} - S_{or}} \frac{\partial f_w}{\partial S^*} \qquad (4.119)$$

where S^* represents the normalized water saturation.

4.4.4 Numerical Examples

Based on above numerical simulation theory and method, the MATLAB is used to compile corresponding full tensor two-phase flow numerical simulator. Before giving the flow simulation examples of two complex fractured-vuggy media at coarse scale, we have initially tested the validity of equivalent method and numerical program in a simple fractured medium.

(1) Numerical validation
We consider a single fracture in the matrix block. Water-flooding simulations are carried out for two different orientations of the fracture ($\theta = 0, \pi/4$). Figure 4.23a is the geometrical configuration. We consider a fracture thickness $e = 100\ \mu m$ and

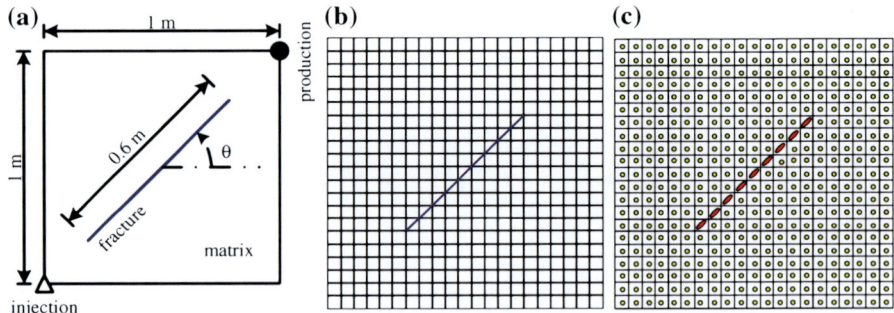

Fig. 4.23 Geometrical configuration of the fractured media with a single fracture (*left*) and a single fracture (*left*) and a mesh of grid blocks (*medium*) and its corresponding permeability tensor map (*right*)

$(k_f = 8.37 \times 10^5 \ \mu m^2)$. The porosity and the permeability of the matrix are $\phi = 1.0$ and $k_m = 1 \ \mu m^2$, respectively. The medium is initially filled with oil. Residual oil saturation and irreducible water saturation are both zero. We inject water at the bottom left corner at the rate of $q = 0.01$ PV/day. Liquid is produced from the top right corner at the same rate of injector. For simplicity, we neglect the gravity and capillary effects and the original matrix and fracture relative permeability curves are both X shape. The pseudo-relative permeability curve in coarse grid is X shape too, as discussed in Sect. 4.3.2.

First, we generated a mesh of grid blocks for the region, by uniformly subdividing it into 21×21 grid blocks, as illustrated in Fig. 4.23b. Here we just show the inclined fracture cases with $\theta = \pi/4$. Figure 4.23c illustrates the corresponding effective permeability tensor ellipses.

Then let us evaluate the validation and accuracy of the present equivalent continuum method by comparing the results with those obtained by the discrete-fracture model. Two different meshes of grid blocks are considered, one is 21×21 and the other is 31×31. Figure 4.24 presents the water saturation distribution at 0.5 PV water injection. As can be seen, the results from the equivalent continuum simulation are in excellent agreement with the discrete-fracture model. It also implies that the numerical results will be more satisfied with the refining of the mesh.

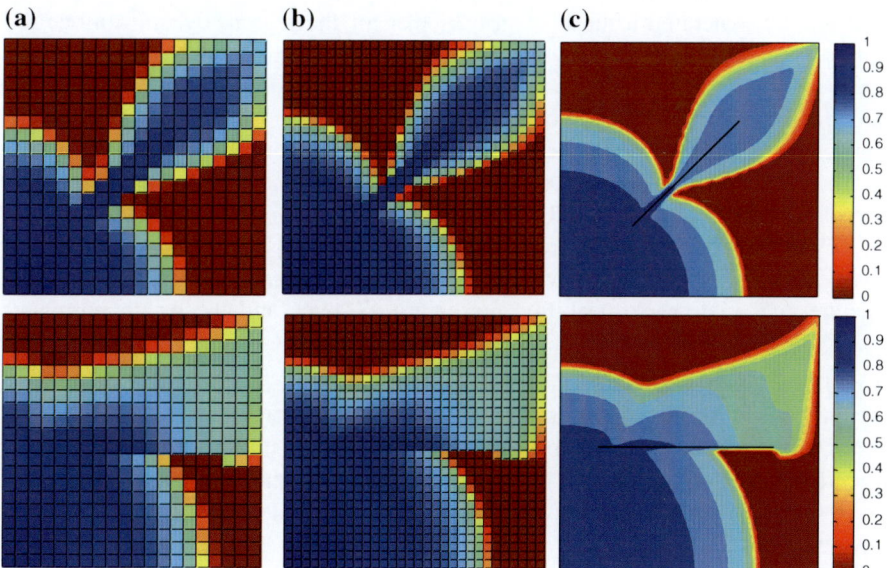

Fig. 4.24 Water saturation profiles at 0.5 PV water injection: simple fractures media with a single fracture, one with a tilted fracture (*top*) and one with horizontal fracture (*bottom*). **a** 21×21 grid block. **b** 31×31 grid block. **c** discrete fracture model

Table 4.2 Macro-fractures' statistic data

Characteristic parameter	Minimum value	Maximum value	Average value
Length (m)	20	160	65.2
Orientation (°)	45	45	45
Intensity (1/m)	0.14	0.58	0.33

Table 4.3 Macro-cavities statistic data

Characteristic parameter	Minimum value	Maximum value	Average value
Axis length (m)	2.1	8.3	6.5
Orientation (°)	0	15	5.0
Density (1/km^2)	1026	2100	1750

(2) Complex fractured karst reservoir 1

Based on statistical data from fractured karst reservoir outcrop in the TAHE oilfield in west China, we applied random modeling technology to generate corresponding discrete fractured-vuggy model. Some of the fracture statistics for fractures system are presented in Table 4.2. And the cavities are simplified into some ellipses with some statistics characteristics, which are presented in Table 4.3. Using these data, we generated the realization of the fractured karst system depicted in Fig. 4.25a. The size of this region is 100 m × 200 m ($x \times y$).

Then we generated a mesh of grid blocks for the region, by uniformly subdividing it into 10 × 20 grid blocks, as illustrated in Fig. 4.25b. The permeability tensor of the whole mesh system can be obtained by homogenization theory. The permeability map along y-direction is presented in Fig. 4.25c. From this map, we can see that the fracture-vug networks have an important influence in the effective permeability. We chose a matrix permeability of $k_m = 1$ μm^2 and a uniform fracture aperture of 100 μm ($k_f = 8.37 \times 10^5$ μm^2). The corresponding porosity and effective permeability tensor are calculated by using Eqs. 4.67 and 4.73. For simplicity, we also neglect the gravity and capillary effects and the original normalized fracture-vug relative permeability curves are X shape and the origin normalized matrix relative permeability curves are $k_{rw,m} = \left(S_{w,m}^*\right)^2$ and $k_{ro,m} = \left(1 - S_{w,m}^*\right)^2$. Both the connate water saturation and residual saturation of matrix and fracture-vug system are zero. The pseudo-relative permeability curves and corresponding parameter distributions for coarse grid blocks are shown in Fig. 4.26b. The medium is initially filled with oil. We inject water at the bottom left corner at the rate of $q = 0.004$ PV/day. Liquid is produced from the top right corner at the same rate of injector.

Figure 4.27 shows the influence of variations in the effective parameters on the motion of the water through the fractured karst region. Three snapshots of the subsequent evolution of the water flooding are presented in the figure. They help to illustrate how fluid moves through the homogenized grid blocks. In the figure, we can see that the variations in the effective permeability and pseudo-relative

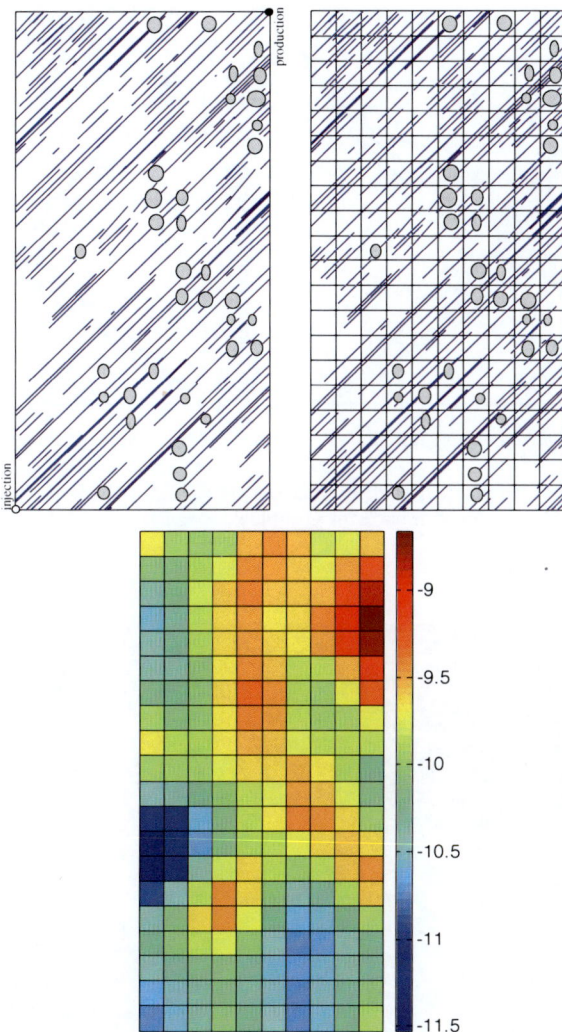

Fig. 4.25 Realization of a fracture-vug media geometric model generated with the statistics corresponding to the TAHE outcrop from Tables 4.2 and 4.3 (*left*); the mesh of grid blocks (*medium*); the permeability logarithm map along *y*-direction (*right*)

permeability curves have had a pronounced and cumulative effect on the flow through the region. In Fig. 4.27, we also superimpose the fracture-cavity system onto a plot of the water saturation profile at 0.5 PV water injection. We can see that the fluid flow is primarily determined by the orientation and intensity of fracture-cavity system. The figure shows that the preferred direction of motion is primarily determined by the properties of the fracture-cavity system. And the corresponding effective parameters of the homogenized grid blocks honor these properties.

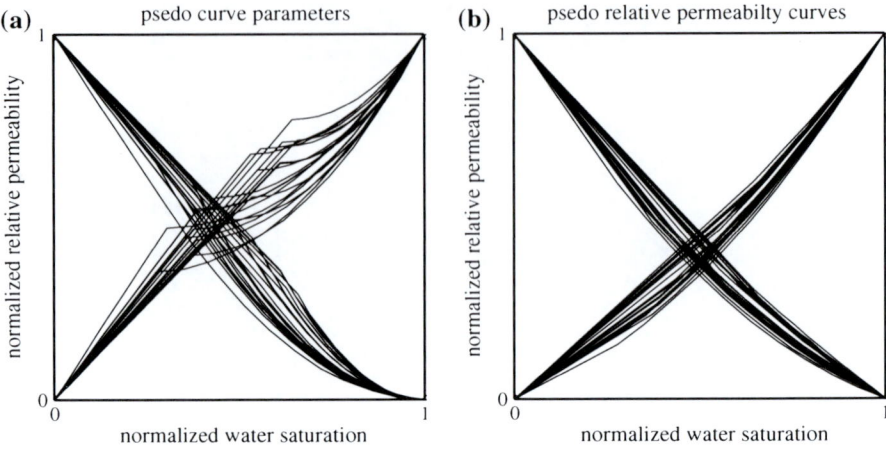

Fig. 4.26 Pseudo-curve parameters determination for each gird blocks (*left*) and the corresponding pseudo-relative permeability curves of grid blocks (*right*)

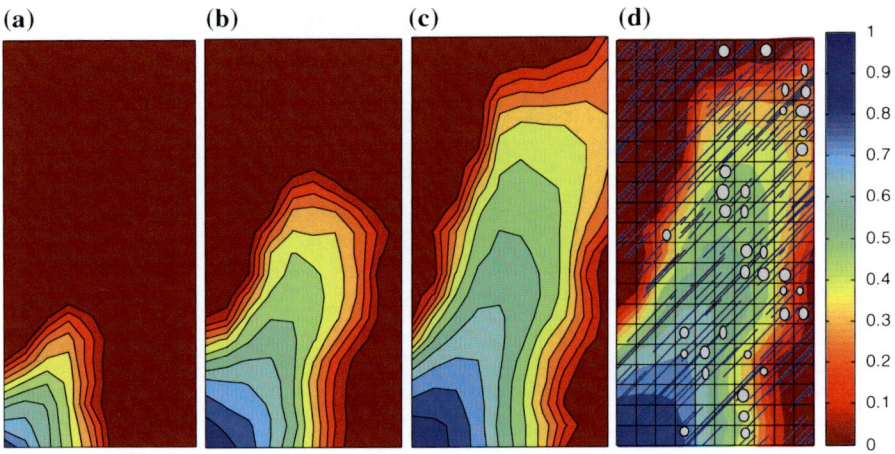

Fig. 4.27 Three water saturation profiles at different times and the superposition of the fracture-vug system on the evolved water saturation map. **a** 25 d. **b** 75 d. **c** 125 d. **d** 125 d

(3) Complex fractured karst reservoir 2

In the above example, the fractures are aligned with the same direction. In this section, we will give a more general case with fractures in multiple directions. As illustrated in Fig. 4.28a, the size of the study domain is 100 m × 100 m ($x \times y$), where the coordinate system is as same as that of the complex fractured karst reservoir 1 depicted in Fig. 4.25. The permeability of matrix $k_m = 11\ \mu m^2$, and a uniform fracture aperture of 100 μm ($k_f = 8.37 \times 10^5\ \mu m^2$). The medium is initially filled with oil. We inject water at the bottom left corner at the rate of $q = 0.01$ PV/day. Liquid is produced from the top right corner at the same rate of injector. The other parameters are the same as those given in above example.

First, we generated a mesh of grid blocks for the region. Two different meshes of grid blocks are considered, one is 20×20 and the other is 10×10, as shown in

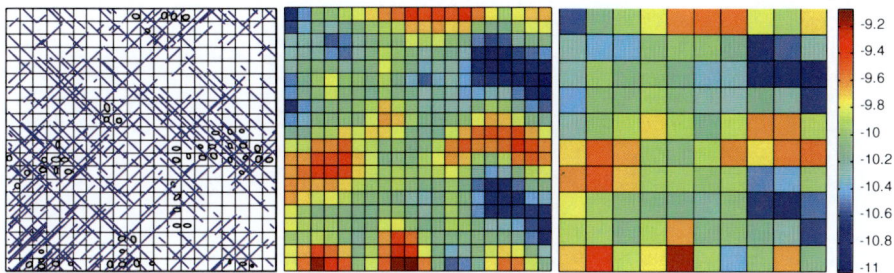

Fig. 4.28 A conceptual fractures system (*left*); the permeability logarithm map along x-direction at the fine grid (*medium*); the corresponding permeability logarithm map along $x =$ direction at the coarse grid (*right*)

(a)

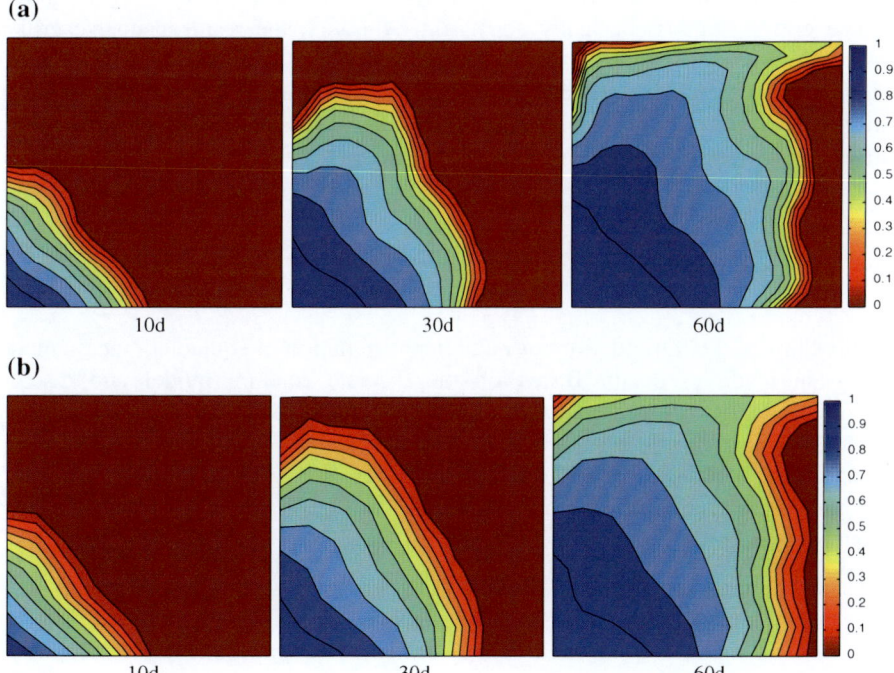

(b)

Fig. 4.29 Comparison with water saturation profiles between the fine grid (*top*) and the coarse grid (*bottom*) at different times. **a** 20×20 grid block. **b** 10×10 grid block

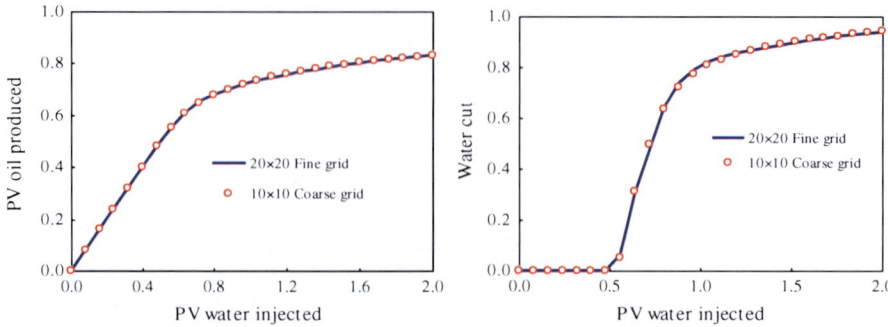

Fig. 4.30 Comparison with cumulative oil production between the fine grid and coarse grid (*left*); and water cut curves for the fine grid and the coarse grid (*right*)

Fig. 4.28. Then we obtain the absolute permeability tensor (as shown in Fig. 4.28b, c) and the pseudo-relative permeability curve (similar to Fig. 4.26).

Figure 4.29 shows the water saturation map at different times. As can be seen from the figure, the existence of fracture and cavity results in the in-homogeneity of media, and coarse-scale equivalent flow simulation method can adapt to the condition very well. The guidance effect of fracture-vug system can be seen from water saturation map. Simultaneously, the numerical computation result of two different coarse grid systems is basically consistent. Although the result at coarse grid is more homogeneous, the strong heterogeneity of reservoir can be reflected. Figure 4.30 presents the cumulative oil production curve and water cut for both two grid systems until 2PV water injection. We observe very close agreement between the two grid systems, which again demonstrate the validity of our approach.

4.5 Summary

In this chapter, the introduction for numerical simulation of equivalent media model has been made in detail. Based on the discrete fracture model and discrete fracture-vug network model, the equivalent theory and numerical simulation methods, which can be used to simulate two-phase flow in fractured-vuggy reservoirs at reservoir scale, have been established. It provides an efficient method for the numerical simulation of fractured-vuggy reservoirs. Specific works and conclusions as follows:

(1) For the coarse grids with fractures, the equivalent permeability tensors have been obtained based on oversample technology, DFM method and Galerkin finite element, and the accuracy is verified by several numerical examples.

(2) For the coarse grids with fractures and vugs, the equivalent permeability tensors have been obtained based on homogenization theory, DFVN method

and Galerkin finite element, and the accuracy is verified by several numerical examples.

(3) On the basis of the preferential flow assumption, an analytical calculation for pseudo-relative permeability curves of grid blocks has been conducted easily. And these effective parameters are used in equivalent continuum simulations of naturally fractured karst reservoirs. The simulation results indicate that the pseudo-relative permeability can efficiently represent the motion characteristic of water-oil front in fracture-vuggy media. However, it is not suitable for the coarse grids with little fractures and vugs, which means a new computation method is needed and it will be the next research points in the future.

(4) Based on the mixed finite element and finite volume method, an efficient numerical simulation method is formed, which can be used to full tensor permeability. The validity of numerical method has been verified by several numerical examples.

References

Aarnes JE, Gimse T, Lie KA (2007) An introduction to the numerics of flow in porous media using Matlab. In: Geometric modelling, numerical simulation, and optimization, pp 265–306

Aarnes JE, Lie KA, Kippe V, Krogstad S (2009) Multiscale methods for subsurface flow. In: Multiscale Modeling and Simulation in Science, pp 3–48

Abdel-Ghani R (2009) Single porosity simulation of fractures with low to medium fracture-to-matrix permeability contrast. In: Paper presented at the SPE/EAGE reservoir characterization & simulation conference

Afif M, Amaziane B (2002a) Convergence of finite volume schemes for a degenerate convection–diffusion equation arising in flow in porous media. Comput Methods Appl Mech Eng 191 (46):5265–5286

Afif M, Amaziane B (2002b) On convergence of finite volume schemes for one-dimensional two-phase flow in porous media. J Comput Appl Math 145(1):31–48

Afif M, Amaziane B (2003) Numerical simulation of two-phase flow through heterogeneous porous media. Numer Algorithms 34(2–4):117–125

Allaire G (1992) Homogenization and two-scale convergence. SIAM J Math Anal 23(6):1482–1518

Arbogast T, Gomez MSM (2009) A discretization and multigrid solver for a Darcy-Stokes system of three dimensional vuggy porous media. Comput Geosci 13(3):331–348

Arbogast T, Lehr HL (2006) Homogenization of a Darcy-Stokes system modeling vuggy porous media. Comput Geosci 10(3):291–302

Arbogast T, Brunson DS, Bryant SL et al (2004). A preliminary computational investigation of a macro-model for vuggy porous medium. In: Paper presented at the computational methods in water resources XV, New York

Auriault J (1991) Heterogeneous medium. Is an equivalent macroscopic description possible? Int J Eng Sci 29(7):785–795

Bensoussan A, Lions J-L, Papanicolaou G (2011) Asymptotic analysis for periodic structures, vol 374. American Mathematical Society

Durlofsky LJ (1993) A triangle based mixed finite element—finite volume technique for modeling two phase flow through porous media. J Comput Phys 105(2):252–266

Efendiev Y, Durlofsky L (2002) Numerical modeling of subgrid heterogeneity in two phase flow simulations. Water Resour Res 38(8):1128

Feng J, Chen L, Li C, Liu L (2007) Equivalent continuous medium model for fractured low-permeability reservoir. Pet Drill Tech 35(5):94–97

Hearn C (1971) Simulation of stratified waterflooding by pseudo relative permeability curves. J Petrol Technol 23(07):805–813

Hornung U (1997) Homogenization and porous media, vol 6. Springer

Huang Z, Yao J, Li Y, Wang C, Lü X (2010) Permeability analysis of fractured vuggy porous media based on homogenization theory. Sci China Technol Sci 53(3):839–847

Huang Z, Yao J, Li Y, Wang C, Lv X (2011a) Numerical calculation of equivalent permeability tensor for fractured vuggy porous media based on homogenization theory. Commun. Comput. Phys. 9(1):180–204

Huang ZQ, Yao J, Wang YY (2011b) An efficient numerical model for immiscible two-phase flow in fractured karst reservoirs. Submitted

Huang Z-Q, Yao J, Wang Y-Y (2013) An efficient numerical model for immiscible two-phase flow in fractured karst reservoirs. Commun Comput Phys 13(2):540–558

Liu J, Liu X, Hu Y, Zhang S (2000) The equivalent continuum media model of fracture sand stone reservoir. J Chongqing Univ (Nat Sci Ed)(z1), 158–161

Long J (2012) Porous media equivalents for networks of discontinuous fractures. Water Resour Res

Long J, Witherspoon PA (1985) The relationship of the degree of interconnection to permeability in fracture networks. J Geophys Res: Solid Earth (1978–2012), 90(B4):3087–3098

Louis C (1972) Rock hydraulics. In: Rock mechanics. Springer, pp 299–387

Neale G, Nader W (1973) The permeability of a uniformly vuggy porous medium. Old SPE J 13 (2):69–74

Oda M (1986) An equivalent continuum model for coupled stress and fluid flow analysis in jointed rock masses. Water Resour Res 22(13):1845–1856

Popov P, Efendiev Y, Qin G (2009) Multiscale modeling and simulations of flows in naturally fractured karst reservoirs. Commun. Comput. Phys. 6(1):162–184

Pruess K, Wang J, Tsang Y (1990) On thermohydrologic conditions near high-level nuclear wastes emplaced in partially saturated fractured tuff: 2. Effective continuum approximation. Water Resources Research 26(6):1249–1261

Qin G, Bi L, Popov P, Efendiev Y, Espedal M (2010) An efficient upscaling procedure based on stokes-brinkman model and discrete fracture network method for naturally fractured carbonate karst reservoirs. In: Paper presented at the CPS/SPE international oil and gas conference and exhibition, Beijing, China

Snow DT (1970) The frequency and apertures of fractures in rock. In: Paper presented at the International journal of Rock mechanics and Mining sciences & Geomechanics Abstracts

Telleria M, Virues C, Crotti M (1999) Pseudo relative permeability functions. Limitations in the use of the frontal advance theory for 2-Dimensional systems. In: Paper presented at the Latin American and Caribbean petroleum engineering conference

Tian K (1984) The discussion on hydrogeology model of fractured rock. Site Invest Sci Technol 4:27–34

van Golf-Racht TD (1982) Fundamentals of fractured reservoir engineering. Elsevier

van Lingen P, Sengul M, Daniel JM, Cosentino L (2001) Single medium simulation of reservoirs with conductive faults and fractures. SPE Middle East Oil Show

Wilson C, Witherspoon P, Long J, Galbraith R, Dubois A, McPherson M (1983) Large-scale hydraulic conductivity measurements in fractured granite. In: Paper presented at the International Journal of Rock Mechanics and Mining Sciences & Geomechanics Abstracts

Yotov I (1996) Mixed finite element methods for flow in porous media. Citeseer

Zhou D, Jiao F, Ge J (2004) Investigating progress for fluid flowing in fractured media. Offshore Pet 24(2):34–38

Chapter 5
Hybrid Fracture Model

Abstract This chapter introduces the numerical simulation of mixed models for fractured medium. It starts with presenting the rules for dealing with the fractures during the modeling process. It then introduces the concept models that are usually used to describe the flowing characteristic of the fluids in fractured medium. However, the fractured medium is too complicated, so several kinds of mixed models, like coupled model and combined model, are introduced to realize better modeling. Then the categories, characteristics, and realization process of the coupled model are introduced. Also, the categories and realization process of combined model are introduced briefly.

Keywords Combined model · Coupled model · Numerical simulation · Fractured medium · Seepage concept model

5.1 Development Characteristics of Fractured Medium

Under the effect of geological tectonic stress or artificially applied stress, rock medium will be damaged and fractures will come into being. There are many factors influencing the development of rock fractures, including the size and direction of stress, lithology, formation thickness, and the position of the fracture on the geological structure. Owing to these different influencing factors, fractures in rock medium tend to develop in a rather complex way, which is manifested as a diversity of fractures' length, aperture, strike, dip, density, etc. Thus, fractured medium tends to have obvious multi-scale characteristic and strong heterogeneity (shown in Fig. 5.1). Multi-scale characteristic means there are fractures of various lengths in the same area. And strong heterogeneity means there is obvious difference between the development degrees of the fractures in different areas, for example, there are fractures of various sizes near the fault area or the fracturing area, and with the distance from the fault area or the fracturing area increases, the development degree of the fractures decreases (most of the fractures are of middle or small sizes), and the number of the fractures decreases significantly.

© Petroleum Industry Press and Springer-Verlag Berlin Heidelberg 2017
J. Yao and Z.-Q. Huang, *Fractured Vuggy Carbonate Reservoir Simulation*,
Springer Geophysics, DOI 10.1007/978-3-662-55032-8_5

Fig. 5.1 Complex fractured reservoirs with fractures of different scales

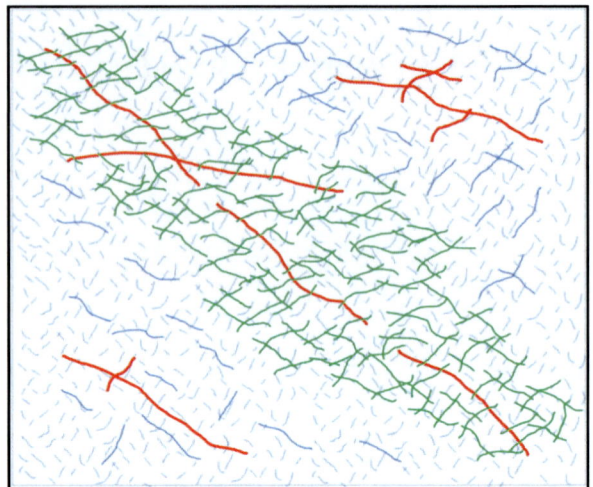

Fractures of different development degrees have different permeabilities. And for the fracture areas that have different fracture development degrees, the connectivity and permeability of the fracture network in these areas be different. Thus the fracture development degree can influence and change the seepage path. The description of complex fracture and the measurement of its connectivity and permeability are especially important for the research of the seepage in fractured medium.

When describing a fracture, we need to describe its length, aperture, dip, strike, density, etc. Underground fractures develop in a rather complex way, however its development has certain laws we can find. Research shows that, the development of underground fractures conforms to geo-statistic law. Generally, fracture length, l conforms to certain exponential function law:

$$n(l, L) = d_c L^2 \cdot (a - 1) \cdot \frac{l^{-a}}{l_{min}^{-a+1}}, \quad l \in [l_{min}, l_{max}] \tag{5.1}$$

where $n(l, L)$ is the number of the fractures whose system scale is L and whose length is in the range of $[l, l + dl]$; a is one parameter in the exponential function for fracture distribution, usually $a \in [1.5, 3]$; d_c is the number of fractures that have fracture center point in per unit area; $[l_{min}, l_{max}]$ is the range of fracture length.

The value of a reflects the development status of fracture length in the model, when $a > 3$, the model contains both big and small fractures, and it can be recognized as typical seepage model; when $a < 2$, the fluids mainly flow in the big fractures, so we can adopt the discrete fracture grid model; and when $2 \leq a \leq 3$, the condition will be rather complex, and both big fractures and small fractures both can influence the seepage process of the fluids.

Parameter C indicates the connectivity of the fracture. We usually set a parameter, $C0$, as the threshold value. When $C > C0$, the fracture network is connected. But $C0$ can also be regarded as a parameter that is independent of the change of the fracture scale (Bour and Davy 1997, 1998). The change of parameter a, which is in the exponential function, has so little influence on the $C0$ that this influence can be neglected.

$$\begin{cases} p \sim d \cdot L & a < 2 \\ p \sim d \cdot \frac{L^{3-a}}{l_{min}^{2-a}} & 2 < a < 3 \\ p \sim d \cdot l_{min} & a > 3 \end{cases} \tag{5.2}$$

When $a > 3$, the fracture's permeability and connectivity is independent of the scale. When $a < 3$, its connectivity is related to d_c (fracture's density) and L (fracture's scale). There are four independent variables in the fracture exponential function: a (fracture length index), l_{min} (fracture's minimum length), L (the fracture system length), d_c (fracture density or permeability parameter). There can be three kinds of situations (Dreuzy et al. 2001): when $a < 2$, the fluids mainly flow through long fractures and the fracture network is in a secondary position; for fracture network, characteristic length is the system length, so reducing the size of the grid cannot improve the permeability. When $a > 3$, fracture network is composed of small fractures, so we apply the seepage theory to describing the system. When $2 \leq a \leq 3$, there is no need to classify the fracture length strictly.

5.2 Conceptual Model of Seepage

It is especially important for us to choose a proper seepage model when simulating the flow of the fluids in fractured medium. Owing to the obvious multi-scale characteristic and strong heterogeneity, ordinary seepage models cannot satisfy the requirements of the simulation. There are mainly two categories of seepage models, continuum model and discrete fracture model that are usually applied to describing the flow of fluids in fractured medium. Continuum model assumes that the fractures are uniformly distributed in the matrix rock, and mathematical methods like statistical averaging and volume averaging are used to approximate the parameters of the model. Equivalent medium model and double-porosity model are the main parts of the corresponding conceptual models (Barenblatt et al. 1960; Kazemi et al. 1976; Warren and Root 1963). While the other kind focuses on building a seepage model that accords with the real fractures' form according to the real geological information of fracture development. The corresponding conceptual models contain discrete fracture grid model and discrete fracture model. Those conceptual models all have respective applicable conditions and scope.

5.2.1 *Continuum Model*

Equivalent medium model assumes that fractured medium is an ideal hypothetical continuum, and the physical quantities of each point of the system are in partial equilibrium state. This model will distribute the fracture physical parameters equivalently in the entire medium, and we regard the medium as an anisotropic medium with symmetric tensor. This model ignores the physical structure of each single fracture so it cannot describe or characterize every fracture precisely.

Double-porosity model assumes that the medium is two parallel continuous systems, namely a fracture system and a matrix system. And continuous matrix is assumed that to be divided into a series of rocks by the fracture system, these two systems are coupled together by channeling. This kind of model can be roughly divided into four types: double-porosity model, dual-permeability model, multirole model, and sub-domain model. In order to characterize the change of pressure and saturation in matrix, we use the sub-domain decomposition method to build different models. For example, the MINC model is very common (Gong et al. 2006; Karimi-Fard et al. 2006; Pruess 2013; Wu and Pruess 1988). Research shows that, it is very difficult to obtain an exact solution if we use a double-porosity model to characterize the strongly heterogeneous fractured reservoir that contains large-scale fractures. Thus, double-porosity model is suitable for fractured reservoir where many small-scale fractures developed and the fractures developed richly and have a good connectivity. However, double-porosity is not suitable for the situation where fractures developed limitedly and have a bad connectivity.

5.2.2 *Discrete Fracture Model*

Discrete fracture network model assumes that fluids flow in the fracture network only. The matrix's permeability is ignored while emphasis is laid on the flow of fluids in the fracture network. Discrete fracture network model is suitable for the reservoir in which fracture development degree is high, the connectivity is high and the matrix's permeability can be neglected.

Discrete fracture model takes the matrix's permeability into consideration, namely when building a discrete fracture model we conduct precise explicit expression on the fractures on the basis on taking matrix's permeability into consideration. When compared to double-porosity model, discrete fracture model has explicit expression and calculation for each fracture; so there is no need to calculate the fracture-matrix flow between fractures and matrix; and fracture's dimensionality is reduced so the calculation amount that fine grid decomposition on the fracture aperture brings is reduced. In numerical simulation, most models adopt the finite element method, which is based on unstructured grid. The fractures are located at the boundary of the finite element. Later researchers put forward embedded discrete fracture model, which adopts the regular structured grid, the fractures are embedded in the matrix grid and are treated as the well's source. Embedded discrete fracture model can make the gridding process of complex fractured medium very effective.

5.2.3   *Hybrid Model*

Usually, fractured medium is so complex that single seepage models cannot satisfy our requirements. So the researchers began to combine different seepage models to solve the practical problems. Overall, there are several situations as listed below

(1) Hierarchical model (Lee et al. 2001): classify the fractures according to their scales and choose different methods to calculate their equivalent permeability.
(2) Coupling model (Sarda et al. 2002): in the same area, different kinds of seepage models are coupled together. A representative example is that discrete fracture model is coupled with other kinds of seepage models.
(3) Combined model (Wang et al. 2011): in the simulated space, adopt different suitable seepage models in different areas and obtain a simultaneous solution of the areas.

According to the analysis of the above seepage models, every model has certain applicability. It is difficult to characterize precisely the multi-scale characteristic and the heterogeneity characteristic if we only adopt single seepage model. For this reason, we adopted the coupling model (continuum model and discrete fracture model are coupled together) and combined model (shown in Fig. 5.2). Coupling model is that many seepage models are coupled together in one research area; Hybrid model is that many seepage models in different partitions are combined together to use.

Fig. 5.2 Coupling model and hybrid model for fractured medium

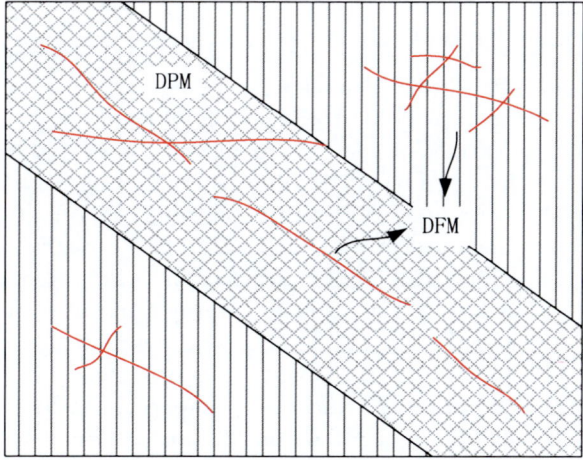

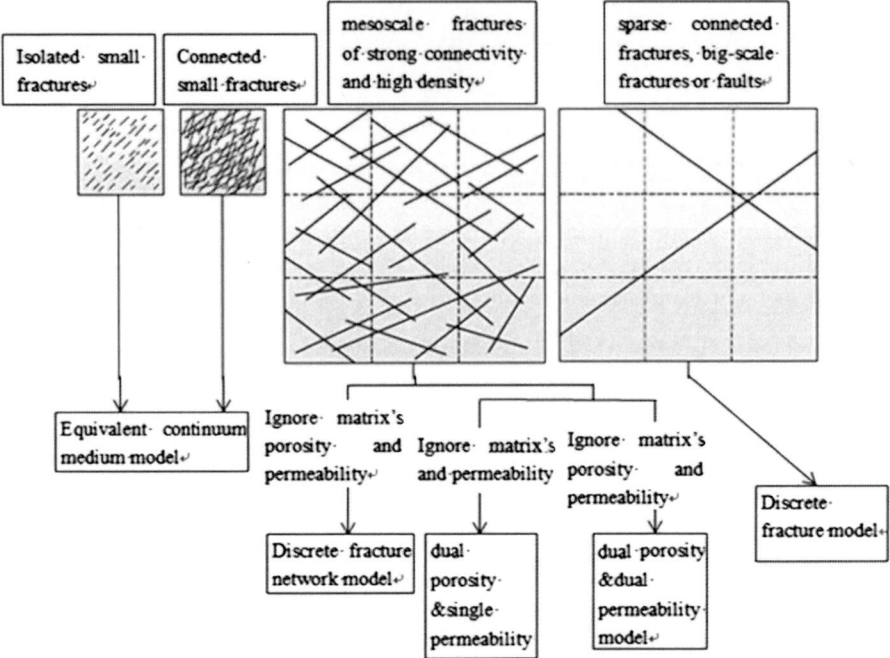

Fig. 5.3 Seepage models applicable for different fractured medium

When building a hybrid model, we have chosen the applicable seepage model according to the fracture development characteristics, in order to characterize the fluids in the fractured medium precisely.

According to the above analysis, fractures can be divided into isolated small fractures, small fractures of strong connectivity, mesoscale fractures of strong connectivity and high density, sparse connected fractures, big-scale fractures or faults, etc. The complex fractures of different lengths and different connectivity have different kinds of applicable seepage models (shown in Fig. 5.3):

(1) Small isolated model can be equivalent to matrix pore, the Representative Element Volume (REV) of the same scale can characterize the seepage characteristics of it and matrix pores. And its flow characteristics can be described by single equivalent anisotropic medium model.

(2) As for the small fractures of strong connectivity, the permeability is obviously stronger than the permeability of the matrix pores, so the porosity and the permeability characteristic of the whole Representative Element Volume is improved. Thus, double-porosity equivalent continuum medium model can be adopted and the porosity and permeability in the Representative Element Volume are the statistical average value of the fracture and matrix.

(3) For mesoscale fractures of strong connectivity and high density, they have different properties from the matrix. And the seepage velocities and fluids

properties are also different. According to the matrix's seepage properties, we can choose from discrete fracture network model, double-porosity and single-permeability model, and double-porosity and dual permeability model.

(4) As for sparse connected fractures, big-scale fractures or faults, the number of fractures is limited and the permeability is strong, and these factors influence the seepage path of the seepage area. Thus, we can choose discrete fracture model.

5.3 Types of Coupling Models and Their Realization

Coupling model means that many seepage models are coupled together in the same area. It is mainly for the multi-scale characteristics of fractured media. That is, choose the applicable seepage model according to the development characteristics of the fractures of different scale and couple the chosen seepage models together.

5.3.1 Coupling Model

The types of coupling model and the fractured medium they are applicable for is listed in Table 5.1.

Now, equivalent medium-double-porosity-discrete fracture model is taken as an example to illustrate the realization of coupling model.

5.3.2 Mathematical Equation

It is assumed that both the rock and the fluids are incompressible, the influence of gravity and capillary pressure is neglected, and the flow of the water phase and the oil phase, both comply with Darcy's law.

$$\phi \frac{\partial S_o}{\partial t} + \nabla \cdot v_o = F_o \tag{5.3}$$

$$\phi \frac{\partial S_w}{\partial t} + \nabla \cdot v_w = F_w \tag{5.4}$$

$$v_o = -\lambda_o(S_o) K \nabla p_o \tag{5.5}$$

$$v_w = -\lambda_w(S_w) K \nabla p_w \tag{5.6}$$

Table 5.1 Types of coupling models and the fractured medium they are applicable for

Types of coupling	The development characteristics of the Fractured medium	Seepage model	The development characteristics of the applicable fracture
Equivalent medium-double-porosity model	There are many isolated or connected small-scale fractures and many mesoscale fractures of good connectivity in the medium	Equivalent medium model	The small fractures whose scales are far smaller than the size of the simulation grid as well as that are isolated or well connected
		Double-porosity model	The well connected fractures whose scale is same with the size of the simulation grid
Double-porosity-discrete fracture model	There are many mesoscale fractures of high connectivity and limited number large-scale fractures in the medium	Double-porosity model	The well connected fractures whose scale are same with the size of simulation grid
		Discrete fracture model	Large-scale fractures, faults whose number is limited and scales are far larger than the simulation grid
Equivalent medium-double-porosity discrete fracture model	There are many isolated or connected small-scale fractures and many mesoscale fractures that are highly connected and limited number of large-scale fractures	Equivalent medium model	The fractures whose scale is far less than the size of the simulation grid and small fractures that are isolated or well connected
		Double-porosity model	The well connected fractures whose scales are same with the size of the simulation grid
		Discrete fracture model	The big fractures, faults whose number is limited and scales are far larger that the simulation grid

where, ϕ is porosity; S_o and S_w are respectively the oil phase saturation and the water phase saturation, and $S_o + S_w = 1$; v_o and v_w are respectively the oil phase flow rate and the water phase flow rate; q_o and q_w are the source terms; K is the permeability tensor; p_o and p_w are respectively the oil phase pressure and water phase pressure; λ_o and λ_w are respectively the oil phase mobility and the water phase mobility, which comply with

$$\lambda_o = \frac{k_{ro}}{\mu_o}, \quad \lambda_w = \frac{k_{rw}}{\mu_w} \tag{5.7}$$

where, k_{ro} and k_{rw} are the relative permeability of oil and water, respectively; μ_o and μ_w are the viscosity of oil and water respectively.

Because the influence of capillary pressure is neglected, so $p_o = p_w$. Let $p = p_o = p_w$, so we can derive the pressure formula from formula (5.3) to formula (5.6):

$$\nabla \cdot \boldsymbol{v} = F \tag{5.8}$$

$$\boldsymbol{v} = -\lambda \boldsymbol{K} \nabla p \tag{5.9}$$

where, $v = v_o + v_w$ is the total flow rate; $\lambda = \lambda_o + \lambda_w$ is the total mobility; $q = q_o + q_w$ is the total source item.

Let the fractional flow be $f_w = \lambda_w / \lambda$, so we can derive the formula to calculate the water saturation by formula (5.4):

$$\boldsymbol{v} = -\lambda \boldsymbol{K} \nabla p \tag{5.10}$$

Formulas from (5.8)–(5.10) form the fractional flow model for the seepage of the incompressible two phases in matrix.

5.3.3   Longitudinal Coupled Model of Equivalent Medium, Dual Medium and Discrete Fracture

The mathematical model of equivalent medium is applicable for the formula (5.6) and (5.7). What is important is calculating the equivalent porosity and equivalent permeability of the Representative Element Volume according to the small fractures.

(1) Double-porosity mathematical model (DPM)
A matrix system and a fracture system are coupled together by channeling function (shown in Fig. 5.4). Fracture-matrix flow q is related to the factors like the permeability, the shape factor a, and the pressure difference between the two systems. While shape factor a is related to the factors like the rock size, the fracture spacing and the fracture density of the matrix system. And big fractures are divided into two parts, and then they are coupled to the matrix system and the fracture system respectively. And these two parts are also coupled together by channeling. In this model, both the porosity and the permeability of matrix are taken into

Fig. 5.4 Schematic diagram
of dual medium (F_m and F_f
are the channeling between
dual medium and big fracture)

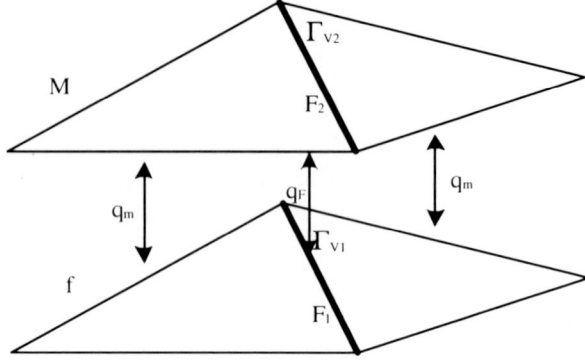

consideration. Thus, we adopt the double porosity-dual permeability model. So the
seepage equation in Ωm and Ωf is

$$\nabla \cdot \boldsymbol{v}^n + (-1)^n q = F^n \tag{5.11}$$

$$\boldsymbol{v}^n = -\lambda^n \boldsymbol{K}^n \nabla p^n \tag{5.12}$$

$$\phi^n \frac{\partial S_w^n}{\partial t} + \nabla \cdot \left(f_w^n \boldsymbol{v}^n \right) + (-1)^n q_w = F_w^n \tag{5.13}$$

$$q_m = \begin{cases} \alpha_m K^m \lambda^m \left(p^m - p^f \right) & \text{for } p^m \geq p^f \\ \alpha_m K^m \lambda^f \left(p^m - p^f \right) & \text{for } p^m < p^f \end{cases} \tag{5.14}$$

$$q_{m,w} = \begin{cases} \alpha_m K^m \lambda_w^m \left(p^m - p^f \right) & \text{for } p^m \geq p^f \\ \alpha_m K^m \lambda_w^m \left(p^m - p^f \right) & \text{for } p^m < p^f \end{cases} \tag{5.15}$$

where, the superscript $n = 1$ or 2. And $n = 1$ represents a fractured system, $n = 2$
represents a matrix system; F and F_w are the total fracture-matrix flow and the water
phase fracture-matrix flow; α is the shape factor of the matrix system; K^m is the
absolute permeability tensor.

Similarly, big fracture also accords with the above equations. And the shape
factor in fracture-matrix flow is α_f. Adopt the absolute permeability K^f of the big
fractures as the absolute permeability.

(2) Discrete fracture mathematical model (DFM)
For big fractures, discrete fracture model is adopted. The method of dimension
reduction is adopted, so the 2-D fractures are seen as 1-D element located at the
interface of the surrounding medium unit. The final equation of the relation between
the big fractures and the surrounding medium is established by the relationship
between the flow rate and the pressure. The seepage equation is

$$\nabla \cdot \boldsymbol{v}^{\mathrm{F}n} + (-1)^n q_{\mathrm{F}} = F^{\mathrm{F}n} \tag{5.16}$$

$$\boldsymbol{v}^{\mathrm{F}n} = -\lambda^{\mathrm{F}n} \boldsymbol{K}^{\mathrm{F}n} \nabla p^{\mathrm{F}n} \tag{5.17}$$

$$\phi^{\mathrm{F}n} \frac{\partial S_{\mathrm{w}}^{\mathrm{F}n}}{\partial t} + \nabla \cdot \left(f_{\mathrm{w}}^{\mathrm{F}n} \boldsymbol{v}^{\mathrm{F}n} \right) + (-1)^n q_{\mathrm{F,w}} = Q_{\mathrm{w}}^{\mathrm{F}n} + F_{\mathrm{w}}^{\mathrm{F}n} \tag{5.18}$$

where, the superscript $n = 1$ or 2. And $n = 1$ represents the big fractures that are coupled with a fractured system, $n = 2$ represents the big fractures that are coupled with a matrix system; $\phi^{\mathrm{F}n}$ is the porosity of the big-scale fractures. $S_{\mathrm{w}}^{\mathrm{F}n}$ is the saturation of big-scale fractures; $q_{\mathrm{w}}^{\mathrm{F}n}$ is the effect that source sink term has on the big-scale fractures; $Q_{\mathrm{w}}^{\mathrm{F}n}$ is the flow rate between the big-scale fractures and their surrounding medium.

(3) Coupling conditions

The longitudinal coupling between the seepage models is realized according to certain coupling conditions. The equivalent media part can be seen as the matrix system of the dual media. We focus on the longitudinal coupling between the discrete fracture and the dual medium. $\Omega f \cup \Omega F1$ and $\Omega m \cup \Omega F2$ can be seen as two parallel systems. Like conventional dual medium, the two systems are coupled together by fracture-matrix flow. In each system, the big fractures are coupled with the surrounding medium according to the relationship of flow rate and pressure. That is, on $\Gamma V1$, the surface pressure of the large-scale fracture is equal to the pressure that the fracture system unit has on this side, and the amount of inflow/outflow of the large-scale fracture is same with the amount of outflow/inflow of the fracture system (Fig. 5.5)

$$\begin{cases} q_{e,l}^{\mathrm{f}} + q_{e',l}^{\mathrm{f}} = Q_l^{\mathrm{F}1} & \text{on } \Gamma_{\mathrm{V}1} \\ p_{l,e}^{\mathrm{f}} = p_{l,e'}^{\mathrm{f}} = p_l^{\mathrm{F}1} & \text{on } \Gamma_{\mathrm{V}1} \end{cases} \tag{5.19}$$

Fig. 5.5 Schematic diagram of discrete fracture model

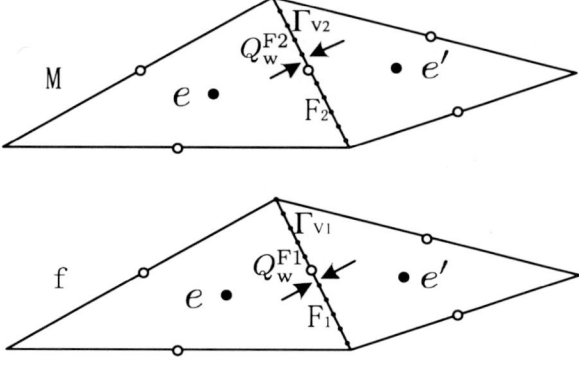

where, $q^f_{e,l}$ and $q^f_{e',l}$ represent respectively the flow rate of the inflow from $-e$ and e' (the micro-fracture system units on both sides of the big fracture l) into l; Q^{F1}_l is the flow rate of the inflow from the micro-fracture system units on both sides of the big fracture 1 into l; $p^f_{e,l}$ and $p^f_{e',l}$ represent respectively the pressure that the micro-fracture units on both sides of the big fracture have on the side l; $p^{F1}_{e,l}$ is the pressure on the large-scale fracture l.

Similarly, on Γ_{V2}, the surface pressure of the big-scale fracture equals to the pressure the matrix system unit has on this side. And the amount of inflow/outflow of the large-scale fracture is same with the amount of outflow/inflow of the matrix system

$$\begin{cases} q^m_{e,l} + q^m_{e',l} = Q^{F2}_l & \text{on } \Gamma_{V2} \\ p^m_{l,e} = p^m_{l,e'} = p^{F2}_l & \text{on } \Gamma_{V2} \end{cases} \tag{5.20}$$

5.3.4 Numerical Simulation for Coupling Model

(1) The comparison between coupling model and the dual medium model that contain large-scale fractures

In order to test the accuracy of coupling model, dual medium model is chosen to be compared with the coupling model of dual medium and discrete fracture. It is assumed that the size of the complex medium model that contains one big fracture and many fractures of middle or small scale is 70 cm × 10 cm, the physical parameters of the model is shown in Table 5.2. And point A, which is 10 cm from the right side, is chosen as the reference point. When the big fracture has different lengths (shown in Fig. 5.6), the calculation results of point A are shown in Fig. 5.7. The maximum size of the triangular grids in the model is 2.0 cm. And the grids near the big fracture are in-filled and the maximum of the grids is 0.2 cm. The model contains 807 nodes and 1532 finite elements (shown in Fig. 5.6).

Table 5.2 The relative physical parameters of the model

Matrix system properties	$\phi^m = 0.1$, $K^m = 1 \times 10^3\ \mu m^2$
Fracture system properties	$\phi^f = 0.01$, $K^f = 6 \times 10^3\ \mu m^2$
Large-scale fracture properties	$K^F = 1 \times 10^6\ \mu m^2$, $\varepsilon = 0.001$ m
Fluid properties	$\mu_w = \mu_o = 1$ mPa S, $\rho_w = \rho_o = 1000$ kg/m^3
Type of the relative permeability curve	Linear
Capillary pressure	Neglected
Irreducible water saturation and residual oil saturation	$S_{rw} = 0.0$, $S_{ro} = 0.0$

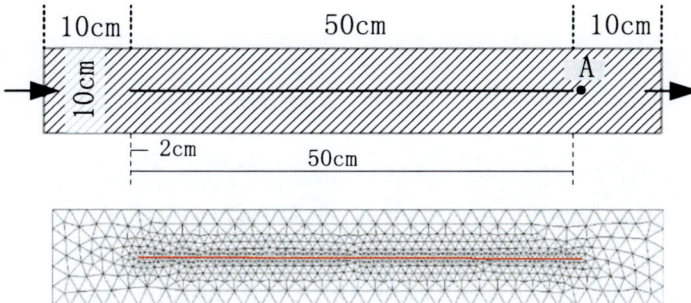

Fig. 5.6 Mesh division of the model

As we can see from the Fig. 5.7, when the length of the large scale is 2 cm, there is only slight difference between the water saturation of point A from the two models. However, if the length becomes 50 cm, there would be an obvious difference between the water saturation of point A from the two models. For the coupling model, the water breakthrough time is early, and the fluid seeps through the fractures quickly into point A. And for dual medium model, this process lags

Fig. 5.7 0–1 The water saturation curve of point A. **a** The length of the bigfracture is 2 cm. **b** The length of the big fracture is 50 cm

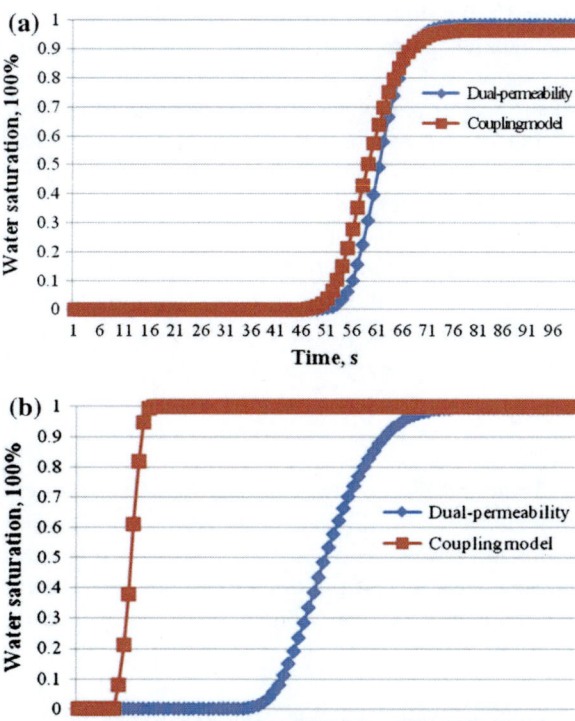

(a)

(b)

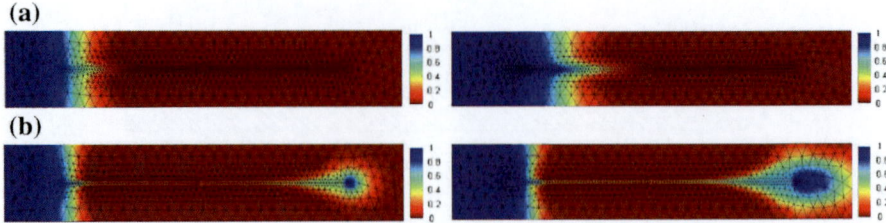

Fig. 5.8 The comparison of the water saturation distribution map at different time (the length of the fracture is 50 cm). **a** The water saturation distribution in the dual medium model at the time of 13 and 20 s. **b** The water saturation distribution in the coupling model at the time of 13 and 20 s

obviously. The results of the numerical simulation are shown in Fig. 5.8. Thus, coupling model can realize the detailed characterization of the seepage characteristics of large-scale fractures.

(2)   The pressure of single phase flow in fractured reservoir

Design a 1×1 two-dimension unit area fractured medium model, in which many micro fractures and large-scale fractures developed. In the model, for the large-scale fracture, the permeability $K^f = 100{,}000 \ \mu m^2$, $\phi^m = 0.1$, $\phi^f = 0.01$, $\phi^F = 1$, the aperture $a = 0.001$ m, $\mu_w = 1$ mPa s, $\mu_o = 1$ mPa s, the shape factor $\alpha_m = 20$, $\alpha_F = 10$, the fluid density $\rho_w = \rho_o = 1000 \ kg/m^3$. Linear relationship is adopted for the relative permeability curve, capillary pressure is neglected, and residual oil saturation and irreducible water saturation are assumed to be zero. The bottom left of the model is a water injection well and the up right is a production well.

Let us analyze the pressure change characteristic in the model of single phase flow seepage. In the simulation for single phase fluids seepage, the compression properties of rock are taken into consideration. The compressibility Cm = 0.02, Cf = 0.1, CF = 0.2. Produce under constant pressure and the pressure of the injection well is 1×106 Pa, and the pressure of production well is 1×105 Pa. The absolute permeability tensor Km (unit: μm^2) of the matrix system and the absolute permeability tensor K^f (unit: μm^2) of the fracture system are respectively

$$\boldsymbol{K}^m = \begin{pmatrix} 0.01 & 0.0 \\ 0.0 & 0.01 \end{pmatrix} \quad \boldsymbol{K}^f = \begin{pmatrix} 1.0 & 0.0 \\ 0.0 & 1.0 \end{pmatrix} \qquad (5.21)$$

From the Fig. 5.9 we can find that the pressure in the fractured system changes dramatically, the differential pressure between the water injection well and the production well is about 1×106 Pa; while the pressure in the matrix system changes gently, and the differential pressure between the water injection well and the production well is about 1×105 Pa. In the fracture system, the pressure field near the big fracture changes obviously, and the pressure drops along the big fracture. The fracture-matrix flow rate between the systems changes obviously. Around the water injection well, fluids flow into the matrix system from the fracture system; while around the production well, fluids flow into the fracture system from

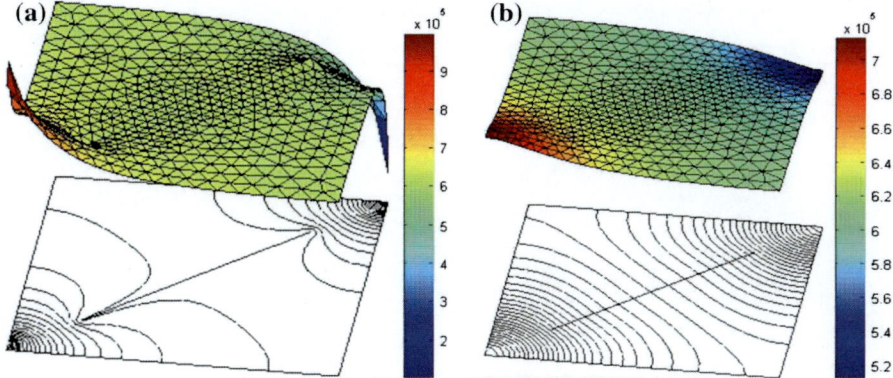

Fig. 5.9 The pressure distribution in the coupling models of single phase flow for fractured reservoir **a** pressure distribution in the fracture system **b** pressure distribution in the matrix system

the matrix system. That shows, the characteristics of the pressure distribution of the numerical simulation and the change of the fracture-matrix flow rate are in accordance with the theory. The porosity of the fracture system is low, the permeability is high, the pressure changes obviously. And a high pressure zone around the water injection is well formed in a short time. The porosity of the matrix system is high, permeability is low, and the pressure changes slowly. Owing to the difference between the properties of the two systems, the pressures change in a different way, which makes the channeling directions of the systems be different.

(3)   The two-phase fluids in fractured reservoir

For dual medium seepage model, it is required that the permeability of the fracture system is far larger than the permeability of the matrix system and the porosity of the fracture system is far less than the porosity of the matrix system. In the numerical simulation of water flooding oil, the speed of the process of water flooding oil in fracture system is far larger than the speed of matrix system. In order to analyze the displacement result, the difference of the two speed is reduced so as to analyze the characteristics of the oil-water two-phase seepage in the coupling model. The absolute permeability K^f (unit: 103 μm^2) of the fracture system and the absolute permeability Km (unit: 103 μm^2) of the fracture system are respectively

$$K^f = \begin{pmatrix} 6.0 & 0.0 \\ 0.0 & 8.0 \end{pmatrix} \quad K^m = \begin{pmatrix} 1.0 & 0.0 \\ 0.0 & 1.0 \end{pmatrix} \tag{5.22}$$

The rate of oil production and water injection are both $q = 10$ m^3/d. And the physical properties of the model are shown in Table 5.3.

As we can see from the Figs. 5.10 and 5.11, the isotropic matrix system exhibits an anisotropic pressure distribution characteristic, which indicates that the seepage characteristic influences the pressure change of the matrix system directly. On the

Table 5.3 The relative physical properties of the model

Matrix system	$\phi^m = 0.2$
Porosity of the fracture system	$\phi^f = 0.02$
Large-scale fracture properties	$K^F = 1 \times 106\ \mu m^2$, a = 0.001 m
Fluid properties	$\mu_w = 1$ mPa S, $\rho_w = 1000$ kg/m^3; $\mu_o = 2$ mPa S, $\rho_o = 800$ kg/m^3
Type of the relative permeability curve of the matrix system	Linear
Type of the relative permeability curve of the fracture system	Quadric form
Capillary pressure	Neglected
Irreducible water saturation and residual oil saturation	$S_{rw} = 0.0$, $S_{ro} = 0.0$

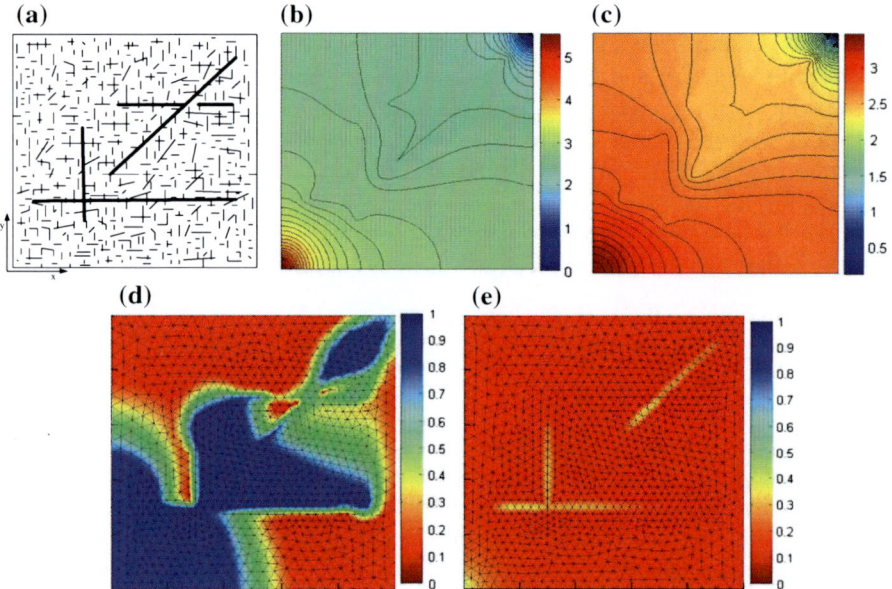

Fig. 5.10 The distribution of pressure and water saturation distribution when PV = 0.1. **a** Complex fractured medium model. **b** Pressure of the fracture system. **c** Pressure of the bed rock system. **d** Saturation of the fracture system. **e** Saturation of the bed rock system

water saturation distribution map, the large-scale fracture can directly influence the change of the water saturation of the fracture system around the fracture and the matrix system. The fluids around the fracture are obviously being displaced.

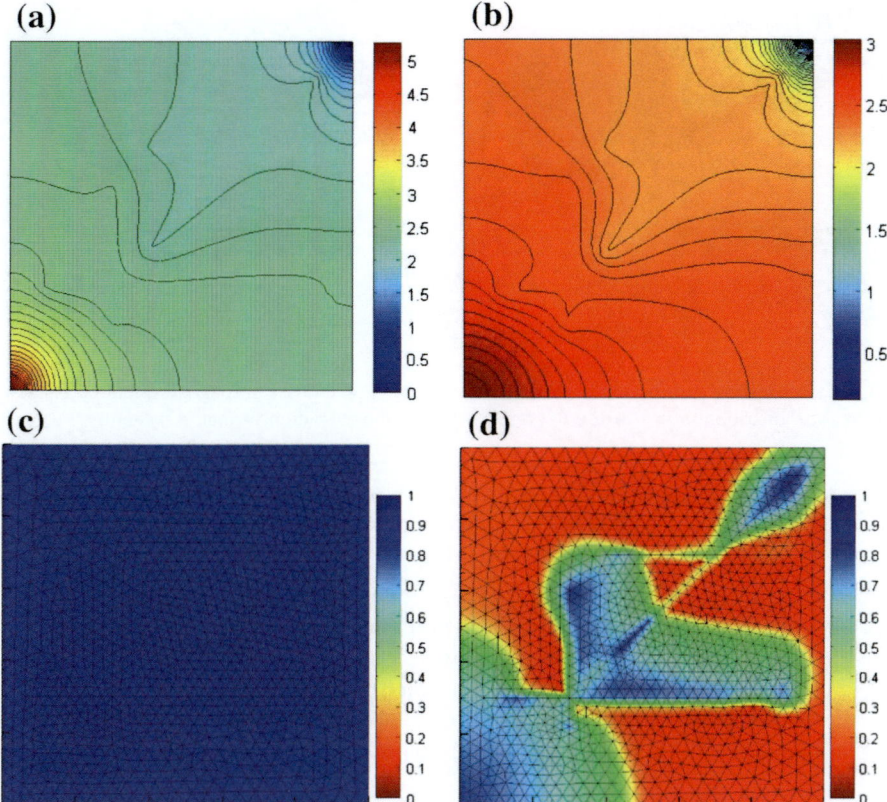

Fig. 5.11 The distribution of pressure and water saturation distribution when PV = 2.3. **a** Pressure of the fracture system. **b** Pressure of the bed rock system. **c** Saturation of the fracture system. **d** Saturation of the bed rock system

5.4 The Types of Hybrid Model and Its Realization

5.4.1 The Types of Hybrid Model

The mathematical equations of equivalent medium model and discrete fracture model are applicable to formulas from (5.3) to (5.10). But the fractures in discrete fracture model are upgraded: the fracture in a 2-D space is treated as a line with a certain aperture; the fracture in a 3-D space is treated as a plane with certain thickness. The seepage equations of dual medium model are same with the formulas from (5.11) to (5.15).

Realize the coupling between the seepage models according to the conditions for coupling. The equivalent medium portion can be seen as the matrix system of the dual medium or the matrix portion of the discrete fracture model.

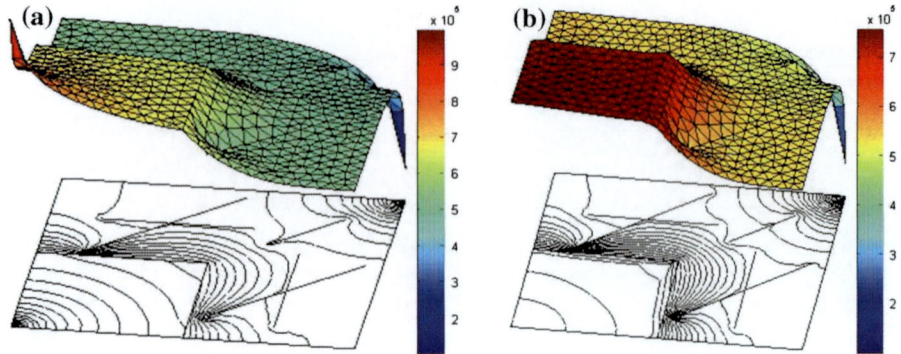

Fig. 5.12 pressure distribution in the hybrid model of single phase seepage in fractured reservoir.
a Pressure distribution in fractured system. **b** Pressure distribution in matrix system

(1) When equivalent medium and discrete fracture are going to be coupled, the matrix boundary pressure is required to be equal to the pressure within the fracture, and the flow rate of the outflow of the matrix units near the fracture boundary is required to be equal to the inflow of the fracture at this area.
(2) When discrete fracture and dual medium are going to be coupled, the discrete fracture-matrix portion is coupled with the dual medium matrix system. Make them have the same pressure and comply with conservation of mass. And discrete fracture is coupled with dual medium fracture system, make them have the same pressure and comply with conservation of mass.

5.4.2   Result Analysis for the Numerical Simulation

The pressure distribution in the hybrid model of single phase seepage in fractured reservoir is shown in Fig. 5.12.

We can see from the Fig. 5.12, the pressure variation of the dual medium part is different from the discrete fracture portion. The pressure of the dual medium fracture system dropped obviously, while the pressure of the matrix system changes gently. In the discrete fracture portion, the discrete fracture exhibits isobaric feature. The pressure around the discrete fracture drops obviously.

5.5   Summary

Because of the strong heterogeneity and the multi-scale property of fracture medium, single flow model cannot realize the detailed characterization. But we can adopt hybrid model. That means choosing applicable models according to the fracture development characteristic and using several models at the same time. Thus, the advantages of different models can be made good use of. The numerical simulation of hybrid model can broaden the scope of application of single model and realize the detailed characterization of the heterogeneity and multi-scale characteristic of the fracture development in complicate fractured medium.

References

Barenblatt GI, Zheltov IP, Kochina IN (1960) Basic concepts in the theory of seepage of homogeneous liquids in fissured rocks [strata]. J Appl Math Mech 24:1286–1303

Bour O, Davy P (1997) Connectivity of random fault networks following a power law fault length distribution. Water Resour Res 33:1567–1583

Bour O, Davy P (1998) On the connectivity of three-dimensional fault networks. Water Resour Res 34:2611–2622

Dreuzy J, Davy P, Bour O (2001) Hydraulic properties of two-dimensional random fracture networks following a power law length distribution: 1. Effective connectivity. Water Resour Res 37:2065–2078

Gong B, Karimi-Fard M, Durlofsky L, Gong B, Karimi-Fard M, Durlofsky L (2006) An upscaling procedure for constructing generalized dual-porosity/dual-permeability models from discrete fracture characterizations. Spe Annu Tech Conf, Exhib

Karimi-Fard M, Gong B, Durlofsky LJ (2006) Generation of coarse-scale continuum flow models from detailed fracture characterizations. Water Resour Res 42:382–385

Kazemi H, Merrill LS Jr, Porterfield KL, Zeman PR (1976) Numerical simulation of water-oil flow in naturally fractured reservoirs. Soc Pet Eng J 16:317–326

Lee SH, Lough MF, Jensen CL (2001) Hierarchical modeling of flow in naturally fractured formations with multiple length scales. Water Resour Res 37:443–455

Pruess K (2013) A practical method for modeling fluid and heat flow in fractured porous media. Soc Pet Eng J. 25:14–26

Sarda S, Jeannin L, Bourbiaux B (2002) Hydraulic characterization of fractured reservoirs: simulation on discrete fracture models. Spe Reserv Eval Eng 5:154–162

Wang Y, Yao J, Huang Z (2011) Review on fluid flow models through fractured rock. J Daqing Pet Inst 35:42–48

Warren JE, Root PJ (1963) The behavior of naturally fractured reservoirs. Soc Pet Eng J 3:245–255

Wu Y-S, Pruess K (1988) A multiple-porosity method for simulation of naturally fractured petroleum reservoirs. SPE Reserv Eng 3:327–336

Chapter 6
Multiscale Numerical Simulation

Abstract The difficulty in analyzing multi-phase fluid flow in real reservoirs is mainly caused by the strong heterogeneity of the reservoirs. The multiple scales in reservoirs may span several orders of magnitude. It takes a long time to calculate multi-scale problem by utilizing conventional numerical method. Multi-scale method incorporates the small-scale information into the base functions; therefore, multiple scale method has exclusive advantages when it is applied to reservoir numerical simulation. The multi-scale methods only need to carry out the coarse mesh on the macro scale. The multi-scale basis function, constructed by solving the partial differential equations on the coarse mesh, could capture the small scale information. It aims at reducing the computational amount and capturing the small scale characteristics. Besides, the efficiency can be further improved by applying parallel computation. In this chapter, we present some applications of multi-scale methods to fluid flows in carbonate reservoirs. We discuss multi-scale methods for transport equations and their coupling to flow equations which are solved using MsFEMs.

Keywords Multi-scale finite element methods · Numerical simulation · Fractured vuggy carbonate reservoir · Discrete fracture model · Discrete fracture-vug network model

6.1 Background and the State of Art of Multi-Scale Methods

A broad range of scientific and engineering problems involve multiple scales. Traditional approaches have been known to be valid for limited spatial and temporal scales. Multiple scales dominate simulation efforts wherever large disparities in spatial and temporal scales are encountered. Such disparities appear in virtually all areas of modern science and engineering, for example, composite materials, porous media, turbulent transport in high Reynold's number flows, and so on. A complete analysis of these problems is extremely difficult. For example, the

© Petroleum Industry Press and Springer-Verlag Berlin Heidelberg 2017 209
J. Yao and Z.-Q. Huang, *Fractured Vuggy Carbonate Reservoir Simulation*,
Springer Geophysics, DOI 10.1007/978-3-662-55032-8_6

difficulty in analyzing groundwater transport is mainly caused by the heterogeneity of subsurface formations spanning over many scales. This heterogeneity is often represented by the multi-scale fluctuations in the permeability (hydraulic conductivity) of the media.

When the traditional numerical methods are applied to solve the multiple scale problems, we need to partition the area firstly, and solve the problem on the fine grid. The amount of calculation is very huge when the grid partition is fine. The direct numerical solution of multiple scale problems is difficult even with the advent of supercomputers. The major difficulty of direct solutions is the size of the computation. A tremendous amount of computer memory and CPU time are required, and this can easily exceed the limit of today's computing resources. The situation can be relieved to some degree by parallel computing; however, the size of the discrete problem is not reduced. Whenever one can afford to resolve all the small-scale features of a physical problem, direct solutions provide quantitative information of the physical processes at all scales. On the other hand, from an application perspective, it is often sufficient to predict the macroscopic properties of the multi-scale systems. Therefore, it is desirable to develop a method that captures the small-scale effect on the large scales, but does not require resolving all the small-scale features.

When dealing with multi-scale processes, it is often the case that input information about processes or material properties is not available everywhere. For example, if one would like to study the fluid flows in a subsurface, then the subsurface properties at the pore-scale are not available everywhere in the reservoir. In this case, one can use Representative Volume Element (RVE) which contains essential information about the heterogeneities. Assuming that such information is available over the entire domain in macroscopic regions (see Fig. 6.1 for illustration), one can perform up-scaling (or averaging) and simulate a process over the entire region. Multi-scale methods can easily handle such cases. For example, when we study the fluid flow in the subsurface, the permeability field is a spatial field varying over multiple scales. It is possible that the full description of permeability at

Fig. 6.1 Schematic description of Representative Volume Element and macroscopic elements

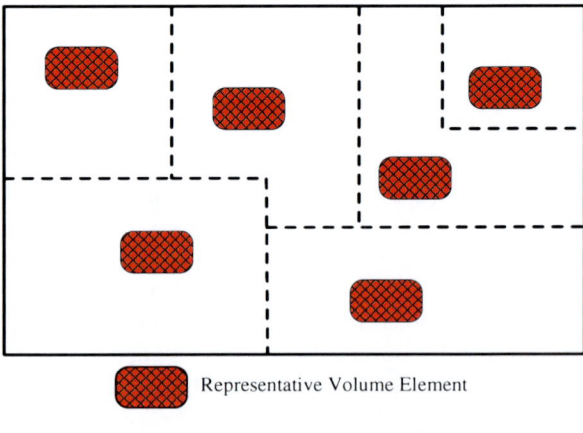

◼ Representative Volume Element

- - - - Macroscopic region boundaries

the finest resolution is not available, and we can only access it in small portions of the domain (RVE). One can attempt to simulate the macroscopic behavior of the material or subsurface processes based on RVE information. However, we must assume that the material has some type of scale separation.

In Fig. 6.2, geological variation over multiple scales is shown (Efendiev and Hou 2009). Here, one can observe faults (in Fig. 6.2a) with complicated geometry, thin but laterally extensive compaction bands that represent low-conductivity regions (in Fig. 6.2b) as well as other features at different scales. A blowup of the fault zone is shown in Fig. 6.2b. The fault rock is of low conductivity and the slip

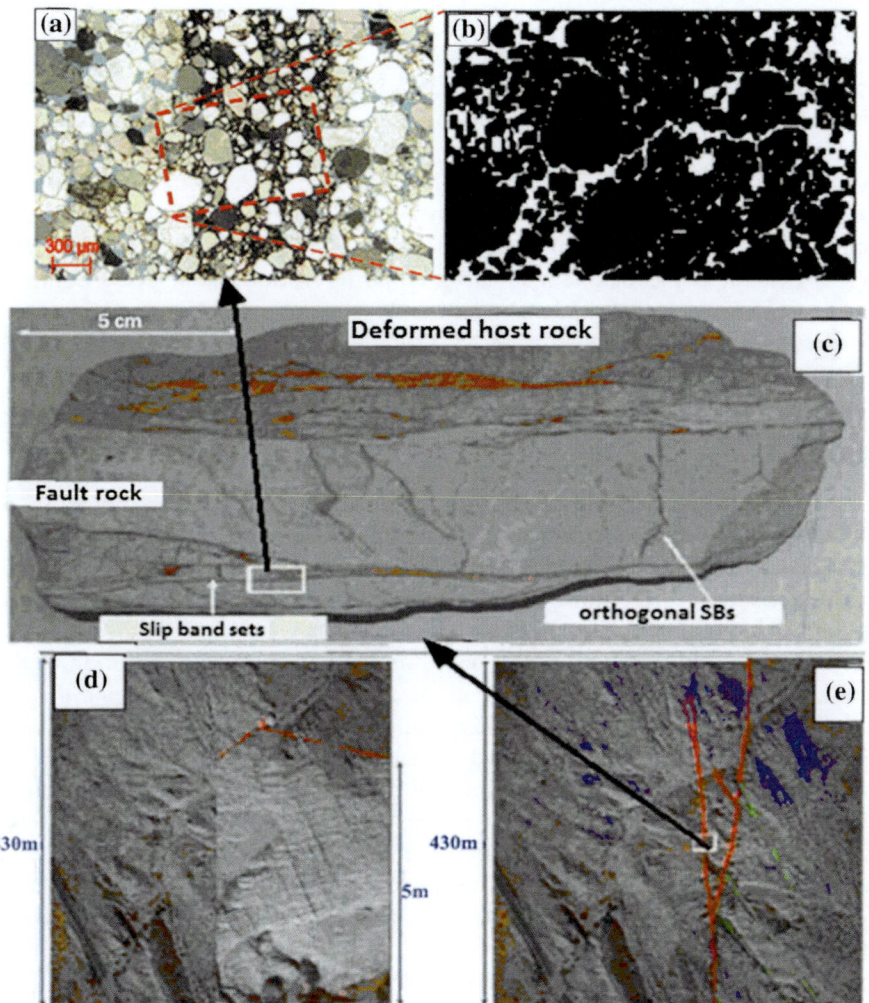

Fig. 6.2 Schematic description of hierarchy of heterogeneities in subsurface formations (modified from Efendiev and Hou 2009)

band sets consist of fractures that are filled (fully or partially) with cement. Pore-scale views of portions of a slip band set are shown in Fig. 6.2c, d. When simulating based only on RVE information as discussed before, the large-scale nonlocal information is disregarded, and this can lead to large errors. Thus, it is crucial to incorporate the multi-scale structure of the solution at all scales that are important for simulations.

With the development of the multiple scale simulation, the multiple scale problems have caused the widespread attention and promote the development of multiple scale computing method. There are many disadvantages when the traditional Finite Element Method is applied to deal with the multiple scale problems. However, the Multi-scale Finite Element Method can solve the problems. Especially the development of computer and the advanced theory has supported the development of the multi-scale simulation.

With the process of technology, the spatial scale has been developed to micro-level in the scientific research and engineering application. Meanwhile the multi-scale modeling and multi-scale calculation are also the hot topics of the world. The direct numerical solution of multiple scale problems is difficult even with the advent of supercomputers. Furthermore, it is not necessary to obtain all the microcosmic information when we analyze multiple scale problems. Therefore, researchers at home and abroad have proposed various multi-scale computing methods, divided into traditional multi-scale computing method and multi-scale computing method developed in recent years.

6.1.1 Traditional Finite Element Method

When traditional finite element method is applied to perform numerical simulation, we need to partition the area in order to form a series of finite elements, and suppose that the media in these finite elements is homogeneous. The polynomial interpolation is adopted to represent the characteristics of physical field in these elements. The finite element equation is built based on the Ritz method or Galerkin method. The characteristics of physical field in the area are obtained by solving the finite element equation in the end. The interpolation functions are linear polynomial and this is an approximation of the real physical field in these elements. When the porous media is heterogenic, the error will be enlarged. Therefore, we assume that the porous media is homogeneous when traditional finite element method is adopted. It is necessary to divide the area into finer finite elements so as to ensure that the parameters in every element are constant if the porous media is heterogeneous. Theoretically, the numerical solution of finite element method will be closer to the theoretical solution if the interpolation function is closer to the physical field function, the grids are finer and the points are more intensive. In general, interpolation function is linear interpolation. However, the physical field is nonlinear when the porous media is heterogeneous, and this will lead to large error. A tremendous amount of computer memory and CUP time are required when the finer grids are adopted. As a result, there are some disadvantages when we apply FEM to solve multiple scale problems related to fractured reservoir, one of the main reasons is the basis functions could not describe the physical field accurately. For MsFEM method, the basis functions satisfy the governing equation, therefore the effect of small scales on the large scales is correctly captured.

MsFEM is proposed for solving multiple scale problems, for example, the heterogeneous porous media involves multiple scales, this heterogeneity is often represented by the multi-scale fluctuations in the permeability (hydraulic conductivity) of the media. The difficulty in analyzing groundwater transport is mainly caused by the heterogeneity of subsurface formations spanning over many scales. MsFEM solves the multiple scale problems on coarse grid, and incorporates the small-scale information into the base functions. For the problems of simulation on large region, MsFEM applies great to the porous media with strong heterogeneity. The parameters in every element could change when MsFEM is adopted, the variation of pressure distribution will be captured by the basis functions. Therefore, we can depict the change of parameters and the distribution of flow field on the coarse grid with very few calculations. The key of the MsFEM is to build the basis functions that can capture the fine-scale features of physical field; this is the essential difference with traditional finite element method.

6.1.2   Multi-Scale Methods

Traditional multi-scale methods include multi-grid method, domain decomposition method and adaptive finite element method. These three methods will be briefly described as follows.

Multi-grid Method is proposed by Brandt (1977), which is widely used in fluid mechanics and other application fields. The basic idea is to form coarse and fine meshes, and then to establish the corresponding differential or finite element discretization equation. Then, the equation is solved on the coarse and fine meshes iteratively. The use of coarse mesh is to eliminate the low-frequency smooth part, and the use of fine mesh is to eliminate the high frequency error section. However, this method is still difficult to solve the multi-scale problem.

Adaptive Finite Element (AFE) method is proposed by Babuska and Rheinboldt (1978). AFE method is divided into h, p, and h-p types (Guo and Babuška 1986). When keep the degree of basis function of h type AFE method constant, decreasing element size h to obtain the desired accuracy. P type method is to keep h constant and increase p to improve the accuracy of approximation. h-p type is the combination of h type and p type.

Domain Decomposition Method (DDM) can be divided into two types, i.e., overlapping and non-overlapping. Overlapping DDM is derived from Schwarz alternating method in (Schwarz 1890). Non-overlapping DDM is more intuitive than DDM, which is suitable for the complex problem in different regions.

For the above several multi-scale methods, these methods are computed on the small scale. Thus, there are still some difficulties when we solve practical engineering problems. Therefore, it is necessary to find a more effective method to solve the multi-scale problem. In recent years, many scholars put forward many multi-scale calculation method, in addition to the multi-scale finite element method, including homogenization method, wavelet numerical uniform method, variational multi-scale method, non uniform multi scales method, multi-scale finite volume method, adaptive multi-scale finite element method, scale lifting method and the method of Equation-Free (Kevrekidis et al. 2004) and some other methods.

Homogenization Method (Gr et al. 1992; Cui and Cao 1999) is analyzed to study the small scales on periodic unit, and then the information of small scale is mapped to large scale. Thus, we can derive the homogenization equation on large scale. In engineering and other applications, this method has been successful used. But it is based on the assumption of the periodic structure of the micro structure; its application range is limited.

Wavelet-based Numerical Homogenization Method is proposed to solve the elliptic problems by Dorobantu and Engquist (1996). The method is based on multi-resolution analysis, and the discrete operator of the original equation is established on a small scale. Then the discrete operator is imposed on the wavelet transform. The method greatly reduces the computation time, but the process of wavelet transform is very complex.

The method is based on the multi-scale model analysis and the posteriori error estimates. The scalar field is decomposed into large-scale and small-scale solutions. The decomposition of the solution is based on the assumption that the small scale information of the solution cannot be captured on a given grid. Generally, the scale solution can be determined by analyzing or numerical method. More accurate numerical approximation can be obtained by the modified variational form.

Heterogeneous multi-scale method (HMM) is a general framework of the multi-scale calculation method, proposed by Weinan (2003), Weinan and Yue (2004), Ren and Weinan (2005). The method firstly constructs the coarse grid format with the unknown coefficients, and then solves local small scale unit to estimate macroscopic coefficients. Finally in the whole region, the macroscopic equation is solved. This method is composed of two important parts: coarse grid format and macroscopic coefficients. High-order finite element can be used as macro algorithm and it can be applied to high-order cases, as shown in Fig. 6.3 (Weinan et al. 2005).

Multi-scale Finite Volume Method (MsFVM) is proposed by Jenny et al. (2003, 2005). The method is to divide the region on the large scale and then determine the control volume on these elements; the differential equations are integrated to obtain cell balance equation. MsFVM is suitable to solve the problems with complex boundaries. The basic feature of this method is able to maintain physical local conservation.

Adaptive Multi-scale Finite Element Method (Adaptive MsFEM) is first proposed by He and Ren (2009a, b). In this method, the multi-scale basis functions

Fig. 6.3 Schematic
description of HMM

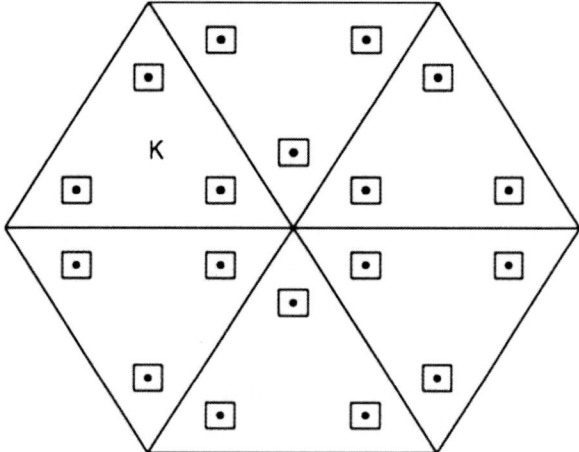

with adaptive characteristics in time domain are coupled to deal with the local problem defined in small scale. The idea of this method is to use modified iterative format as the framework of multi-scale method, and build adaptive multi-scale base functions which could capture the small scale information in the coefficients of the equation.

The basic idea of up-scaling method is to build large-scale equations with a known analytical form, which may be different from the basic small scale equations, and then solve these up-scaled equations in the large-scale grid (Desbarats 1998; Neuweiler and Cirpka 2005). The method has been successfully applied in some ways, but it depends on some specific assumptions of the porous media and this makes its application limited. The up-scaling methods proposed by Durlofsky (1991), Mccarthy (1995) is more general. Durlofsky used up-scaling method to calculate the permeability tensor of the porous media, which is shown in Fig. 6.4.

Fig. 6.4 Space periodic
porous media

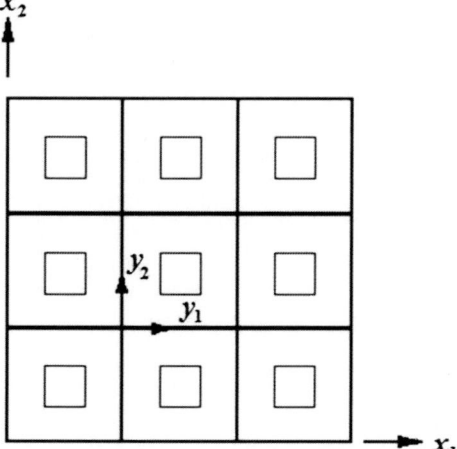

Durlofsky assumed that permeability tensor in small scale y is changed but in large scale x is constant, which simplify the determination of the multi-scale variation of permeability tensor to solving small scale y equivalent permeability tensor, but it is only for heterogeneity in two different scales or the equivalent permeability tensor. Up-scaling method is based on the homogenization theory, so it may be not suitable for the non periodic problem. When medium correlation scale is larger, the application is also very difficult.

Multi-scale finite element method is derived from the pioneering work of Babuška et al. In 1983, (Babuška and Osborn 1983) applied the multi-scale base functions to solve elliptic equation with a special scale parameter for the one-dimensional problem. In 1994, he (Babuška and Osborn 1994) analyzed the two-dimensional problems. In 1997, Hou and Wu (1997), Hou and Cai (1999) extended it to solve the general two-dimensional problem with oscillating coefficients, and then proposed multi-scale finite element method. They used multi-scale finite element method for solving elliptic problems.

Hou and Wu (1997) simulated the single phase stable flow in two-dimensional porous media. They proposed two types of boundary conditions for basis functions: one choice is linear boundary conditions, similar with the basis function of standard finite element method of which is only related to the node coordinates, along the boundary changing from 1 to 0. The other one is the oscillatory boundary condition. The basis function on the boundary satisfies simplified elliptic equation, and it can reflect the changes of boundary parameters caused by pressure changes. For example, the basis function ϕ_i of element E (Fig. 6.5) satisfies the following simplified elliptic equation:

Fig. 6.5 Schematic description of element E

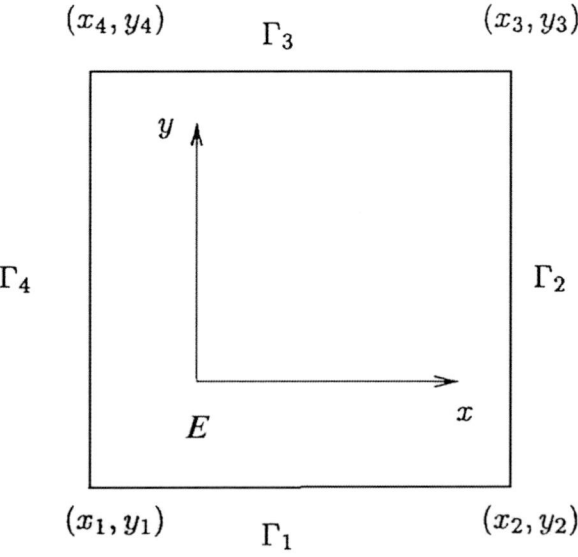

$$\nabla K(x) \nabla \phi_i = 0, \quad \mathbf{x} \in E \tag{6.1}$$

where $K(\mathbf{x})$ is permeability tensor, and $\phi_i(\mathbf{x}_j) = \delta_{ij}$. When $K(\mathbf{x})$ is separable in space, i.e., $K(\mathbf{x}) = K_1(x)K_2(y)$, ϕ_i can be computed analytically. For example, on Γ_1 we have

$$\phi_i(x)|_{\Gamma_1} = \frac{\int_x^{x_2} \dfrac{dt}{K(t)}}{\int_{x_1}^{x_2} \dfrac{dt}{K(t)}} \tag{6.2}$$

If K is a constant, then $\phi_i(x) = (x_2 - x)/(x_2 - x_1)$ is linear. If K is not a constant, ϕ_i is oscillatory due to the oscillations in K. Combining with (6.1), we can get the multi-scale basis functions.

Also, with both types of boundary condition, one has $\sum_{i=1}^{d} \phi_i = 1$.

Hou and Wu proposed an oversampling method to overcome the difficulty due to scale resonance. Let E be the original domain, S be a oversampled domain (see Fig. 6.6). First, the basis functions $\psi_i (i = 1, \ldots, d)$ of sampled domain S is obtained, we then form the actual basis $\phi_i (i = 1, \ldots, d)$ by linear combination of $\psi_i (i = 1, \ldots, d)$

$$\phi_i = \sum_{i=1}^{d} c_{ij} \psi_j, \quad i = 1, \ldots d \tag{6.3}$$

The coefficients c_{ij} are determined by condition $\phi_i(x_j) = \delta_{ij}$.

Hou and Wu also applied MsFEM to solve the nonlinear problem Efendiev et al. (2004) and Efendiev and Pankov (2004). Multi-scale basis function is the key of multi-scale finite element method. To eliminate the adverse effects of element scale,

Fig. 6.6 Schematic description of oversampled domain

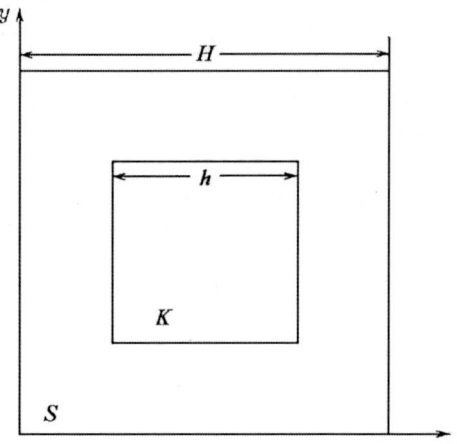

basis function can better reflect the small scale information. But Hou only studied the two-dimensional steady flow.

In 2003, Chen and Hou (2003) proposed mixed multi-scale finite element method for severely heterogeneous porous media, combining with oversampling technique to solve the local Neumann boundary value problem, to obtain more accurate the multi-scale basis functions.

In 2002, Chen and Yue (2002) studied steady flow problems in heterogeneous porous media based on MsFEM. They proposed a scale technology, which makes the small scale information integrated into macro information on large scale.

Many kinds of multi-scale methods are designed to build a general computational framework for multi-scale simulation. Equation-Free (Kevrekidis et al. 2002), HMM and other methods are used to solve the macro equations in RVE, and these methods are widely used. When solving partial differential equations, MsFEM is similar to these methods. For these problems, the basis function is approximated by RVE. For MsFEM, the local problem can be expressed by different global equations. The important step is to determine the form of the macro equations and the variables of the basis function. However, many general numerical methods are not described the way of determining the variables that affect the macro-scale basis functions in multi-scale simulation.

The multi-scale methods considered pre-compute the effective parameters that are repeatedly used for different sources and boundary conditions. In this regard, these methods can be classified as up-scaling methods where the up-scaled parameters are precomputed. In multi-scale approaches, one can reuse precomputed quantities to form coarse-scale equations for different source terms, boundary conditions and so on. Moreover, adaptive and parallel computations can be carried out with these methods where one can downscale the computed coarse-scale solution in the regions of interest. These features of up-scaling methods and MsFEMs are exploited in subsurface applications. The multi-scale methods differ from domain decomposition methods where the local problems are solved many times. Domain decomposition methods are powerful techniques for solving multi-physics problems; however, the cost of iterations can be high, in particular, for multi-scale problems. These iterations guarantee the convergence of domain decomposition methods under suitable assumptions. Multi-scale methods with up-scaling concepts in mind, on the other hand, attempt to find accurate sub-grid capturing resolution and avoid the iterations. This may not be always possible, and for that reason, some type of hybrid methods with accurate sub-grid modeling can be considered in the future.

In recent years, one of the research directions of multi-scale simulation is the use of some limited global information. Limited global information is often used in the up-scaling method. It uses some simplified models to extract important information about the nonlocal physical processes. One example is the two-phase flow in heterogeneous medium. Chen and Durlofsky (2006) studied the two-phase immiscible flow in heterogeneous media, using single phase flow information to upscale the governing equations of two-phase flow. In particular, the flow and transport of two-phase flow in the coarse grid are calculated by solving the global single phase flow equation with the calculation of the scale of the penetration coefficient. Similar with the basic idea of up-scaling using limited global data, multi-scale finite element method using limited global information is also proposed Aarnes (2004), Owhadi and Zhang (2007). Owhadi and Zhang (2007) provides a theoretical basis for the use of the limited global information. MsFEM using limited global information is to establish a multi-scale basis function using limited global information.

6.2   Multi-Scale Finite Element Method

The difference between multi-scale finite element method and traditional finite element method is the choice of basis functions. Basis functions of traditional finite element are usually linear that could not reflect the physical field. Multi-scale basis functions of MsFEM are obtained by solving local differential equations based on the governing equation. MsFEM is especially suitable for large-scale heterogeneous reservoir problem, which overcomes the above-mentioned shortcomings of FEM. Multi-scale basis functions of MsFEM can reflect the fine features of the unit parameters through computing on the coarse meshes, which could reduce the amount of calculation.

6.2.1   Multi-Scale Basis Functions

Multi-scale basis functions are important ingredients of MsFEM. Basis functions are designed to capture the multi-scale features of the results. Consider

$$-\nabla \cdot \left(\frac{\mathbf{K}(x, y)}{\mu} \nabla p \right) = f(x, y) \in \Omega \tag{6.4}$$

where $\mathbf{K}$ is a permeability tensor; p is pressure; f is source/sink term. Let $\mathcal{K}^h$ be a coarse partition of Ω. We consider an element $E \in \mathcal{K}^h$ that has d vertices, the multi-scale basis functions $\{\phi_E^i, i = 1 \ldots d\}$ are given by:

$$-\nabla \cdot \left(\frac{\mathbf{K}}{\mu} \nabla \phi_E^i \right) = 0 \tag{6.5}$$

where $\phi_E^i(\mathbf{x}_j) = \delta_{ij}$ and $1 \sum_{i=1}^d \phi_E^i = 1$. We would like to note that the solution of (6.5) is usually not analytic. We need to divide the coarse element into fine elements. Thus, numerical solution of (6.5) can be got through finite element methods. The small features of the local domain are incorporated into multi-scale basis functions (see Fig. 6.7).

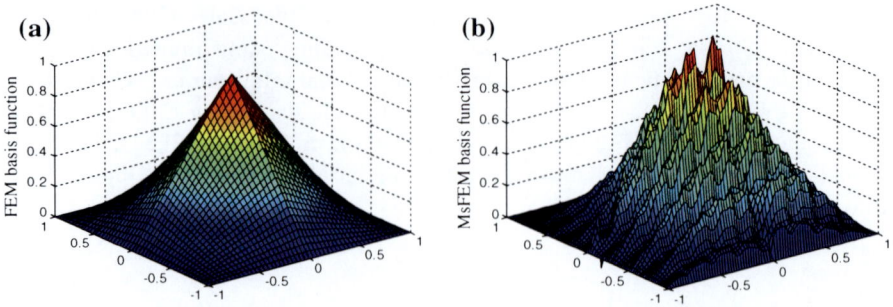

Fig. 6.7   2D basis functions of MsFEM and FEM

6.2.2   *Boundary Conditions of Multi-Scale Basis Functions*

The choice of boundary conditions in defining the multi-scale basis functions plays a crucial role in approximating the multi-scale solution. The boundary conditions are generally divided into linear boundary conditions and oscillatory boundary conditions.

Linear boundary conditions are similar with the variation of the basis function of finite element method that cannot reflect the changed of the boundary parameters. The linear conditions are only related to the coordinates of points on the boundary. The oscillatory boundary conditions can reflect the changes of parameters. When solving multi-scale basis functions on a coarse grid, if the division of fine mesh of coarse grid resolves heterogeneity of cell medium, the solutions of multi-scale finite element method have the same convergence with that of finite element method. If the division of fine mesh could not resolve heterogeneity, finite element method will not get accurate results, the results of multi-scale finite element method are much better than that of finite element method.

6.2.3   *Oversampling Technique*

In order to avoid the scale resonance, Hou proposed an oversampling method to overcome the difficulty due to scale resonance.

Let Δ_{ijk} be a coarse grid, Δ_{abc} be a sampled domain, as shown in Fig. 6.8. The sampled domain may slightly, or many times, be larger than the original domain, but too much times will cause extra computation of solving the basic function. It should be noted that the sampled domain must be similar to the shape of original domain.

We denote ϕ_K^i, ϕ_K^j, ϕ_K^k be the actual basis functions of points i, j, k; ψ_1, ψ_2, ψ_3 be the basis functions of points a, b, c. Take, for example, ψ_1 of point a in Δ_{abc}. ψ_1 is set to satisfy the following equations

Fig. 6.8 Schematic description of oversampled domain

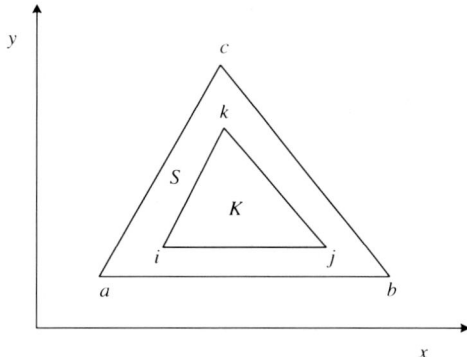

$$\begin{cases} L^{\varepsilon}\psi_1 = 0 & \text{in} \quad \mathbf{x} \in \Delta_{abc} \\ \psi_1 = \omega_1 & \text{on} \quad \partial\Delta_{abc} \end{cases} \tag{6.6}$$

where $\omega_1|_a = 1$, $\omega_1|_b = 0$, $\omega_1|_c = 0$. Similar to the computation of ψ_1, we can get ψ_2, ψ_3.

Because $K \subset S$, the actual basis functions ϕ_K^i, ϕ_K^j, ϕ_K^k can be formed by linear combination of $\psi_j (j = 1, 2, 3)$

$$\phi_K^m = \sum_{l=1}^{3} C_{m,l}\psi_l, \quad m = i, j, k$$

where $C_{m,l}$ is given by $\phi_K^m(\mathbf{x}_l) = \delta_{lm}$.

6.3   MsFEM for Fluid Flow in Heterogeneous Reservoir

Many problems of fundamental and practical importance in science and engineering have multi-scale solutions; for example, porous media, composite materials and so on. The direct numerical simulation of problems involving multi-scale solutions is difficult, due to the requisite of tremendous amount of computer memory and CPU time, which can easily exceed the limit of today's computer resources. On the other hand, in practice, it is often sufficient to predict the large-scale solutions to a certain accuracy. Thus, a number of multi-scale numerical methods have been presented, such as heterogeneous multi-scale methods, multi-scale finite element methods (MsFEM) and variational multi-scale method.

In this part we will apply MsMFEM to numerical computation of incompressible, immiscible two-phase flow in heterogeneous reservoir. The main idea of MsMFEM is to model fine-scale patterns in the velocity field by computing multi-scale base functions that reflect the impact of the fine-scale heterogeneous structures. It allows the pressure equation to be solved on a coarse grid and it produces velocity fields that are mass conservative on a sub-grid scale so that we can use upstream finite volume method to compute the saturation equation on a fine grid. So, large-scale effects are accounted for by the degrees of freedom on a coarse grid, and fine-scale effects are accounted for by subresolution in the basis functions.

6.3.1 Mathematical Model

Incompressible flow in strong heterogeneous reservoir obeys Darcy's law. Neglecting the effect of gravity, compressibility, capillary pressure, and the governing equations for two-phase immiscible flow can be described by the fractional flow model.

$$\phi \frac{\partial S_w}{\partial t} + \nabla \cdot (f_w \mathbf{v}) = q_w \tag{6.7}$$

$$\mathbf{v} = -(\lambda_w + \lambda_o)\mathbf{K}\nabla p, \quad \nabla \cdot \mathbf{v} = q \tag{6.8}$$

where ϕ is porosity; S_w is water saturation, and $S_o + S_w = 1$; $q = q_o + q_w$ is the total volumetric source term; $\mathbf{v} = \mathbf{v}_o + \mathbf{v}_w$ is the total velocity; $\mathbf{K}$ is a permeability tensor; p is the pressure; $f_w = \lambda_w / \lambda$ is the fractional flow of water; λ_o, λ_w are the oil and water mobility, that satisfy

$$\lambda_o = \frac{K_{ro}}{\mu_o}, \quad \lambda_w = \frac{K_{rw}}{\mu_w} \tag{6.9}$$

where K_{ro} and K_{rw} are the relative permeability of phase oil and water; μ_o, μ_w denote the viscosity of phase oil and water.

In the following, we will study (6.7) on the coarse and fine-scale and (6.8) is solved separately, using the sequential fully implicit method. For the model, we will assume no-flow boundary conditions and neglect body forces.

6.3.2 MsMFEM Method

First, we state some notations to be used in the paper. Let Ω be the reservoir domain and $\mathbf{n}$ be the outward-pointing unit normal on $\partial \Omega$. To solve the pressure equations on the coarse scale, we will use a mixed finite element formulation. Then, for (6.7) we find $(\tilde{p}, \tilde{\mathbf{v}}) \in L_0^2(\Omega) \times H_0^d(\Omega)$ such that,

$$\int_\Omega \mathbf{u} \cdot (\lambda \mathbf{K})^{-1} \tilde{\mathbf{v}} d\Omega - \int_\Omega \tilde{p} \nabla \cdot \mathbf{u} \, d\Omega = 0 \quad \forall \mathbf{u} \in L_0^2(\Omega) \tag{6.10}$$

$$\int_\Omega l \nabla \cdot \tilde{\mathbf{v}} d\Omega = \int_\Omega q l d\Omega, \quad \forall l \in H_0^d(\Omega) \tag{6.11}$$

Let $\mathcal{T}$ be a coarse-grid partition of Ω. U and V are finite dimensional subspace of $H_0^d(\Omega)$ and $L^2(\Omega)$. To derive a discretization of (6.10) and (6.11), we find $(p, \mathbf{v}) \in U \times V$ such that

$$\int_\Omega \mathbf{u} \cdot (\lambda \mathbf{K})^{-1} \mathbf{v} d\Omega - \int_\Omega p \nabla \cdot \mathbf{u} d\Omega = 0 \quad \forall \mathbf{u} \in U \tag{6.12}$$

$$\int_\Omega l \nabla \cdot \mathbf{v} d\Omega = \int_\Omega q l d\Omega, \quad \forall l \in V \tag{6.13}$$

Let $\{\psi_i\}$ and $\{\phi_k\}$ be the basis functions of U and V. The local equations can be assembled to form a hybrid system

$$\begin{bmatrix} \mathbf{B} & \mathbf{C} \\ \mathbf{C}^\mathsf{T} & \mathbf{0} \end{bmatrix} \begin{bmatrix} \mathbf{v} \\ -\mathbf{p} \end{bmatrix} = \begin{bmatrix} \mathbf{0} \\ \mathbf{q} \end{bmatrix} \tag{6.14}$$

where $v = \sum v_i \psi_i$, $p = \sum p_k \phi_k$, $\mathbf{B} = \{b_{ij}\}$, $b_{ij} = \int_\Omega \psi_i \cdot (\lambda K)^{-1} \psi_j dx$; $\mathbf{C} = \{c_{ik}\}$, $c_{ik} = \int_\Omega \phi_k \nabla \cdot \psi_i dx$; $\mathbf{q} = \{q_k\}$, $q_k = \int_\Omega \phi_k q dx$; $\mathbf{v} = \{v_i\}$; $\mathbf{p} = \{p_k\}$.

Let $T_h = \{\Omega_i\}$ be a coarse-grid partition of Ω by a collection of polyhedral elements and $\Omega_{ij} = \Omega_i \cup \Gamma_{ij} \cup \Omega_j$ contains two neighboring grid blocks Ω_i and Ω_j (see Fig. 6.9). For each $\Gamma_{ij} = \partial\Omega_i \cap \partial\Omega_j$ and Ω_i, we assign a multi-scale basis function $\psi_{ij} \in U_{Ms}$ and a basis function $\phi_i \in V$ separately.

(1) Basis function for velocity
In the MsMFEM, the basis function associated with the interface $\Gamma_{ij} = \partial\Omega_i \cap \partial\Omega_j$ is constructed by solving

Fig. 6.9 Multi-scale grid

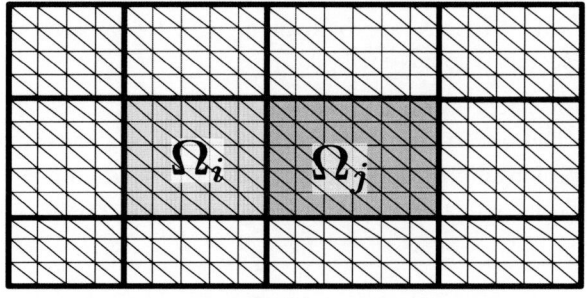

$$\psi_{ij}(x) = -\lambda K \nabla \phi_{ij} \qquad (6.15)$$

$$\nabla \cdot \psi_{ij} = \begin{cases} \omega_i(x), & x \in \Omega_i \\ -\omega_j(x), & x \in \Omega_j \end{cases} \qquad (6.16)$$

$$\psi_{ij}(x) \cdot n = 0 \quad \forall x \in \partial \Omega_{ij} \qquad (6.17)$$

$$\psi_{ij}(x) \cdot n_{ij} = v_{ij} \quad \forall x \in \Gamma_{ij} \qquad (6.18)$$

here, **n** is the outward unit normal on $\partial \Omega_{ij}$, n_{ij} is the unit normal to Γ_{ij} pointing from Ω_i to Ω_j, $\omega(x)_i$ is a source distribution function, which is to produce a flow with unit average from Ω_i to Ω_j. For all Ω_i such that $\int_{\Omega_i} q dx \neq 0$, we choose $\omega(x)_i$ to be

$$\omega(x)_i = \begin{cases} 1/|\Omega_i|, & \int\limits_{\Omega_i} q dx = 0 \\ q(x) \big/ \int_{\Omega_i} q(\xi) d\xi, & \int\limits_{\Omega_i} q dx \neq 0; \end{cases} \qquad (6.19)$$

To ensure a conservative approximation on the fine grid, for block-grids with no source, the simplest choice is $\omega(x)_i = 1/|\Omega_i|$. The local boundary condition v_{ij} should reflect heterogeneous structures. So we define v_{ij} according to

$$v_{ij}(x) = \frac{v_{ij}^0(x)}{\int_{\Gamma_{ij}} v_{ij}^0(s) ds}, \quad x \in \Gamma_{ij} \qquad (6.20)$$

where

$$v_{ij}^0 = n_{ij} \cdot (K\lambda) \cdot n_{ij}. \qquad (6.21)$$

(2) Basis Functions for Pressure
To approximate the pressure we will use functions that are constant on each coarse grid block with no source. Thus, for each Ω_i we assign a basis function $\phi_i \in V$, such that

$$\phi_i(x) = \begin{cases} 1, & x \in \Omega_i \\ 0, & x \notin \Omega_i \end{cases} \qquad (6.22)$$

This type of approximation space for pressure is also used in the lowest order Raviart–Thomas method.

(3) Coarse-scale Hybrid System
Then we arrange all the basis functions ψ_{ij} as columns in a matrix $\mathbf{\Psi}$. Let I is the prolongation from blocks to cells. If block number i contains cell number j, then $I_{ij} = 1$, otherwise $I_{ij} = 0$. The coarse grid system can then be obtained in the form

$$\begin{bmatrix} \mathbf{B}^c & \widetilde{\mathbf{C}} \\ \widetilde{\mathbf{C}}^{\mathrm{T}} & \mathbf{0} \end{bmatrix} \begin{bmatrix} \mathbf{v}^c \\ -\mathbf{p}^c \end{bmatrix} = \begin{bmatrix} \mathbf{0} \\ \mathbf{q}^c \end{bmatrix} \tag{6.23}$$

where $\mathbf{B}^c = \mathbf{\Psi}^{\mathrm{T}} \mathbf{B}^f \mathbf{\Psi}$; $\widetilde{\mathbf{C}} = \mathbf{\Psi}^{\mathrm{T}} \mathbf{C}^f \mathrm{I}$; $\mathbf{q}^c = \mathrm{I}^{\mathrm{T}} \mathbf{q}^f \mathrm{I}$.

6.3.3 Numerical Experiments

In this section, we apply the MsMFEM and the up-scaling method to model incompressible and immiscible two-phase flow. The system is considered to be one of the layers of the benchmark test, the SPE comparative project. The log-permeability of the layer number 85 is given in Fig. 6.10. The fine field is 220×60. Assuming that the reservoir is initially fully oil-saturated. The water and oil mobilities are defined by

$$\lambda_{\mathrm{w}} = \frac{S_{\mathrm{w}}^2}{\mu_{\mathrm{w}}} \quad \text{and} \quad \lambda_{\mathrm{o}} = \frac{(1 - S_{\mathrm{w}})^2}{\mu_{\mathrm{o}}}.$$

We consider a quarter-five spot problem. A source and a sink are introduced in bottom left corner and top right corner, respectively.

Figure 6.11 shows saturation profiles after an injection of water corresponding to 100 % of the total pore volume in the reservoir. The grids are up-scaled by a factor four and ten in each coordinate direction, respectively, so that the coarse grid

Fig. 6.10 Permeability
(logarithm)

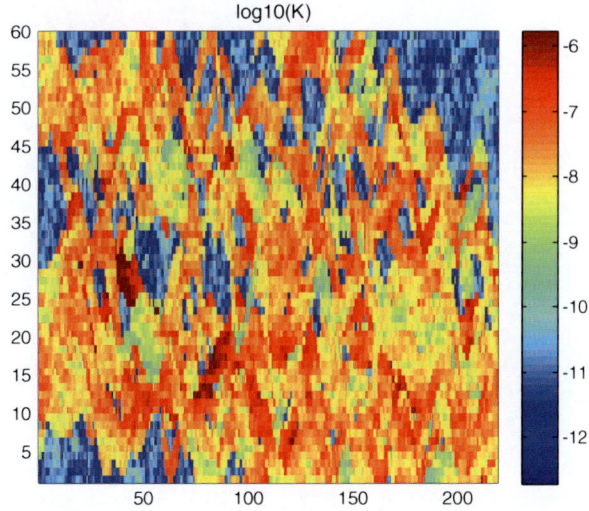

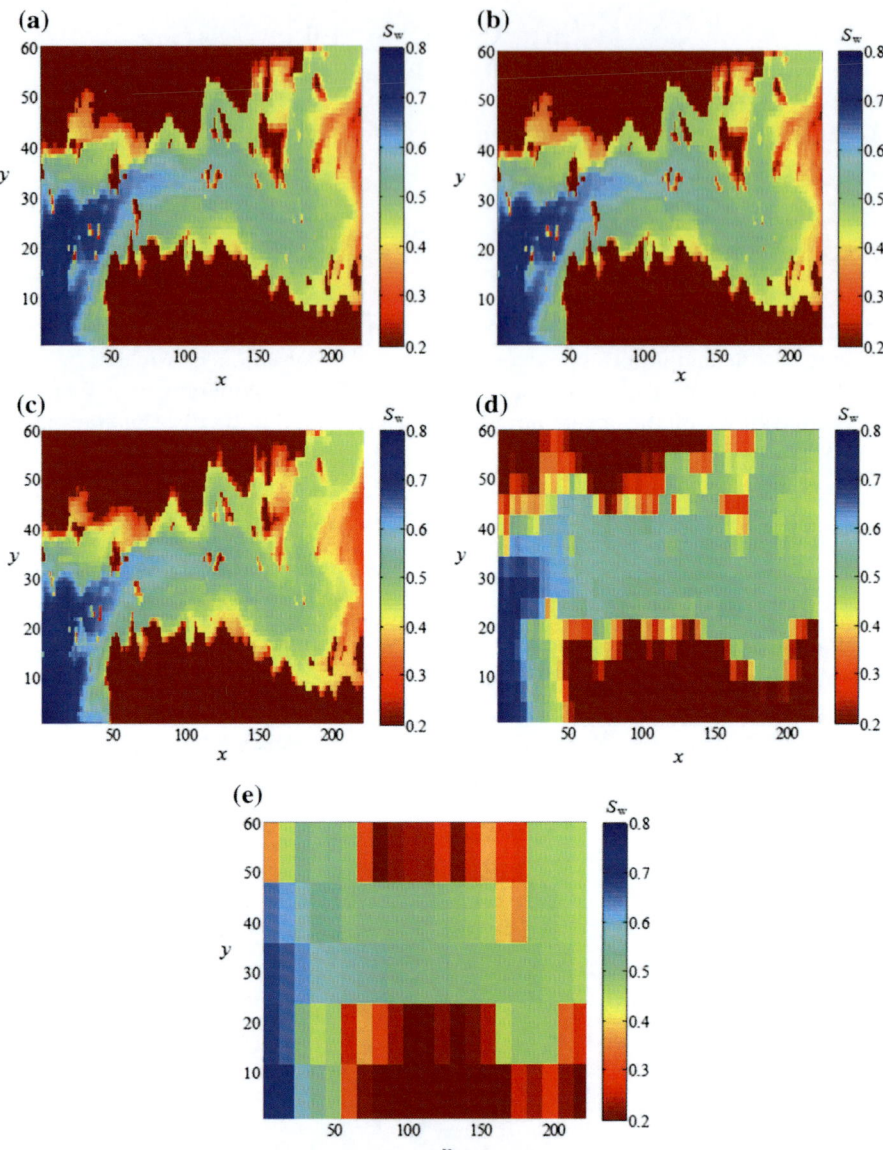

Fig. 6.11 Water saturation profiles at 1.0PVI. **a** Fine-scale solution (220 × 60), **b** MsMFEM (55 × 15), **c** MsMFEM (22 × 6), **d** upascaling method (55 × 15), **e** upascaling method (22 × 6)

consists of 55 × 15 blocks and 22 × 6 blocks. The figure indicates that the water distributions obtained using MsMFEM are in excellent agreement with the reference fine-scale solutions. The up-scaling method has lost all fine-scale information and only gives a crude approximation to the global flow pattern.

6.4   MsFEM for Fluid Flow in Fractured Reservoir

Numerical simulation in naturally fractured media is challenging because of the coexistence of porous media and fractures on multiple scales that need to be coupled. We present a new approach to reservoir simulation that gives accurate resolution of both large-scale and fine-scale flow patterns. Multi-scale methods are suitable for this type of modeling, because it enables capturing the large-scale behavior of the solution without solving all the small features. Double-porosity models in view of their strength and simplicity can be mainly used for sugar-cube representation of fractured media. In such a representation, the transfer function between the fracture and the matrix block can be readily calculated for water-wet media. For a mixed-wet system, the evaluation of the transfer function becomes complicated due to the effect of gravity.

In this part, we use a multi-scale finite element method (MsFEM) for two-phase flow in fractured media using the discrete fracture model. By combining MsFEM with the discrete fracture model, we aim toward a numerical scheme that facilitates fractured reservoir simulation without up-scaling. MsFEM uses a standard Darcy model to approximate the pressure and saturation on a coarse grid, whereas fine-scale effects are captured through basis functions constructed by solving local flow problems using the discrete fracture model.

6.4.1   Governing Equations

We illustrate our method for the case where two immiscible fluid phases, oil and water for example, are flowing in fractured porous medium. The basic equations of incompressible two-phase flow include mass conservation equation, saturation equation, and capillary pressure relationship,

$$\frac{\partial(\phi S_n)}{\partial t} = \nabla \cdot (K\lambda_n(\nabla p_n - \rho_n g \nabla z)) + q_n \tag{6.24}$$

$$\frac{\partial(\phi S_w)}{\partial t} = \nabla \cdot (K\lambda_w(\nabla p_w - \rho_w g \nabla z)) + q_w \tag{6.25}$$

$$S_n + S_w = 1 \tag{6.26}$$

$$p_n - p_w = p_c(S_w) \tag{6.27}$$

where the subscripts n and w denote the non-wetting and wetting phase, respectively. S_i, p_i, and q_i denote the saturation, pressure and source/sink term of phase i, respectively. p_c is capillary pressure. K is permeability tensor, ϕ is the porosity of the medium. g is the acceleration of gravity, z is the vertical coordinate (positive in the upward direction) and t denotes the time. The relative mobility of each phase is defined by

$$\lambda_n = -\frac{k_{rn}}{\mu_n}, \quad \lambda_w = -\frac{k_{rw}}{\mu_w} \tag{6.28}$$

where k_{ri} and μ_i denote the relative permeability and viscosity of phase i respectively. Due to the incompressible fluids assumption, the viscosity of a particular phase is constant. We define a flow potential for each phase

$$\Phi_n = p_n + \rho_n g z \tag{6.29}$$

$$\Phi_w = p_w + \rho_w g z \tag{6.30}$$

In addition, we define the capillary flow potential Φ_c to include the gravity and the capillary effects.

$$\Phi_c = \Phi_n - \Phi_w = p_c + (\rho_n - \rho_w) g z$$

Equations (6.24)–(6.27) can be combined and formulated in terms of only two dependent variables, Φ_w and S_w, based on these definitions. With this procedure, the flow equations for two-phase flow are written as

$$\nabla \cdot (K(\lambda_w + \lambda_n) \nabla \Phi_w) = \nabla \cdot (K \lambda_n (\nabla \Phi_c)) - (q_n + q_w) \tag{6.31}$$

$$\frac{\partial(\phi S_w)}{\partial t} = \nabla \cdot (K \lambda_w (\nabla \Phi_w)) + q_w \tag{6.32}$$

The flow equations can be written in the form of matrix:

$$\begin{bmatrix} 0 & 0 \\ 0 & \phi \end{bmatrix} \frac{\partial}{\partial t} \begin{bmatrix} \Phi_w \\ S_w \end{bmatrix} + \nabla \cdot \left(-\begin{bmatrix} K(\lambda_m + \lambda_n) & K\lambda_n p_c' \\ K\lambda_m & 0 \end{bmatrix} \nabla \begin{bmatrix} \Phi_w^m \\ S_w^m \end{bmatrix} \right) = \begin{bmatrix} q_n + q_w \\ q_w \end{bmatrix} \tag{6.33}$$

where

$$\nabla \Phi_c = \frac{d\Phi_c}{dS_w} \nabla S_w = \frac{dp_c}{dS_w} \nabla S_w = p_c' \nabla S_w$$

In the following, we will study (6.33) as our flow models on the coarse and fine-scale. We will assume no-flow conditions for both models. The initial and boundary conditions are taken to be given in the form:

① Initial conditions:

$$\Phi_i(x, t = 0) = \Phi_{i,0}(x), \ S_i(x, t = 0) = S_{i,0}(x) \quad \forall x \in \Omega$$

② Dirichlet boundary conditions:

$$\Phi_i(x, t) = \widetilde{\Phi}_i, \; S_i(x, t) = \widetilde{S}_i \text{ on } \Gamma_D$$

③ Neumann boundary conditions, the boundary conditions in this paper are assumed to be impervious:

$$v_i \cdot n = -(\lambda_i \nabla \Phi_i) \cdot n = 0, \; \nabla S_i \cdot n = 0 \text{ on } \Gamma_N$$

where n is the outer normal unit vector of outer boundary, the subscript i represents the non-wetting or wetting phase.

For 2D problems, 1D element is employed to represent fracture in the discrete fracture model. Thus, the system of Eqs. (6.31) and (6.32) will be discretized in two-dimensional form for the matrix and in one-dimensional form for the fractures. The whole domain Ω is $\Omega = \Omega_m \sum_i d_i \times \Omega_{f,i}$, where m and f represent the matrix and the fracture, and d_i is the aperture of the ith fracture. Let FEQ represent the flow equations. Therefore, the integral form of these equations can be written as

$$\int_\Omega FEQ d\Omega = \int_{\Omega_m} FEQ d\Omega_m + \sum_i d_i \times \int_{\Omega_{f,i}} FEQ d\Omega_{f,i}. \tag{6.34}$$

To reduce the dimension of fractures, we assume the wetting potential should be continuous at the matrix–fracture interface, that is, $\Phi_w^m = \Phi_w^f$. This implies that the capillary potential and capillary pressure are also continuous at the interface. However, at the matrix–fracture interface have the same capillary pressure p_c^*, wetting saturation may be discontinuous at the interface. So, special measures should be taken toward the saturation equation of wetting phase in fractures. Corresponding to the continuity of capillary potential at the matrix–fracture interface, there is a relation between S_w^m and S_w^f:

$$S_w^f = \frac{p_c^m(S_w^m)}{p_c^f} \tag{6.35}$$

Equation (6.33) for the fracture domain can be written as

$$\begin{bmatrix} 0 & 0 \\ 0 & a_w \phi^f \end{bmatrix} \frac{\partial}{\partial t} \begin{bmatrix} \Phi_w^m \\ S_w^m \end{bmatrix} + \nabla \cdot \left(- \begin{bmatrix} K(\lambda_m^f + \lambda_n^f) & a_w K \lambda_n^f (p_c^f)' \\ K \lambda_m^f & 0 \end{bmatrix} \nabla \begin{bmatrix} \Phi_w^m \\ S_w^m \end{bmatrix} \right)$$
$$= \begin{bmatrix} q_n^m + q_w^m \\ q_w^m \end{bmatrix} \tag{6.36}$$

where $a_w = \dfrac{ds_w^f}{ds_w^m}$.

6.4.2 Multi-scale Finite Element Method

(1) Discretization and Hybrid System
To solve (6.36) on the fine-scale and the coarse scale, we will use the finite element formulation. To simplify the presentation of the finite element formulation, we assume the boundary conditions are impervious. Here, the flow potential equation and the saturation equation in (6.36) are derived separately. The variational problems of the equations are to seek $\tilde{\Phi}_w$ and $\tilde{S}_w$ from suitable discrete approximation spaces defined over Ω, such that

$$\int_\Omega K(\lambda_w + \lambda_n)\nabla\tilde{\Phi}_w\nabla\Psi d\Omega = -\int_\Omega K\lambda_n\nabla\tilde{\Phi}_c\nabla\Psi d\Omega + \int_\Omega (q_n + q_w)\Psi d\Omega \quad (6.37)$$

$$\int_\Omega \phi\frac{\partial\tilde{S}_w}{\partial t}\nabla\Psi d\Omega = -\int_\Omega K\lambda_w\nabla\tilde{\Phi}_w\nabla\Psi d\Omega + \int_\Omega q_w\Psi d\Omega \quad (6.38)$$

where Ψ is the weight function. Let T_h be a coarse-grid partition of Ω by a collection of polyhedral elements, and for each element $E \in T_h$, the region is discretized using triangular elements for the matrix and line elements for the fractures, as shown in Fig. 6.12. In each element E, we define a set of local basis functions $\{\Psi_i^E, i = 1,\ldots d\}$ with d being the number of nodes of the element. The superscript E will be neglected if the basis functions are considered in the same element. Thus, the flow potential and saturation are approximated as

$$\Phi_w \approx \tilde{\Phi}_w = \sum_{i=1}^d \Psi_i\Phi_{w,i}, \ S_w \approx \tilde{S}_w = \sum_{i=1}^d \Psi_i S_{w,i} \quad (6.39)$$

Substituting (6.39) into (6.37) and (6.38) yield a hybrid system of the form:

$$\begin{bmatrix} \mathbf{0} & \mathbf{0} \\ \mathbf{0} & \mathbf{A} \end{bmatrix}\begin{bmatrix} \dot{\Phi}_w^f \\ \dot{S}_w^f \end{bmatrix} + \begin{bmatrix} \mathbf{B} & \mathbf{C}_1 \\ \mathbf{C}_2 & \mathbf{0} \end{bmatrix}\begin{bmatrix} \Phi_w^f \\ S_w^f \end{bmatrix} = \begin{bmatrix} \mathbf{Q}_1 \\ \mathbf{Q}_2 \end{bmatrix} \quad (6.40)$$

Fig. 6.12 Mesh schematics of discrete fractured model

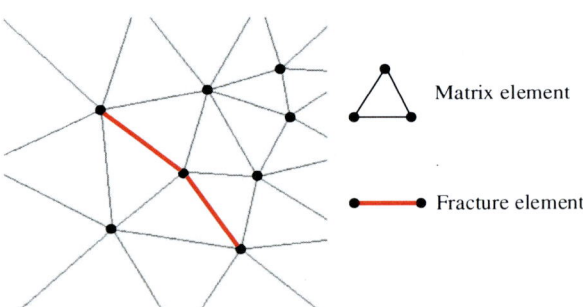

Matrix element

Fracture element

where Φ_w, $\mathbf{S}_w$ are the vectors of flow potential values $\Phi_{w,i}$ and saturation values $S_{w,i}$, respectively. The entries in the matrices are

$$A_{ij} = \int_\Omega \Psi_i \phi \, \Psi_j d\Omega, \; B_{ij} = \int_\Omega \nabla^T \Psi_i [K(\lambda_w + \lambda_n)] \nabla \Psi_j d\Omega,$$

$$C_{ij,1} = \int_\Omega \nabla^T \Psi_i (K\lambda_n p') \nabla \Psi_j d\Omega, \; C_{ij,2} = \int_\Omega \nabla^T \Psi_i (K\lambda_w) \nabla \Psi_j d\Omega$$

$$Q_{i,1} = \int_\Omega \nabla^T \Psi_i (q_n + q_w) d\Omega, \; Q_{i,2} = \int_\Omega \nabla^T \Psi_i q_w d\Omega.$$

The mass matrix A is not diagonal. To be able to update the saturation explicitly, we use a lumped matrix. The lumped matrix A^E in each element E is defined as

$$A_{ii}^E = \sum_{j=1}^d \int_\Omega \Psi_i \phi \Psi_j d\Omega = \int_\Omega \Psi_i d\Omega, \quad A_{ij}^E = 0$$

(2) Basis Functions and the Boundary Condition

Based on the discrete-fracture model, in coarse element E, the basis functions $\{\Psi_i\}$ are set to satisfy the following problem:

$$-\nabla \cdot (K\nabla\Psi_i) = 0 \tag{6.41}$$

for some function g_i defined on the boundary of the coarse element E.

Although the final multi-scale results are not sensitive to the accuracy of the multi-scale basis function, the boundary condition of the basis functions can have a big influence on the accuracy of MsFEM, in other words, a good choice of boundary conditions would significantly improve the accuracy of the multi-scale method. In most previous literatures, one choice of the function g_i for each i is to let g_i vary linearly along ∂E. Another way is to choose g_i to be the solution of the reduced elliptic problems on each side of ∂E. For example, on Γ_1 in Fig. 6.13, Ψ_i satisfies the reduced elliptic problem:

$$\frac{\partial}{\partial y}\left(K(y)\frac{\partial\Psi_i}{\partial y}\right) = 0 \tag{6.42}$$

The boundary condition of the 1D elliptic equation is given by $\Psi_i(\mathbf{x}_j) = \delta_{ij}$ (δ_{ij} is the Kronecker delta, i.e., $\delta_{ii} = 1$, while $\delta_{ij} = 0$ for $i \neq j$). The (6.43) can be solved analytically, that is,

$$\Psi_i|_{\Gamma_1} = \int_{y_{i-1}}^y \frac{dt}{K(t)} \Big/ \int_{y_{i-1}}^{y_i} \frac{dt}{K(t)} \tag{6.43}$$

If K is constant, then Ψ_i is linear.

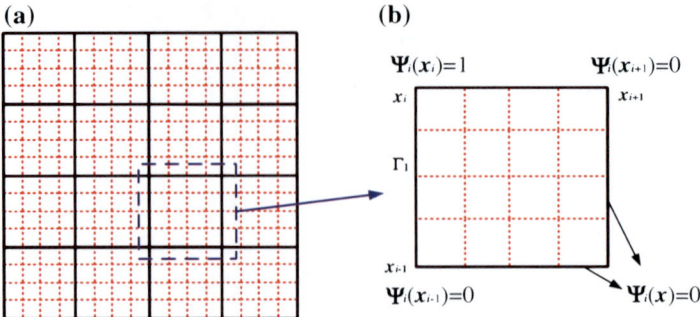

Fig. 6.13 Schematic showing **a** fine and coarse scale grid and **b** nodal points

In the coarse-grid partitioning model, there may exist fractures (line segments) in coarse grids. Generally, the fracture distributes in a coarse grid with three patterns (see Fig. 6.14). In one case, no fracture intersects the boundaries (see Fig. 6.14a), then the boundary condition of the basis functions is the same as (6.43). In the second case, there are some fractures intersected the boundaries of the coarse grid. To motivate the construction of the boundary condition, let us consider a simple case that only one fracture intersects the boundary (see Fig. 6.14b), the boundary condition is given by

$$\Psi_i\big|_{\Gamma_1} = \begin{cases} \int_{y_{i-1}}^{y} \frac{\mathrm{d}_t}{K^m(t)} \Big/ \left(\int_{y_{i-1}}^{y_i} \frac{\mathrm{d}_t}{K^m(t)} + \frac{d}{K^f} \right), & \text{if } y < y^f \\ \left(\int_{y_{i-1}}^{y} \frac{\mathrm{d}_t}{K^m(t)} + \frac{d}{K^f} \right) \Big/ \left(\int_{y_{i-1}}^{y_i} \frac{\mathrm{d}_t}{K^m(t)} + \frac{d}{K^f} \right), & \text{if } y \geq y^f \end{cases} \tag{6.44}$$

where y^f is the node intersected by the fracture and the boundary, d is the aperture of the fracture. In the third case, the fracture coincides with the boundary (see Fig. 6.14c). We have

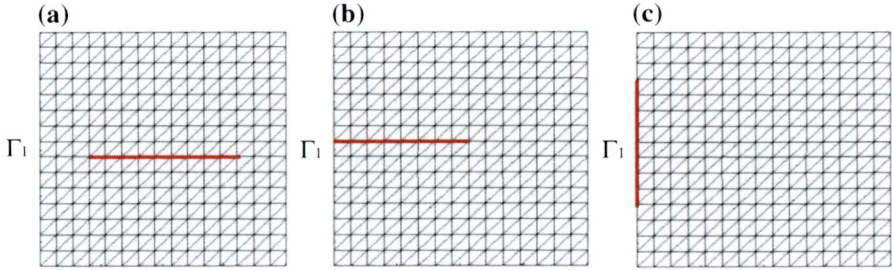

Fig. 6.14 Fracture distributes in a coarse grid with three patterns: **a** the fracture does not intersect the boundary; **b** the fracture intersect the boundary; **c** the fracture coincides with the boundary

$$\Psi_i\big|_{\Gamma_i} = \begin{cases} \int_{y_{i-1}}^{y} \frac{dt}{K^m(t)}\Big/A^{fm}, & \text{if } y < y^{f_1} \\[2mm] \left(\int_{y_{i-1}}^{y^{f_1}} \frac{dt}{K^m(t)} + \int_{y^{f_1}}^{y} \frac{dt}{(K^m(t)+K^f)} + \right)\Big/A^{fm}, & \text{if } y^{f_1} \leq y \leq y^{f_2} \\[2mm] \left(A^{fm} - \int_{y}^{y_i} \frac{dt}{K^m(t)}\right)\Big/A^{fm}, & \text{if } y > y^{f_2} \end{cases} \qquad (6.45)$$

where y^{f_1}, y^{f_2} are the nodes intersected by the fracture and the boundary, and

$$A^{fm} = \int_{y_{i-1}}^{y^{f_1}} \frac{dt}{K^m(t)} + \int_{y^{f_2}}^{y^{f_1}} \frac{dt}{K^f(t)+K^m(t)} + \int_{y^{f_2}}^{y_i} \frac{dt}{K^m(t)}.$$

Figure 6.15 displays basis functions for three domains with a fracture. We see that the basis function reflects the fine-scale details in the fractured domain.

(3) Coarse-scale Hybrid System

Here, we arrange all the multi-scale basis function Ψ_i as columns in a matrix Ψ. Then, the multi-scale system is now obtained by summing the fine grid equations as follows:

$$\begin{bmatrix} \Psi^T & 0 \\ 0 & \Psi^T \end{bmatrix} \begin{bmatrix} 0 & 0 \\ 0 & A \end{bmatrix} \begin{bmatrix} \dot{\Phi}_w^f \\ \dot{S}_w^f \end{bmatrix} + \begin{bmatrix} \Psi^T & 0 \\ 0 & \Psi^T \end{bmatrix} \begin{bmatrix} B & C_1 \\ C_2 & 0 \end{bmatrix} \begin{bmatrix} \Phi_w^f \\ S_w^f \end{bmatrix} = \begin{bmatrix} Q_1^f \\ Q_2^f \end{bmatrix} \qquad (6.46)$$

For the fine-scale potential and saturation, we have

$$\Phi^f \approx \Psi\Phi^c, \; S^f \approx \Psi S^c$$

With the above approximation of coarse grid-block potential, the coarse-scale hybrid system reads

$$\begin{bmatrix} 0 & 0 \\ 0 & \widetilde{A} \end{bmatrix} \begin{bmatrix} \dot{\Phi}_w^c \\ \dot{S}_w^c \end{bmatrix} + \begin{bmatrix} \widetilde{B} & \widetilde{C}_1 \\ \widetilde{C}_2 & 0 \end{bmatrix} \begin{bmatrix} \Phi_w^c \\ S_w^c \end{bmatrix} = \begin{bmatrix} \widetilde{Q}_1^c \\ \widetilde{Q}_2^c \end{bmatrix} \qquad (6.47)$$

where $\widetilde{A} = \Psi A^f \Psi$, $\widetilde{B} = \Psi B^f \Psi$, $\widetilde{C}_i = \Psi C_i^f \Psi$, $\widetilde{Q}_i = \Psi^T Q_i$ $(i = 1, 2)$.

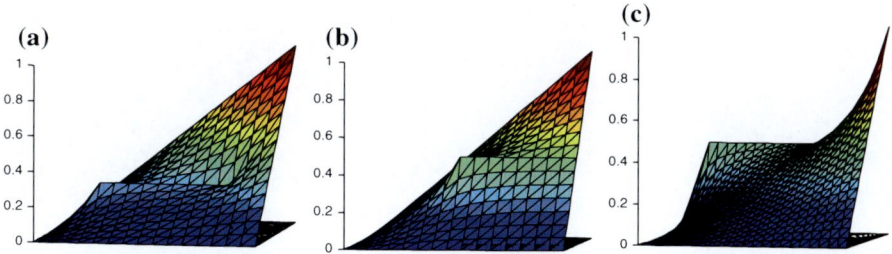

Fig. 6.15 Three MsFEM basis functions for a homogeneous domain

6.4.3   Numerical Experiments

The model has three small fractures, with apertures of 1 mm and lengths of 280 cm (see Fig. 6.16). The fracture permeability is $K_f = d^2/12 = 8.33 \times 10^5 \ \mu m^2$. The model has 2009 fine-scale grids, and the multi-scale discretization is illustrated in Fig. 6.17. The injection well was placed at the lower left corner and production well was placed at the upper right corner. The density and viscosity of each phase are shown in Table 6.1. The porosity ϕ of the matrix is 0.2 and the permeability k_m is 1 md.

Fig. 6.16 Fractured media with complex fractures

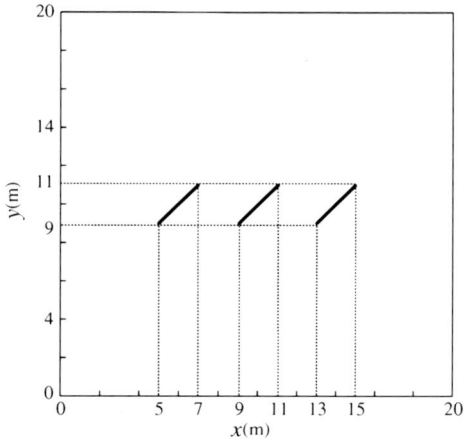

Fig. 6.17 Multiscale mesh for discrete fractured media

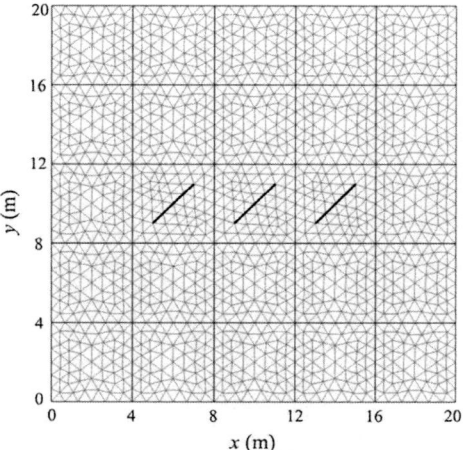

Table 6.1 Fluid parameters

Property	Water	Oil
Density, kg/m^3	1000	800
Viscosity, Pa s	1.0×10^{-3}	0.5×10^{-2}

We measure the relative permeability functions are specified as

$$K_{rw} = (S^*)^2, \ K_{ro} = (1 - S^*)^2, \ S^* = \frac{S - S_{wc}}{1 - S_{wc} - S_{or}}, \qquad (6.48)$$

with $S_{wc} = S_{or} = 0.1$, and initial saturation $S_o = S_{wc}$. The water flow rate was set to 0.2 PV/d. Ignore the effect of capillary pressure.

In Fig. 6.18, the saturation profiles at 0.25 and 0.45 PV are compared. To provide a comparison of the accuracy of the saturation solution, the saturation

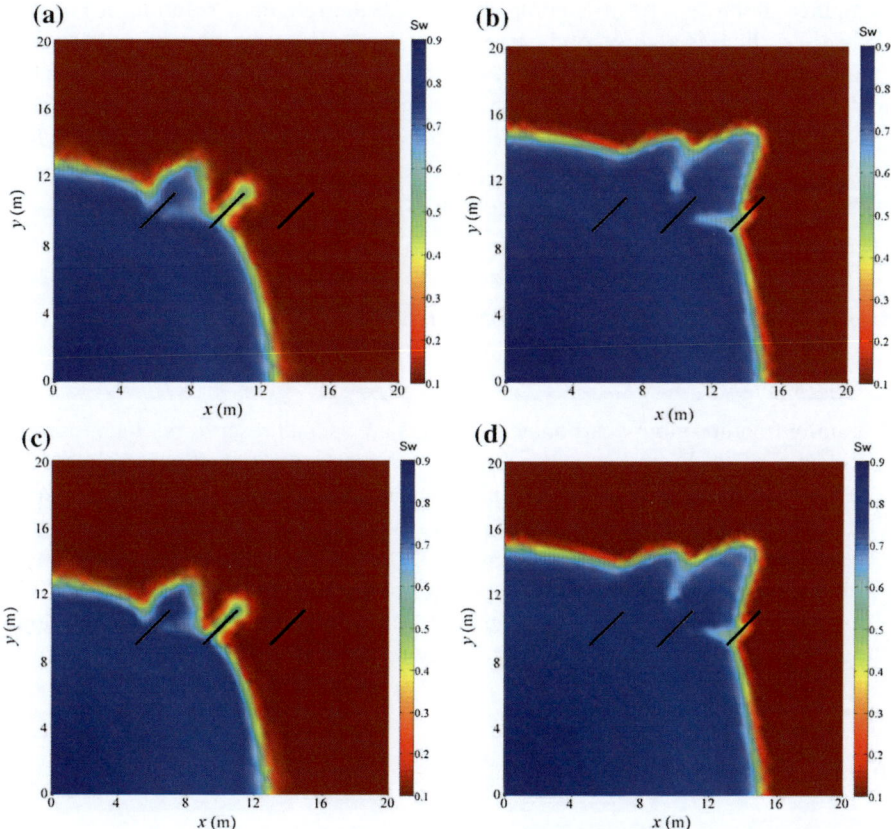

Fig. 6.18 Saturation maps for reference solution and MsFEM. **a** Reference saturation at 0.25 PV, **b** reference saturation at 0.45 PV, **c** MsFEM saturation at 0.25 PV, **d** MsFEM saturation at 0.45 PV

Fig. 6.19 Comparison of saturation distribution along $y = 10$ m

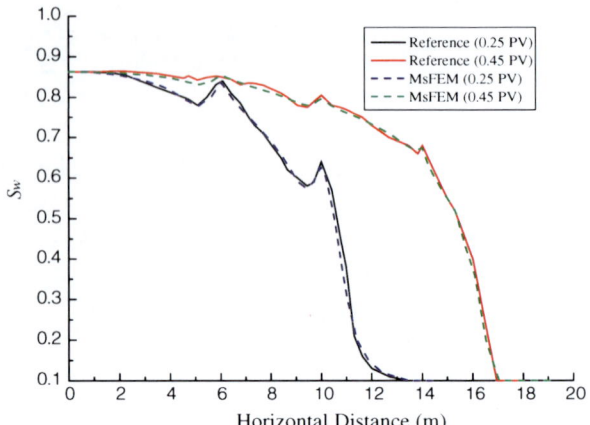

computed along $y = 10$ m is compared against the reference solution in Fig. 6.19. As we see, there is a very good agreement between the reference saturation and the saturation field obtained by the MsFEM. Saturation errors for MsFEM are 0.0272 and 0.0158. The results show that the computations are accurate. The total CPU time of solving the problem by MsFEM requires about 40 % less CPU time than the reference solution. It does not greatly reduce the total CPU time, as the MsFEM needs an extra basis function calculation step.

6.5 Multi-Scale Numerical Simulation of Discrete Fracture-Vug Model

Naturally fracture-vuggy carbonate reservoirs are special reservoirs. Such reservoirs are characterized by the presence of fractures, vugs, and caves at multiple scales. The main difficulty in numerical simulations in such reservoirs is the coexistence of porous and free flow regions, typically at several scales that require coupling. We need to analyze the fine-scale information if the traditional finite element method is applied to solve the problem, and this requires a tremendous amount of CPU time. This section presents a multi-scale mixed finite element method for simulation of fluid flow in naturally fractured and vuggy reservoirs.

6.5.1 Mathematical Models

In this paper, we consider the fluid flow is isothermal, single phase, and incompressible with constant fluid viscosity. In the DFVN conceptual model, the flow in matrix and fractures follows Darcy's law, and incompressible flow in the vugs obeys the Stokes equations.

The mathematical model of matrix and fracture system Ω_{pf} is

$$\frac{\partial(\rho\phi)}{\partial t} + \nabla \cdot \mathbf{v}_D = q, \quad \mathbf{v}_D = -\frac{\mathbf{K}}{\mu}\nabla p_D \qquad (6.49)$$

The mathematical model of matrix and vugs system Ω_{pf} is

$$\frac{\partial(\rho\phi)}{\partial t} + \nabla \cdot \mathbf{v}_S = q, \quad \mu\nabla \cdot \left(\nabla\mathbf{v}_S + \nabla\mathbf{v}_S^T\right) - \nabla p_S = \rho\frac{\partial\mathbf{v}_S}{\partial t} \qquad (6.50)$$

The equations of the free-flow and the porous media domains are unified into a single system of equations

$$\frac{\partial(\rho\phi)}{\partial t} + \nabla \cdot \mathbf{v} = q, \quad -\mu\mathbf{K}^{-1}\mathbf{v} - \nabla p + \tilde{\mu}\Delta\mathbf{v} = C\rho\frac{\partial\mathbf{v}}{\partial t} \qquad (6.51)$$

where $\mathbf{v}$ is velocity; q is source term; $\mathbf{K}$ is a permeability tensor; p is pressure; μ is fluid viscosity; $\tilde{\mu}$ is an effective viscosity. $\mathbf{K}$, $\tilde{\mu}$ and C are determined by the type of area. Equation (6.51) is Stokes-Brinkman equation.

In vugs system Ω_v, we let

$$\tilde{\mu} = \mu, \quad \mathbf{K} = \infty, \quad C = 1$$

then the Stocks–Brinkman equation can be simplified to the Stokes equations. In matrix and fracture system Ω_{pf}, let $\mathbf{K}$ equal to porous media permeability, $C = 0$, (6.51) becomes

$$\nabla p = -\mu\mathbf{K}^{-1}\mathbf{v} + \tilde{\mu}\Delta\mathbf{v} \qquad (6.52)$$

If $\tilde{\mu} = 0$, Eq. (6.52) simplifies to the coupled Darcy–Stokes equations, which reintroduces the requirement for the interface conditions and computational intractability. If we set $\tilde{\mu} = \mu$, so $\nabla p \approx -\mu\mathbf{K}^{-1}\mathbf{v}$, and Eq. (6.52) can be seen as Darcy's equation with a small viscosity perturbation. Therefore, we set $\tilde{\mu} = \mu$.

For 2D problems, 1D element is employed to represent fracture. The vugs are simplified by (d-1)-dimensional elements, then the whole domain Ω is

$$\int_{\Omega} FEQd\Omega = \int_{\Omega_m} FEQd\Omega_m + \sum_i a_i \times \int_{\Omega_{f,i}} FEQd\Omega_{f,i} + \int_{\Omega_v} FEQd\Omega_v \qquad (6.53)$$

where, a_i is the aperture of the ith fracture.

We use a Darcy model to approximate pressure and velocity on a coarse grid and uses Stokes–Brinkman model, based on DFVN model to solve multi-scale basis function.

6.5.2   The Multi-Scale Mixed Finite Element Method

(1) Discretization and Hybrid System

Let Ω be a polyhedral domain in $\subset \mathbb{R}^d (d = 2, 3)$, with boundary $\partial\Omega$ whose unit outer normal is denoted by n. Then the variational problems of (6.51) are to seek a pair of functions $(\tilde{p}, \tilde{\mathbf{v}}) \in W \times V$, such that

$$\int_\Omega \mathbf{u} \cdot (\lambda\mathbf{K})^{-1}\tilde{\mathbf{v}}\mathrm{d}\Omega - \int_\Omega \tilde{p}\nabla \cdot \mathbf{u}\mathrm{d}\Omega = 0, \quad \forall \mathbf{u} \in W \tag{6.54}$$

$$\int_\Omega l\nabla \cdot \tilde{\mathbf{v}}\mathrm{d}\Omega = \int_\Omega ql\mathrm{d}\Omega, \quad \forall l \in V \tag{6.55}$$

where W and V are finite dimensional subspaces of $H^d(\Omega)$ and $L^2(\Omega)$.

For Darcy problems, we will use a set of generalized RT0 basis functions. In order to obtain the normal velocity that is continuous across cell face, continuity of the normal component is reintroduced using Lagrange multipliers in which the pressure λ at the element faces plays the role of the Lagrange multipliers. Let $\mathbf{v} = \sum Q_i\psi_i$, $p = \sum p_k\delta_k$, the Darcy equations can now be assembled to form a hybrid system of the form

$$\begin{bmatrix} \mathbf{B} & \mathbf{C} & \mathbf{D} \\ \mathbf{C}^T & 0 & 0 \\ \mathbf{D}^T & 0 & 0 \end{bmatrix} \begin{bmatrix} \mathbf{Q}^c \\ -\mathbf{p}^c \\ \lambda^c \end{bmatrix} = \begin{bmatrix} 0 \\ \mathbf{q}^c \\ 0 \end{bmatrix} \tag{6.56}$$

where $\mathbf{Q}^c$ is the vector of the outward fluxes ordered cell-wise; $\mathbf{p}^c$ is vector of cell pressures; λ^c is vector of face pressures; $\mathbf{B}_{ij} = \int_\Omega \psi_i \cdot (\mu\mathbf{K})^{-1}\psi_j\mathrm{d}\Omega$; $C_{ij} = \int_\Omega \delta_j\nabla \cdot \psi_i\mathrm{d}\Omega$; $\mathbf{D}_{ij} = \int_{\partial\Omega} |\psi_i \cdot \mathbf{n}_j|\mathrm{d}s$ ψ_i and ψ_j are outward-pointing velocity basis functions, $\mathbf{n}_j$ is the normal of cell face j, δ_j satisfies

$$\delta_i = \begin{cases} 1, & x \in \Omega_i \\ 0, & x \notin \Omega_j \end{cases}$$

For the Stokes–Brinkman problem, its variational forms are similar to the variational forms of Darcy equation: seek $(\tilde{p}, \tilde{\mathbf{v}}) \in Q \times V$, such that

$$\int_\Omega \mathbf{u} \cdot (\mu\mathbf{K})^{-1}\tilde{\mathbf{v}}\mathrm{d}\Omega - \int_\Omega \tilde{p}\nabla \cdot \mathbf{u}\mathrm{d}\Omega + \int_\Omega \tilde{\mu}\nabla\mathbf{u} \cdot \nabla\tilde{\mathbf{v}}\mathrm{d}\Omega = 0, \quad \forall \mathbf{u} \in Q \tag{6.57}$$

$$\int_\Omega l\nabla \cdot \tilde{\mathbf{v}}\mathrm{d}\Omega = \int_\Omega ql\mathrm{d}\Omega, \quad \forall l \in V \tag{6.58}$$

We split the velocity v into its two spatial components v_1 and v_2. Let $v_k = \sum v_{ik}\psi_i (k = 1, 2)$, $p = \sum p_k \phi_k$, the mixed system can then be assembled in the form

$$
\begin{bmatrix}
\mathbf{B_1} & 0 & \mathbf{C}_1 \\
0 & \mathbf{B_2} & \mathbf{C}_2 \\
\mathbf{C}_1^T & \mathbf{C}_1^T & 0
\end{bmatrix}
\begin{bmatrix}
\mathbf{v}_1 \\
\mathbf{v}_2 \\
-\mathbf{p}
\end{bmatrix}
=
\begin{bmatrix}
0 \\
0 \\
\mathbf{q}
\end{bmatrix}
\tag{6.59}
$$

where $\mathbf{v}_1$, $\mathbf{v}_2$ are vectors of the two velocity components v_{i1} and v_{i2}, respectively, p is the vector of pressure values p_i

$$\mathbf{B}_k = (\mathbf{B}_k)_m + (\mathbf{B}_k)_f + (\mathbf{B}_k)_v$$

$$(\mathbf{B}_{ij,k})_m = \int_{\Omega_m} \psi_i \cdot (\mu \mathbf{K}_{m,k})^{-1} \psi_j d\Omega_m + \int_{\Omega_m} \tilde{\mu} \left(\frac{\partial \psi_i}{\partial x_1} \frac{\partial \psi_j}{\partial x_1} + \frac{\partial \psi_i}{\partial x_2} \frac{\partial \psi_j}{\partial x_2} \right) d\Omega_m$$

$$\mathbf{K}_m = \begin{bmatrix} K_{m,1} & 0 \\ 0 & K_{m,2} \end{bmatrix}$$

$$(\mathbf{B}_{ij,k})_f = d \int_{\Omega_f} \psi_i \cdot \frac{\mu}{K_f} \psi_j d\Omega_f + \int_{\Omega_f} \tilde{\mu} \left(\frac{\partial \psi_i}{\partial x_1} \frac{\partial \psi_j}{\partial x_1} + \frac{\partial \psi_i}{\partial x_2} \frac{\partial \psi_j}{\partial x_2} \right) d\Omega_f$$

$$(\mathbf{B}_{ij,k})_v = \int_{\Omega_v} \psi_i \cdot \frac{\mu}{K_v} \psi_j d\Omega_v + \int_{\Omega_v} \tilde{\mu} \left(\frac{\partial \psi_i}{\partial x_1} \frac{\partial \psi_j}{\partial x_1} + \frac{\partial \psi_i}{\partial x_2} \frac{\partial \psi_j}{\partial x_2} \right) d\Omega_v$$

$$\mathbf{C}_k = (\mathbf{C}_k)_m + (\mathbf{C}_k)_f + (\mathbf{C}_k)_v$$

$$(\mathbf{C}_{ij,k})_w = \int_{\Omega_w} \frac{\partial \psi_i}{\partial x_k} \phi_j d\Omega_w, \quad (\mathbf{C}_{ij,k})_f = d \int_{\Omega_f} \frac{\partial \psi_i}{\partial x_k} \phi_j d\Omega_f$$

$$\mathbf{q} = \mathbf{q}_m + \mathbf{q}_f + \mathbf{q}_v$$

$$(\mathbf{q}_l)_w = \int_{\Omega_w} \phi_l q d\Omega_w, \quad (\mathbf{q}_l)_v = \int_{\Omega_v} \phi_l q d\Omega_v, \quad (\mathbf{q}_l)_f = d \int_{\Omega_f} \phi_l q d\Omega_f$$

where subscript m, f, v denote matrix system, fracture system and vug system, respectively; $w = m, v$; $k = 1, 2$ denotes the spatial dimension. d is the aperture of the fracture.

(2) MsMFEM basis Functions
Let $\mathcal{T}_h = \{\Omega_i\}$ is partition of Ω, $\Gamma_{ij} = \partial\Omega_i \cap \partial\Omega_j$, we seek $\boldsymbol{\psi}_{ij}$ on Γ_{ij} and seek ϕ_i on Ω_i.

In $\Omega_{ij} = \Omega_i \cup \Gamma_{ij} \cup \Omega_j$, the basis function associated with the interface is constructed by solving the following flow problem over

$$\mu\mathbf{K}^{-1}\boldsymbol{\psi}_{ij} + \nabla\phi_{ij} - \tilde{\mu}\Delta\boldsymbol{\psi}_{ij} = 0 \tag{6.60}$$

$$\nabla \cdot \boldsymbol{\psi}_{ij} = \begin{cases} \omega_i(x), & x \in \Omega_i \\ -\omega_i(x), & x \in \Omega_j \\ 0, & x \notin \Omega_{ij} \end{cases} \tag{6.61}$$

$$\boldsymbol{\psi}_{ij}(x) \cdot \mathbf{n} = 0 \quad \forall x \in \partial\Omega_{ij} \tag{6.62}$$

Here, $\omega_i(x)$ is a weight function on Ω_i, satisfies $\int_{\Omega_i} \omega_i(x)dx = 1$, To obtain a conservative method, we choose:

$$\omega(x)_i = \begin{cases} \sigma(x)/\int_{\Omega_i} \sigma(\xi)d\xi, & \int_{\Omega_i} qdx = 0 \\ q(x)/\int_{\Omega_i} q(\xi)d\xi, & \int_{\Omega_i} qdx \neq 0 \end{cases} \tag{6.63}$$

where $\sigma(x) = \mathrm{trace}(\mathbf{K})/d$, $\mathrm{trace}(\mathbf{A})$ denotes the sum of Eigen value of matrix A.

To obtain conservative multi-scale basis functions, we should apply a conservative method to solve fine-scale problem, like finite volume method. The method is also depends on the local grid structure.

So for a cell Ω_i, the corresponding pressure basis function ϕ_i is given by

$$\phi_i(x) = \begin{cases} 1, & x \in \Omega_i \\ 0, & x \notin \Omega_i \end{cases}$$

(3) Coarse-scale Hybrid System
We split the basis functions into two parts:

$$\boldsymbol{\psi}_{ij} = \boldsymbol{\psi}_{ij}^H - \boldsymbol{\psi}_{ji}^H$$

where

$$\boldsymbol{\psi}_{ij}^H(E) = \begin{cases} \boldsymbol{\psi}_{ij}(E), & E \in \Omega_{ij}\backslash\Omega_j \\ 0, & E \notin \Omega_i \end{cases}$$

$$\boldsymbol{\psi}_{ji}^H(E) = \begin{cases} -\boldsymbol{\psi}_{ij}(E), & E \in \Omega_j \\ 0, & E \notin \Omega_j \end{cases}$$

$\boldsymbol{\psi}$ is a matrix which all the basis functions $\boldsymbol{\psi}_{ij}^H$ arranged as columns.

The starting point in developing the multi-scale method is the assumption that the fine-scale fields can approximately be expanded in the corresponding spaces spanned by the basis functions. For the fine-scale pressure and velocity, we have

$$\mathbf{v}^f = \mathbf{\psi} \mathbf{A}^{-1} \mathbf{q}^c, \quad \mathbf{p}^f = \mathbf{I} \mathbf{p}^c$$

Here, $\mathbf{A}$ is a matrix; $\mathbf{I}$ is the prolongation from blocks to cells, if block j contains cell i, $\mathbf{I}_{ij} = 1$, otherwise $\mathbf{I}_{ij} = 0$. The coarse-scale face pressure λ^c is given as

$$\lambda_i^c = \int_{\Gamma_{ij}} \lambda^f \mathbf{\psi}_{ij} \cdot \mathbf{n} ds$$

Let $\mathbf{J}$ is the prolongation from coarse to fine face such that $\mathbf{J}_{ij} = 1$ if coarse face j contains fine face i and is zero otherwise. Then for the fine-scale face pressure is given as

$$\lambda^f = \mathbf{J} \lambda^c$$

The multi-scale system is obtained by summing all the fine-scale equations, the coarse-scale system is

$$\begin{bmatrix} \mathbf{B}^c & \mathbf{C}^c & \mathbf{D}^c \\ \mathbf{C}^{cT} & 0 & 0 \\ \mathbf{D}^{cT} & 0 & 0 \end{bmatrix} \begin{bmatrix} \mathbf{v}^c \\ -\mathbf{p}^c \\ \lambda^c \end{bmatrix} = \begin{bmatrix} 0 \\ \mathbf{q}^c \\ 0 \end{bmatrix} \tag{6.64}$$

where $\mathbf{B}^c = \mathbf{\psi}^T \mathbf{B}^f \mathbf{\psi}$; $\mathbf{C}^c = \mathbf{\psi}^T \mathbf{C}^f \mathbf{I}$; $\mathbf{D}^c = \mathbf{\psi}^T \mathbf{C}^f \mathbf{J}$; $q^c = \mathcal{I}^T q^f$.

6.5.3   Numerical Experiments

The fracture-vugs model shown in Fig. 6.20 is constructed. The model is calculated by MsMFEM and finite element method, respectively. The sample is 25 m/25 m². A pressure gradient of 1.0 MPa/m along the x-direction is created by imposing pressure on the left and right vertical boundary, respectively. We consider no flow at top and bottom sides of the domain. The permeability is 1×10^{-6} m². The aperture of fractures equals 10 cm.

Figure 6.21 illustrates the pressure and velocity solutions from FEM and MsMFEM, respectively. The pressure and velocity distributions along x-direction on $y = 12$ m is shown in Fig. 6.22. Numerical results have shown that, the results

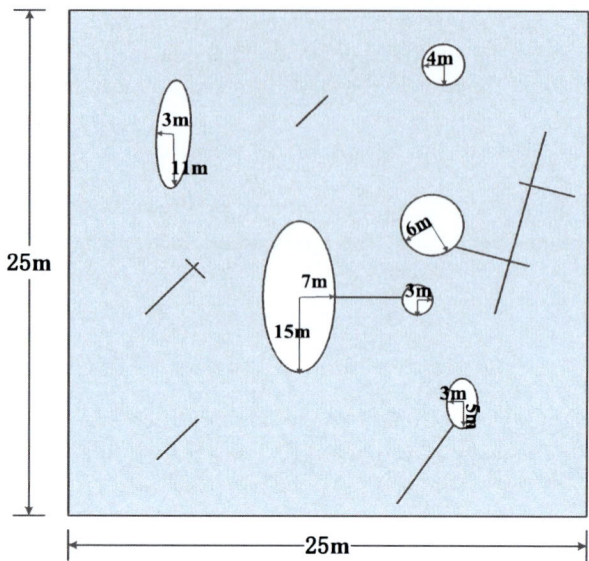

Fig. 6.20 Schematic of three sample models

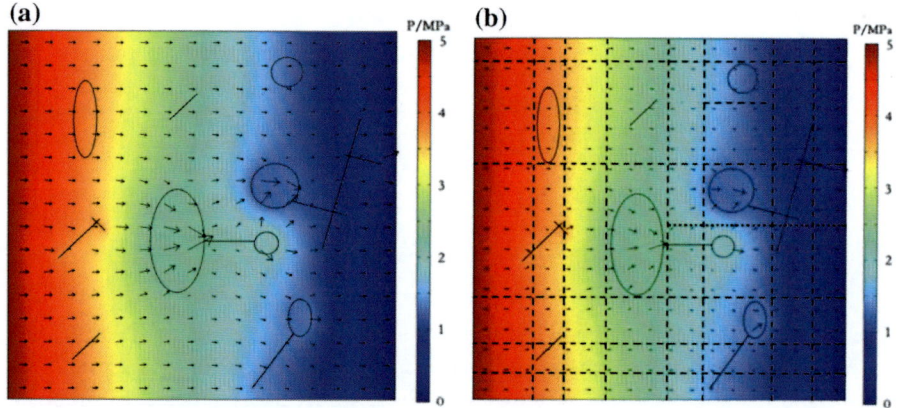

Fig. 6.21 Pressure and velocity distributions. **a** FEM, **b** MsMFEM

from MsMFEM are in close agreement with fine-scale solutions; the pressure and velocity discrepancies are less than 6 %. The calculation speed of MsMFEM is faster than fine-scale solution. The simulation results illustrates that the influence of $\bar{\mu}\Delta v$ is insignificant.

Fig. 6.22 Pressure
distribution for fine-scale
solution and MsMFEM

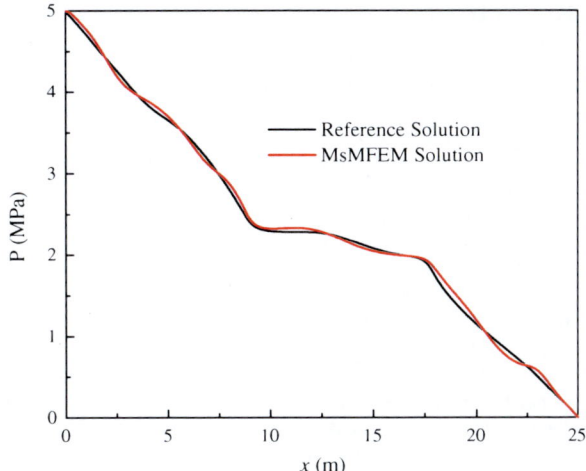

x (m)

6.6   Concluding Remarks

We have presented a first attempt to extend the multi-scale mixed finite element method (MsMFEM) to fluid flow in fractured porous media using DFVN model. This has been successfully achieved. In addition to implementation of the method, several examples were provided to demonstrate the accuracy and robustness of MsMFEM.

In this chapter, we introduce multi-scale simulation method for carbonate reservoirs, different multi-scale mathematical models are built, the theory and method is developed for multi-scale simulation of carbonate reservoirs, the conclusions are follows:

(1) 2D multi-scale model is built for heterogeneous reservoirs, the multi-scale simulation method is developed based on the multi-scale mixed finite element method and some calculation results are used to verify the validity and reliability of the numerical method. The results from MsMFEM are in close agreement with fine-scale solutions. MsMFEM saves calculation amount enormously and it has prominent advantages over conventional method when deal with fluid flow problems in heterogeneous reservoirs.

(2) For fracture reservoirs, the multi-scale mathematical model is built based on discrete fracture model. The multi-scale finite method that based on the global information is applied to analyze the model and the validity is verified by the numerical examples.

(3) We extend the Multi-scale mixed finite element method (MsMFEM) to fluid flow in fractured porous media using DFVN model, several examples were provided to demonstrate the accuracy and robustness of MsMFEM.

References

Aarnes JRE (2004) On the use of a mixed multiscale finite element method for greaterflexibility and increased speed or improved accuracy in reservoir simulation. SIAM J Multiscale Model Simul 2(3):421–439

Babuska I, Osborn JE (1983) Generalized finite element methods: their performance and their relation to mixed methods. SIAM J Numer Anal 20(3):510–536

Babuska I, Osborn JE (1994) Special finite element methods for a class of second order elliptic problems with rough coefficients. SIAM J Numer, Anal

Babuska I, Rheinboldt WC (1978) A-posteriori error estimates for the finite element method. Int J Numer Meth Eng 12(10):1597–1615

Brandt A (1977) Multi-level adaptive solutions to boundary-value problems. Math Comput 31:333–390

Chen Y, Durlofsky LJ (2006) Adaptive local-global upscaling for general flow scenarios in heterogeneous formations. Transp Porous Media 62(2):157–185

Chen Z, Hou T (2003) A mixed multiscale finite element method for elliptic problems with oscillating coefficients. Math Comput 72(242):541–576

Chen Z, Yue X (2002) Numerical homogenization of well singularities in the flow transport through heterogeneous porous media. Multiscale Model Simul 1(2):260–303

Cui JZ, Cao LQ (1999) Two-scale asymptotic analysis methods for a class of elliptic boundary value problems with small periodic coefficients. Math Numer Sin 21(1):19–28

Desbarats JA (1998) Scaling of constitutive relationships in unsaturated heterogeneous media: a numerical investigation. Water Resour Res 34(6):1427–1435

Dorobantu M, Engquist B (1996) Wavelet-based numerical homogenization. SIAM J Numer Anal 35:540–559

Durlofsky LJ (1991) Numerical calculation of equivalent grid block permeability tensors for heterogeneous porous media. Water Resour Res 27(5):699–708

Efendiev Y, Hou TY (2009) Multiscale finite element methods: theory and applications (vol. 4). Springer Science & Business Media

Efendiev Y, Pankov AA (2004) Numerical homogenization of nonlinear random parabolic operators. SIAM Multiscale Model Simul 2(2):237–268

Efendiev Y et al (2004) Multiscale finite element methods for nonlinear problems and their applications. Commun Math Sci

Gr et al (1992) Homogenization and two-scale convergence. SIAM J Math Anal (6):1482–1518

Guo B, Babuška I (1986) The h-p version of the finite element method. Comput Mech 1(3):203–220

He XG, Ren L (2009a) Adaptive multi-scale finite element method for unsaturated flow in heterogeneous porous media I. Numerical scheme. Shui Li Xue Bao 40(1):38–46

He XG, Ren L (2009b) Adaptive multi-scale finite element method for unsaturated flow in heterogeneous porous media II. Numerical results. Shui Li Xue Bao 40(2):138–144

Hou TY, Cai Z (1999) Convergence of a multiscale finite element method for elliptic problems with rapidly oscillating coefficients. Math Comput 68(227):913–943

Hou TY, Wu XH (1997) A multiscale finite element method for elliptic problems in composite materials and porous media. J Comput Phys 134(1):169–189

Jenny P et al (2003) Multi-scale finite-volume method for elliptic problems in subsurface flow simulation. J Comput Phys 187(1):47–67

Jenny P et al (2005) Adaptive multiscale finite-volume method for multiphase flow and transport in porous media. Multiscale Model Simul 3(1):50–64

Kevrekidis IG et al (2002) Equation-free multiscale computation: enabling microscopic simulators to perform system-level tasks. Commun Math Sci

Kevrekidis IG et al (2004) Equation-free, coarse-grained multiscale computation: enabling microscopic simulators to perform system-level analysis. Commun Math Sci (4):715

Mccarthy JF (1995) Comparison of fast algorithms for estimating large-scale permeabilities of heterogeneous media. Transp Porous Media 19(2):123–137

Neuweiler I, Cirpka OA (2005) Homogenization of Richards equation in permeability fields with different connectivities. Water Resour Res 41(2):199–207

Owhadi H, Zhang L (2007) Metric based upscaling. Commun Pure Appl Math 60:675–723

Ren W, Weinan E (2005) Heterogeneous multiscale method for the modeling of complex fluids and micro-fluidics. J Comput Phys 204(1):1–26

Schwarz HA (1890) Gesammelte mathematische abhandlungen. Am Math Soc 2

Weinan E (2003) Analysis of the heterogeneous multiscale method for ordinary differential equations. Commun Math Sci 1(3):423–436

Weinan E et al (2005) Analysis of the heterogeneous multiscale method for elliptic homogenization problems. J Am Math Soc 18(1):21–156

Weinan E, Yue XY (2004) Heterogeneous multiscale method for locally self-similar problems. Commun Math Sci(1):137–144

Printed by Books on Demand, Germany